Universitext

Universitext

Universitext is a series of textbooks that presents material from a wide variety of mathematical disciplines at master's level and beyond. The books, often well class-tested by their author, may have an informal, personal even experimental approach to their subject matter. Some of the most successful and established books in the series have evolved through several editions, always following the evolution of teaching curricula, to very polished texts.

Thus as research topics trickle down into graduate-level teaching, first textbooks written for new, cutting-edge courses may make their way into *Universitext*.

More information about this series at http://www.springer.com/series/223

Stephen Bruce Sontz

Principal Bundles

The Quantum Case

Stephen Bruce Sontz
Centro de Investigación en Matemáticas, A.C.
Guanajuato, Mexico

ISSN 0172-5939   ISSN 2191-6675   (electronic)
Universitext
ISBN 978-3-319-15828-0   ISBN 978-3-319-15829-7   (eBook)
DOI 10.1007/978-3-319-15829-7

Library of Congress Control Number: 2015935233

Mathematics Subject Classification (2010): 16W30, 17B37, 46L87, 81R60

Springer Cham Heidelberg New York Dordrecht London

Printed on acid-free paper

Springer International Publishing AG Switzerland is part of Springer Science+Business Media (www.
springer.com)

To Micho

Preface

First, who is the intended audience? I think I am addressing persons who either are active researchers in some area of mathematics or physics or are preparing themselves for that. Also, they are not experts in the topic of the title and want to learn something about it. As a nonexpert myself, I feel qualified in knowing what other nonexperts need to know. But absolutely everyone is welcome since science is open to all. However, experts in noncommutative geometry should find themselves rather bored. Among other things, I am trying to give a quite detailed mathematical presentation so that everything is accessible to physicists as well as mathematicians. It is easier for mathematicians to skip over details they do not need than for physicists to fill in seemingly enormous gaps that mathematicians barely notice even being there. And I do hope to find many physicists in my audience.

What are the prerequisites? Basically, gumption. With enough of that one can assimilate the knowledge base to get up to speed for page 1. What is that knowledge base? That is not so easy to specify since you might want to know what the *minimal* knowledge base is that's just enough to get by. Unfortunately, no one really knows what that might be. Of course, you have to know calculus, but somehow you absolutely need to know much more calculus than will actually be used here. Why? As a scientist, all I can say is I observe this to be the case. A true explanation[1] would seem to require an amazing amount of knowledge about how a human being acquires and maintains control over, well, an amazing amount of knowledge. But that explanation may well have something to do, for example, with how we go about understanding what a derivative is. The point is that learning the definition and perhaps the basic properties of derivatives is not going to do the trick for most folks. One has to continue to learn in greater depth before fully grasping the basics. Or as another saying goes, you only learn what course n is about by taking course $n + 1$.

Besides calculus, here is a short shopping list of prerequisites: linear algebra including tensor products, modern algebra (some keywords: rings, groups, free,

---

[1] I leave the definition of this as the first of many exercises for the reader.

homomorphisms, ideals), category theory up to functors and natural transformations (but as a language written in diagrams and not so much as a theory), and classical differential geometry. For the last topic the outstanding reference is [49] though other presentations are available, such as the companion volume [76] of this volume. Also, it is traditional to throw an escape clause into the deal: The reader should have mathematical maturity. Far be it from me to break from such well-established tradition.

The organization of the specific contents of this book and further relevant information are given in Chapter 1, the introductory chapter. But briefly, let me say that this book itself is just an introduction to quantum principal bundles.

The approach used here is mathematical in some sense that has gradually evolved over time from Cauchy's days, reaching a sort of climax—or anticlimax according to some—in the Bourbaki school. Personally, I am a great admirer of the 20th-century Russian school, which did not make such a big fuss about distinguishing mathematics and physics as other modern schools did. And I do have great respect for the accomplishments of Bourbaki & Co. in the 20th century. I only wish they had been even more ambitious.

As for acknowledgments, they extend way back. To give some context to this complicated question, let me explain a bit about where I come from. You see, I work in Mexico, a country in which I was not born. So once after beginning the very first day of a course here and explaining its general outline, I asked my students if there were any questions. And one student asked (*en español, por cierto*) where I come from. Of course, I meant questions about the course, not about me. So it was not an appropriate question. But after thinking about it a moment, I felt it was an appropriate question about my scientific background. So I answered: I come from Hilbert. It is not a complete answer, but I deserve partial credit since it is partially true. My acknowledgments go back even further, but clearly Hilbert is up there high on the list even though I never knew him. But some of my first teachers did know him or were indirectly influenced. I owe a great debt to them, especially J. Alperin, I. Kaplansky, A. Liulevicius, S. Mac Lane, and R. Welland, all of whom had a clarity of exposition that was thrilling and captivating for me. They set the standard for me early on. More recently, I have benefited greatly from my contacts with many colleagues but especially (again in alphabetical order) from J. Cruz Sampedro, L. Gross, B. Hall, L. Thomas, and C. Villegas-Blas, who have all reinforced the vision that mathematics is much, much more than formal manipulations of symbols—and that physics provides an illuminating beacon for identifying important mathematical research. These lists are not complete by any means. Think of them as my team's starting players. But I also had great strength on the bench, far too many to list. Thanks to all the unnamed heroes, too!

But as far as the project behind this particular book is concerned, my unending gratitude goes to Micho Đurđevich, who opened my eyes to the wonderful world of noncommutative geometry and quantum groups. For many years we tried to find common ground for our differing approaches to the infinite-dimensional world of mathematical physics, always with Micho's probing curiosity and marvelous patience. And eventually we did hit pay dirt. Professionally, Micho has been a colleague and a collaborator *par excellence*. It is more than wonderful to count him as a great friend as well.

As for specific support for this book, I wish to thank Charles Dunkl for providing a copy of [20]. Also, the comments of the anonymous referees helped me to clarify my presentation of various points. The people at Springer, especially my editor, Donna Chernyk, get a lot of my thanks too.

As for errors, omissions, confusions, and other misdemeanors, I claim full innocence based on my present ignorance (= lack of knowledge) of any such offenses but plead guilty to them all and appeal for mercy. *¡Culpa mía! Nadie es perfecto*. My humble request is that all recriminations be limited to a message informing me of my trespasses, which are mine and mine alone. I meant no harm and hope no offense is taken.

Guanajuato, Mexico   Stephen Bruce Sontz
September 2014

Contents

Abbreviations

FODC   First-order differential calculus
HODC   Higher-order differential calculus
LIVF   Left invariant vector field
QCD   Quantum connection displacement
QED   Quantum electrodynamics
QPB   Quantum principal bundle
SR   Structure representation

Chapter 1
Introduction

What I cannot create I do not understand.

— Richard Feynman in [34]
"Feynman's Office: The Last Blackboards"

These books are based on various courses I have given plus a substantial amount
of material I studied to fill in gaps in my own personal knowledge of some themes
in modern geometry. By tradition, the geometry comes in at least two seemingly
distinct forms. First, there is the geometry based on smooth manifolds, which we
will call classical differential geometry. This goes back to the days of Apollonius
and Euclid. Second, there is the more recent noncommutative geometry, sometimes
also known as quantum geometry. Some of the principal founders here are Connes
[13] and Woronowicz [83]. My interests in these two fields developed independently
for some time, but now my intention is to view them as strongly related. Indeed, the
classical theory is simultaneously a primary motivating example for the quantum
theory as well as being a special case of it.

Moreover, from my perspective, an essential aspect of classical differential
geometry is the role of Lie groups as symmetry groups in that theory, and similarly
an essential aspect in the study of noncommutative (or quantum) spaces is the role
of a new sort of symmetry (played by quantum groups) in that theory. Since precise
definitions of both quantum space and quantum group can vary from author to
author, the real point of the above analogy between classical and quantum geometry
is that this should be a guideline for finding adequate definitions for these newer
concepts. In some sense, quantum groups are an extension of F. Klein's Erlangen
program to the quantum theory. In some expositions, quantum groups appear as a
special case of quantum spaces, just as Lie groups can be viewed as a special case
of a smooth manifold. So these two structures, spaces and groups, are intimately
related, and hence neither will be discussed here in total isolation from the other. But
that does not mean that this is an original point of view. Far from it! For example,
have a look at Woronowicz's comments in the introduction of [86] and Manin's
book [57], especially Chapter 4.

A most important relation in classical differential geometry between Lie groups
and smooth manifolds is embodied in the theory of principal fiber bundles. Of
course, this itself is a particular case of G-spaces. But in these books, ultimately I
focus much attention on the particular case of principal bundles, both classical in the

© Springer International Publishing Switzerland 2015
S.B. Sontz, *Principal Bundles*, Universitext, DOI 10.1007/978-3-319-15829-7_1

companion volume [76] and quantum in the present volume. This is due to several reasons, one of which is the rich geometric structure of principal bundles that arises from the connections on them. To clarify the importance of this remark, notice that the study of differential manifolds per se yields a rich and wonderful theory known as differential topology. The word "topology" is well chosen since the standard geometric structures do not appear at this level although topology can be considered (again per the Erlangen program) as a type of geometry. But the geometric structures I am thinking of go back to the etymological origins of "geometry" (Earth measurement), namely, ideas concerning distance, angle, direction, and parallelism. From these arise two basic concepts of calculus: area and tangents. While classical principal bundles do not support all of these structures, there is geometry encoded in the "infinitesimal" structure of the exterior differential calculus and if there is also a connection on the bundle, in parallel transport, curvature, and covariant derivatives. We will see in particular that a quantum principal bundle can have differential calculi on its spaces as well as a quantum connection with its associated covariant derivatives and curvature. All of these are basic geometric structures.

Another reason for focusing on principal bundles is well known today: They are the structure one uses to define and study gauge fields, a central concept in modern physics. Perhaps less well known is that the gauge fields of most interest in physics lately are quantum fields, which do not fit into the classical theory of principal bundles. And in this regard, the story is far from over. A fully developed theory of quantum gauge field theory—that is both mathematically rigorous and physically nontrivial—remains an important challenge for future research.

While the above remarks describe to some extent my point of view, more must be said about the theory presented here on the quantum side. (On the classical side, the dust has long since settled on these matters.) I adopt a viewpoint based on both the theory of compact quantum groups expounded by Woronowicz in [84] as well as the corresponding theory of quantum principal bundles found in some of Đurđevich's papers, especially in [22]. They go together hand in hand, which is important (at least for me!) since I want to exhibit an inseparable interdependence between quantum symmetries and quantum spaces. And this is provided in the papers of these two researchers. It is also noteworthy that both of the authors, together with collaborators, have developed rather rich theories of these structures. Nonetheless, I will not be presenting the full power of either theory.

One motivation for writing this book is that the work of Woronowicz and Đurđevich appears only in research papers and research-level monographs, and I thought it was time to present some of that in one place in a style accessible to nonexperts. Even though this is only an introduction to that work, I think that this book is the first attempt at such an introduction, especially concerning the theory of quantum principal bundles. Other viewpoints on quantum groups and quantum bundles can also be found in the literature, but in all due respect it seems to me that these two researchers have presented a critical mass of results with plenty of geometric (and not simply algebraic or topological) content. Remarks are given in the notes about those other approaches.

Another motivation is closer to home. After writing a paper (see [31]) in collaboration with Micho Đurđevich, in which we extensively use the theory of quantum principal bundles in order to understand Dunkl operators in a new way, namely, as the covariant derivatives of a quantum connection defined in a certain quantum principal bundle, I found much interest expressed by a number of colleagues in our work. But our paper (as almost any research paper) is not self-contained, nor was it ever intended to be. And I fear that the initial interest of my colleagues often flagged when recognition was made of the difficulty in acquiring the rather specific background needed to read that paper. So this book also serves to provide in one place the necessary prerequisites (and quite a bit more, of course) for understanding that paper. I present the major result of that paper in the last chapter.

The organization of the contents starts with a rather direct presentation of almost all of the material in the paper [86] written by S.L. Woronowicz on differential calculus for quantum groups (Chapters 2–11) together with various interludes, for example, Section 10.2, for the presentation of some of the material in [21] written by M. Đurđevich on universal differential algebras and the quantum germs map. The intent is to slow down the pace and explain things in much more detail than one expects (or gets!) in a research paper for experts. After all, experts are not my principal audience, since they already know this stuff!

The *raison d'être* of this book, as expressed in its title, is Chapter 12, where I discuss quantum principal bundles by following mainly M. Đurđevich's paper [22] on this topic. However, even so, I use a variant of the definition of a quantum principal bundle instead of that given in [22]. This again is done at a rather slow pace, and much effort is put into showing how this theory is inspired by the classical case of principal fiber bundles. I expect it to be a rather tedious drudge for the experts. It is important to remark that even though Chapter 12 weighs in at around 100 pages, it is just an introduction.

Even though Chapter 12 is the major goal of this book, the road to it is, in my opinion, even more important. My point is that a key ingredient in the modern approach to noncommutative geometry is the introduction of differential calculi and the appreciation of the fact that these calculi are not unique. This is why I have included the beautiful theory found in [86] of the braided exterior calculus, which is not strictly necessary for the rest of the text. After all, some of the original ideas about noncommutative geometry go back to Murray and von Neumann in [59–61], and [62], but it has been the works by Connes and Woronowicz that have given much of the flavor to current research by showing how differential calculi work in the noncommutative world. Also, the introduction of plenty of examples of quantum groups of interest in physics by Faddeev and his collaborators has been another critically important aspect of the modern approach. If I did not also mention the importance of the works of Drinfeld, Jimbo, and Manin (among others!) at this juncture, I would deserve to be duly chastised.

It is also important to mention what is not to be found in this book. The seminal works of Connes (see [13]), including cyclic cohomology, are not presented. Woronowicz's theory of compact matrix pseudogroups (quantum groups) in [84], and in particular the theory of their co-representations, is mentioned only in passing even though his paper [86] seems to have been written as a sequel to [84]. There is no essential use made of C^*-algebras, which is part of the treatment in [84]. In fact, you will never see a definition of quantum group. The structure corresponding to the classical concept of a group will be a Hopf algebra with an invertible co-inverse. The co-inverse is also called the antipode.

References will be found throughout that give some idea of the relationship of the topics covered here with other approaches and more recent related developments as well as some totally unscholarly comments about the historical development. Most chapters end with notes to give some perspective on that chapter. Unfortunately, completeness is an ideal rarely seen in practice. No doubt even these comments will not be as complete as some (including me) might like. The references are meant to help the reader in further studies, but the author's responsibility for doing this task has diminished in the 21st century as this burden is falling more and more on the reader's shoulders. Not only can the reader search the Internet, but this can be done years after this book is published. This allows the reader access not only to future theoretical developments not available to the author, but also to evaluations and perspectives that do not presently exist.

The chapters should be read basically in sequence. But Chapter 13 on finite groups can be read at any time. It is placed near the end so that it does not interrupt the main flow of the topics. The more adventurous reader should feel free to take a dip at any time, even without knowing all the definitions, to see this commutative material, which motivates much of the noncommutative material. It is the shallow end of the pool (to maintain a metaphor) even though it is almost the last topic treated. Finally, Chapter 14 presents the main results on Dunkl operators from a joint work in [31] with M. Đurđevich. It depends on results from all of the prior chapters, though not on everything previously presented.

I close with two appendices on a minor point and on Hopf algebras. The latter serves as an occasion to read into the record the many identities as well as various basic results that are used freely throughout the book. Other sources can readily serve the same purpose; that is left entirely to the reader's discretion.

None of the material of this book is original. However, I may have been lucky enough to find some clearer ways of explaining the known. And this is not a history. I respect the work of historians, particularly if it is good work, but I am not academically prepared to engage in that sort of endeavor myself. Inevitably, one feels obliged to make some sort of historical comment. When I indulge this "obligation," my remarks are not to be taken as anything other than folk legends. (Of course, some folk legends do turn out to be right!)

The references I give are those that I have found useful for one reason or another. I am never ascribing any reference to be the original source, with the unique exception of the original definition of the braid group. For all I know or care, Gauss intuited it all. He certainly had the smarts to do just that. Sometimes one prefers a certain source because one first learned the material there. I have never read Wigner's famous 1939 paper on representations of the Poincaré group. No doubt it is excellent, though I am not the one to judge that. I learned that topic unexpectedly while reading a rather unknown physics text, which has consequently become a personal favorite. If I were discussing that topic in this book, that text would be my reference for it, not Wigner's paper. So if some researchers feel overlooked, I am sorry, but my mission here is not to hand out such recognition to one and all as is their due. Besides, such persons are in good company with Eugene Wigner!

Exercises? Yes, there are exercises. Some are rather straightforward, some are puzzles, and some are in between. And yes, by a well-established tradition they are an integral part of the book. I have even done most of them or at least convinced myself that I could do them. I would expect the reader to do more or less the same. After all, the banner motto of the book (given at the beginning of this chapter) declares essentially that understanding is not a spectator sport. Some exercises are quite difficult but are not used subsequently in the text, though they may be used in other exercises. The reader should feel challenged by such exercises, but their solution is not indicated, so that the reader can continue without time pressure to ponder them. Some of the exercises are more routine and will be used later on in the text. However, some exercises are quite difficult and will be used later in the text! So it seems that I am requesting the reader to write some of my book for me. Well, yes and no. Yes, it would be best if the reader does write some of the book. But no, this is not absolutely necessary, since requested material will appear on the Internet from time to time, including the inevitable error corrections. However, a habit that many (including me!) have to break is the demand for instant gratification. It is beneficial for researchers to get used to the slowness of the processes of acquiring and producing new knowledge. In this sense, some aspects of the Internet are not desirable.

I guess that all objects (algebras, modules, etc.) are vector spaces over the field $\mathbb{C}$ of complex numbers unless otherwise stated and that all (or at least most) maps are linear over this field. Of course, $*$-operations are antilinear. However, the motivating examples for much of this material come from classical differential geometry, where the vector spaces are usually over the field of real numbers although complex geometry is, of course, an exception. The restriction to the particular field $\mathbb{C}$ has two motivations. First, it is *the* field for doing physics. This was not so obvious until the development of quantum mechanics in the 20th century. It has even been said, half-seriously, that the discovery of the complex numbers was the major achievement in 20th-century physics. Second, it has a $*$-operation, namely, complex conjugation, which gives the material of this book an extra flavor—as well as an extra chapter.

The symbol $\otimes$ always means the tensor product in the category under consideration. The underlying vector space of the tensor product is the tensor product of the underlying vector spaces over the field of complex numbers. Tensor products over other objects will be written with that object as a subscript, such as $\otimes_A$, unless otherwise indicated. We will use one convention that is not completely standard when applying the canonical isomorphisms between $V \otimes \mathbb{C}$ and V, where V is a complex vector space. We will use $\cong$ rather than the more usual $=$ to identify corresponding elements in these two vector spaces. For example, we write $v \otimes 1 \cong v$ for $v \in V$. We also use similar notation, namely, $1 \otimes v \cong v$, to identify elements in $\mathbb{C} \otimes V$ and in V. Here is a useful fact about the tensor product that the reader may not have seen. Suppose that U, V, and W are vector spaces (over any field, actually) and that $T : V \to W$ is linear. Then $\ker(id_U \otimes T) = U \otimes \ker T$, where $id_U : U \to U$ is the identity map. If you have never seen this, prove it! However, this is something the reader may have seen before but does not recognize in this notation. It just says that in the category of vector spaces the functor $U \otimes \cdot$ is left exact.

Throughout this book $\mathcal{A}$ will denote a Hopf algebra over the field $\mathbb{C}$ of complex numbers unless otherwise indicated. We make no restriction on its dimension. We will also always impose the hypothesis that the co-inverse κ (often called in other texts the antipode and denoted by S) is an invertible map $\kappa : \mathcal{A} \to \mathcal{A}$. However, the reader may note that some results do not depend on this hypothesis. While we will not focus our attention on compact matrix quantum groups (see [84]), they are included as a special case of what we do here since in that case κ is invertible (see Proposition 11.3). The basic language, properties, and identities of Hopf algebras are presented in Appendix B. There are many different definitions of "quantum group," and such a Hopf algebra $\mathcal{A}$ has something to do with most of those definitions. (A good reference for this is [80].) However, for our purposes, we need not concern ourselves with a rigorous definition of quantum group. The Hopf algebra $\mathcal{A}$ will serve here as an intuitive analog of a group in the noncommutative setting, and that is all that matters.

I use Sweedler's notation for co-multiplication and the similar notation for co-actions, of which co-multiplication is a special case. See Appendix B for more details. This notation has the virtue of eliminating summations and subindices at the price of a certain queasiness for the uninitiated reader. For those not (yet!) familiar with this notation, take heart. It really is okay. At first, just expand Sweedler's notation into the more standard form and reassure yourself that indeed it all does make sense. Slowly but surely, you will be weaned from your need to do this, and eventually you will even wonder what the fuss was all about. Just as with many other things in life, practice makes perfect. But none of us is perfect! Sometimes I slip back into the standard notation myself when doing rough calculations, just to make sure that everything is all right.

Chapter 2
First-Order Differential Calculus

The construction of an adequate differential calculus for a given quantum space (generalizing the de Rham theory in the exterior algebra associated to a smooth manifold) is a nontrivial problem. And the resulting theory was not as one had anticipated and was even, at first, considered to be defective in some intuitive sense. The first step is to construct a generalization of the de Rham differential $d : C^\infty(M) \to \Gamma(T^*M)$, which will be called *a* first-order differential calculus (FODC). The use of the indefinite article is essential and is really the difference with the classical de Rham theory, which is *functorial* and as such unique. The noncommutative generalization is *axiomatic* and so, in general, allows for many nonisomorphic realizations of a first-order differential calculus for a given quantum space. This situation, which we can now see as a strength of noncommutative theory, was once considered by Woronowicz in [86] to be ". . . in unpleasant contrast with the more satisfactory situation in classical differential geometry . . . ."

2.1 Basic Theory

Definition 2.1. Let $\mathcal{A}$ be an algebra with identity element, denoted by $1 \in \mathcal{A}$. Let Γ be a bimodule over $\mathcal{A}$. [This means that $\mathcal{A}$ acts on Γ on the left and on the right and these two actions are compatible in the sense that $a(\omega b) = (a\omega)b$ for all $a, b \in \mathcal{A}$ and all $\omega \in \Gamma$. Moreover, $1\omega = \omega 1 = \omega$ for all $\omega \in \Gamma$.] Also, let $d : \mathcal{A} \to \Gamma$ be a linear map.

Then we say that (Γ, d) is a *first-order differential calculus* over $\mathcal{A}$ provided that these two properties are satisfied:

- (Leibniz rule) For all $a, b \in \mathcal{A}$, we have $d(ab) = (da)b + a(db)$.

© Springer International Publishing Switzerland 2015
S.B. Sontz, *Principal Bundles*, Universitext, DOI 10.1007/978-3-319-15829-7_2

- (Standard form) Every element $\omega \in \Gamma$ can be expressed as

$$\omega = \sum_{k=1}^{K} a_k (db_k)$$

for some (not necessarily unique) elements $a_k, b_k \in \mathcal{A}$, where K is a positive integer.

We also say that the map d is the *differential* of the FODC.

Remark. For the time being, this algebra need not have a $*$-structure. However, we will discuss that case later.

The bimodule $\Gamma = \{0\}$ can be made in a unique way into an FODC over $\mathcal{A}$, which is called the *trivial* FODC.

Notice that the first property is exactly Leibniz's rule and *not* some generalized or deformed Leibniz rule. The second property is equivalent to saying that Ran d generates Γ algebraically as a left $\mathcal{A}$-module. In other words, Γ is big enough to accommodate the range of d, but no larger.

The first property, Leibniz's rule, uses the bimodule structure of Γ, while the second only uses the left module structure. In fact, it is essential to have a bimodule structure in order to be able to make sense of the right-hand side of Leibniz's rule. So we have an apparent asymmetry in this definition.

The notation Γ reminds us that in classical differential geometry, this object represents the sections of a vector bundle.

Exercise 2.1. *Suppose that* (Γ, d) *satisfies all the properties of an FODC with the possible exception of the standard form property. Prove that* (Γ, d) *satisfies the standard form property if and only if it satisfies*

- *(Right standard form) Every element* $\omega \in \Gamma$ *can be expressed as*

$$\omega = \sum_{k=1}^{K} (db_k) a_k$$

for some (not necessarily unique) elements $a_k, b_k \in \mathcal{A}$, *where* K *is a positive integer. (This says that* Ran d *generates* Γ *algebraically as a right* $\mathcal{A}$-*module.)*

Hint: Use Leibniz's rule.

The motivating example for FODCs comes from classical differential geometry. There for any smooth (i.e., C^∞) manifold M, one has the commutative algebra $C^\infty(M)$, the real smooth functions on M, and the (unique!) exterior derivative $d : C^\infty(M) \to \Gamma(T^*(M))$ from de Rham theory, where $\Gamma(T^*(M))$ is the real vector space and $C^\infty(M)$-bimodule of smooth sections of the cotangent bundle $T^*(M)$ of M. Of course, there are nontrivial examples of FODCs for noncommutative algebras $\mathcal{A}$. But to get this generality, one no longer has a *unique* Γ associated to a given $\mathcal{A}$.

Rather, an FODC is simply any mathematical structure satisfying certain axioms, as specified in the above definition. This gives one more leeway even in the classical case; that is, there can be other nontrivial examples of an FODC over the algebra $C^\infty(M)$ besides the functorially given $d : C^\infty(M) \to \Gamma(T^*(M))$.

Clearly, the collection of all FODCs over a fixed, given algebra $\mathcal{A}$ can be made into the objects in a category by appropriately defining the morphisms of these objects. Specifically, a *morphism of FODCs* is a morphism $\alpha : \Gamma \to \Gamma'$ of bimodules over $\mathcal{A}$ such that $\alpha d = d'$, where (Γ', d') is also an FODC over $\mathcal{A}$. And following the language of categories, an *isomorphism of FODCs* is any morphism that is invertible, while two such calculi are *isomorphic* if there exists some isomorphism between them.

Given an algebra with unit, one would like to be able to construct FODCs over it. We now define at least one such calculus. Let $\otimes$ denote the tensor product of vector spaces over $\mathbb{C}$. Then the multiplication operation of an algebra $\mathcal{A}$ is (or induces, depending on your point of view) a linear map

$$m : \mathcal{A} \otimes \mathcal{A} \to \mathcal{A}.$$

Notice that both $\mathcal{A} \otimes \mathcal{A}$ and $\mathcal{A}$ can be made into $\mathcal{A}$-bimodules by using m itself to define the left and right actions. It then turns out by the associativity of the multiplication that m is an $\mathcal{A}$-bimodule morphism. (This elementary assertion is proved in Appendix A.) So $\mathcal{A}^2 := \operatorname{Ker} m$ is an $\mathcal{A}$-subbimodule of $\mathcal{A} \otimes \mathcal{A}$. In particular, $\mathcal{A}^2$ is an $\mathcal{A}$-bimodule in its own right. For every $b \in \mathcal{A}$, define

$$Db := 1 \otimes b - b \otimes 1 \in \mathcal{A} \otimes \mathcal{A}.$$

Since $m(Db) = 0$, we have that

$$D : \mathcal{A} \to \mathcal{A}^2.$$

According to [7], D is called the *Karoubi differential*. In fact, it is a differential and this relatively simple construction gives us an FODC.

Proposition 2.1. $(\mathcal{A}^2, D)$ *is a first-order differential calculus over* $\mathcal{A}$.

Proof. We have just shown that $\mathcal{A}^2$ is an $\mathcal{A}$-bimodule. Clearly, D is linear. Next, for any $a, b \in \mathcal{A}$, we have that

$$
\begin{aligned}
(Da)b + a(Db) &= (1 \otimes a - a \otimes 1)b + a(1 \otimes b - b \otimes 1) \\
&= 1 \otimes ab - a \otimes b + a \otimes b - ab \otimes 1 \\
&= 1 \otimes ab - ab \otimes 1 = D(ab),
\end{aligned}
$$

which shows Leibniz's rule.

Next, suppose that $\omega = \sum_k a_k \otimes b_k \in \mathcal{A}^2 = \operatorname{Ker} m$ is arbitrary. So $\sum_k a_k b_k = 0$. Then we have

$$\sum_k a_k \, D b_k = \sum_k a_k (1 \otimes b_k - b_k \otimes 1) = \sum_k a_k \otimes b_k - \sum_k a_k b_k \otimes 1$$

$$= \sum_k a_k \otimes b_k = \omega,$$

which shows that ω has the standard form. ∎

Now this innocent-looking little example has an amazing property, namely, that the FODC $(\mathcal{A}^2, D)$ has a universal property given in the next result.

Theorem 2.1. *Let* $\mathcal{N}$ *be an* $\mathcal{A}$-*subbimodule of* $\mathcal{A}^2$. *Define* $\Gamma_{\mathcal{N}} := \mathcal{A}^2/\mathcal{N}$, *the quotient* $\mathcal{A}$-*bimodule. Let* $\pi_{\mathcal{N}} : \mathcal{A}^2 \to \Gamma_{\mathcal{N}}$ *be the quotient map. Define* $d_{\mathcal{N}} = \pi_{\mathcal{N}} \circ D : \mathcal{A} \to \Gamma_{\mathcal{N}}$. *(We drop the subscript* $\mathcal{N}$ *when context indicates which* $\mathcal{A}$-*subbimodule is being used.) Then* $(\Gamma_{\mathcal{N}}, d_{\mathcal{N}})$ *is an FODC over* $\mathcal{A}$.

Moreover, any FODC over $\mathcal{A}$ *is isomorphic to* $(\Gamma_{\mathcal{N}}, d_{\mathcal{N}})$ *for some* $\mathcal{A}$-*subbimodule* $\mathcal{N}$ *of* $\mathcal{A}^2$.

Remark. We note that in the last statement of the proposition, $\mathcal{N}$ need not be unique. So we have a map from the set of $\mathcal{A}$-subbimodules of $\mathcal{A}^2$ *onto* the set of isomorphism classes of FODCs over $\mathcal{A}$, but this map need not be *one-to-one*. (Find an example where this is not one-to-one.) Notice that the trivial FODC results from the choice $\mathcal{N} = \mathcal{A}$. Because of its last statement, this theorem is a big step toward the classification (up to isomorphism) of all the FODCs over a given algebra $\mathcal{A}$ with unit.

Proof. Clearly, $\Gamma \equiv \Gamma_{\mathcal{N}}$ is an $\mathcal{A}$-bimodule by construction. Also, d is the composite of two linear maps and so is itself linear. For Leibniz's rule, we note that

$$d(ab) = \pi(D(ab))$$

$$= \pi\big((Da)b + a(Db)\big)$$

$$= \big(\pi(Da)\big)b + a\big(\pi(Db)\big)$$

$$= (da)b + a(db),$$

where in the next-to-last equality, we used that π is an $\mathcal{A}$-bimodule morphism.

Since π is an epimorphism, any element in Γ can be written as $\pi(\omega)$ for some $\omega = \sum_k a_k \otimes b_k \in \mathcal{A}^2$. But we have already shown in the previous proposition that in this case $\omega = \sum_k a_k \, D b_k$. So it follows that

$$\pi(\omega) = \pi(\sum_k a_k \, D b_k) = \sum_k a_k \pi(D b_k) = \sum_k a_k \, d b_k, \qquad (2.1)$$

which realizes an arbitrary element in Γ in standard form.

All told, we have now shown that $(\Gamma, d) = (\Gamma_{\mathcal{N}}, d_{\mathcal{N}})$ is an FODC over $\mathcal{A}$.

Now to establish the last statement of the proposition, we start with an FODC over $\mathcal{A}$, denoted by (Δ, δ). First, we wish to define an $\mathcal{A}$-bimodule morphism π' : $\mathcal{A}^2 \to \Delta$ by using a variant of the formula (2.1) given in the first half of this proof for the map π. We note that π' will not be a quotient map. So we define

$$\pi' \left(\sum_k a_k \otimes b_k \right) := \sum_k a_k \delta b_k$$

for all $\sum_k a_k \otimes b_k \in \mathcal{A}^2$; so $\sum_k a_k b_k = 0$ holds. By definition, we clearly have that $\sum_k a_k \delta b_k \in \Delta$. To show that this definition does not depend on the choice of the elements a_k and b_k, suppose that $\sum_k a_k \otimes b_k = 0 \in \mathcal{A}^2 \subset \mathcal{A} \otimes \mathcal{A}$. Apply $id \otimes \delta$ to this equality to get $\sum_k a_k \otimes \delta b_k = 0 \in \mathcal{A} \otimes \Delta$. The left action of $\mathcal{A}$ on Δ is an (unnamed) linear map $\mathcal{A} \otimes \Delta \to \Delta$. Applying this map to the previous equation gives $\sum_k a_k \delta b_k = 0 \in \Delta$, as desired.

Next, we will show that π' is an $\mathcal{A}$-bimodule morphism. Clearly, it is linear. To show that it is a right $\mathcal{A}$-module morphism, we let $\omega = \sum_k a_k \otimes b_k \in \mathcal{A}^2$, where $a_k, b_k \in \mathcal{A}$ and $\sum_k a_k b_k = 0$. We also take $c \in \mathcal{A}$. Then we have

$$\pi'(c\omega) = \pi'(c \sum_k a_k \otimes b_k)$$

$$= \pi'(\sum_k c a_k \otimes b_k)$$

$$= \sum_k (c a_k) \delta b_k$$

$$= c \sum_k a_k \delta b_k$$

$$= c \pi'(\sum_k a_k \otimes b_k)$$

$$= c \pi'(\omega),$$

which shows that π' is a left $\mathcal{A}$-module morphism.

We also have

$$\pi'(\omega c) = \pi' \left((\sum_k a_k \otimes b_k) c \right)$$

$$= \pi'(\sum_k a_k \otimes b_k c)$$

$$= \sum_k a_k \delta(b_k c)$$

$$= \sum_k (a_k(\delta b_k)c) + \sum_k (a_k b_k \delta c)$$

$$= (\sum_k a_k(\delta b_k))c + (\sum_k a_k b_k)\delta c$$

$$= (\sum_k a_k(\delta b_k))c$$

$$= (\pi'(\sum_k a_k \otimes b_k))c$$

$$= \pi'(\omega)\,c,$$

which shows that π' is a right $\mathcal{A}$-module morphism.

Next, we show that π' is an epimorphism. First, we note that since (Δ, δ) is an FODC over $\mathcal{A}$, we can write an arbitrary element $\sigma \in \Delta$ is standard form, namely, $\sigma = \sum_k a_k(\delta b_k)$ for some $a_k, b_k \in \mathcal{A}$. Now, we clearly have that

$$\sum_k a_k \otimes b_k - \sum_k a_k b_k \otimes 1 \in \mathcal{A}^2.$$

Using $\delta 1 = 0$ (which follows from Leibniz's rule), we calculate

$$\pi'(\sum_k a_k \otimes b_k - \sum_k a_k b_k \otimes 1) = \sum_k a_k \delta b_k - \sum_k a_k b_k \delta 1 = \sum_k a_k \delta b_k = \sigma.$$

This shows that $\pi' : \mathcal{A}^2 \to \Delta$ is an epimorphism, as claimed.

Now we define $\mathcal{N} := \mathrm{Ker}\,\pi'$. Since π' is an $\mathcal{A}$-bimodule morphism, its kernel $\mathcal{N}$ is an $\mathcal{A}$-subbimodule of $\mathcal{A}^2$. This puts us in the context of the first part of this proof, and so $(\Gamma, d) \equiv (\Gamma_{\mathcal{N}}, d_{\mathcal{N}})$ as defined there is an FODC over $\mathcal{A}$. (We also write π for $\pi_{\mathcal{N}}$.) We want to show that (Δ, δ) is isomorphic to (Γ, d) as an FODC over $\mathcal{A}$. Since π' is an $\mathcal{A}$-bimodule morphism, $\mathcal{N}$ is an $\mathcal{A}$-subbimodule and π' induces an $\mathcal{A}$-bimodule isomorphism

$$\tilde{\pi}' : \mathcal{A}^2/\mathcal{N} = \Gamma \to \Delta.$$

It only remains to prove that the differentials d and δ correspond to each other under this isomorphism; that is, $\tilde{\pi}'d = \delta$. So take any element $a \in \mathcal{A}$. Then

$$\tilde{\pi}'da = \tilde{\pi}'\pi Da = \pi' Da = \pi'(1 \otimes a - a \otimes 1) = 1(\delta a) - a(\delta 1) = \delta a,$$

which shows that δ is identified under the isomorphism $\tilde{\pi}'$ with $d = \pi \circ D$. Here we have used the identity $\tilde{\pi}'\pi = \pi'$, which one can see, for example, by looking at the associated diagram. And this finishes the proof. ∎

Exercise 2.2 (Optional). *Express this universal property in terms of diagrams or adjoint functors. Or both!*

2.2 An example

This rather pretty example comes from [48]. We will spell it out in more detail here. Other examples of FODCs will be presented later.

This is actually a class of examples of FODCs for the algebra $\mathcal{A} = \mathbb{C}[x]$ of polynomials in x. This is a commutative algebra with 1. The bimodule Γ will also be a copy of $\mathcal{A}$ but will be denoted as $\alpha \mathcal{A}$ to indicate that as a right $\mathcal{A}$-module it is free on one nonzero generator α. The left $\mathcal{A}$-module structure on Γ may not be what the reader is expecting. We let $p \in \mathcal{A}$ be any element, that is, any polynomial with complex coefficients. Then the left $\mathcal{A}$-module structure is uniquely determined by the condition that $x\alpha := \alpha p$, where $x \in \mathcal{A}$ is simply a (well-known) polynomial of degree 1. (The reader may have been expecting the particular bimodule structure that results from taking $p = x$. The reader is also invited to prove that this condition determines a *unique* $\mathcal{A}$-bimodule structure on Γ.) To indicate that this bimodule structure depends on the choice of p, we will write Γ_p instead of Γ. To get an FODC, we need also to define a linear map $d_p : \mathcal{A} \to \Gamma_p$ and prove two things: Leibniz's rule and that every element in Γ_p has standard form. Since the monomials x^n are a basis of $\mathcal{A}$, it suffices to define d_p on each of them. But Leibniz's rule forces $d_p 1 = 0$ as well as the definition of $d_p(x^n) \equiv d_p x^n$ for all $n \geq 2$ once we have the definition of $d_p x$. And we simply wish to define $d_p x := \alpha$. So the general definition is

$$d_p x^n := \sum_{j=0}^{n-1} x^j \, \alpha \, x^{n-1-j}$$

for $n \geq 0$. (The reader should understand how the case $n = 0$ gives us $d_p 1 = 0$.) From now on, we will write $d_p x$ instead of α. Note that we already know that $d_p x \neq 0$.

To show Leibniz's rule, it suffices to check it for $f = x^n$ and $g = x^m$. So we calculate

$$d_p x^n \, x^m + x^n \, d_p x^m = (\sum_{j=0}^{n-1} x^j \, d_p x \, x^{n-1-j}) \, x^m + x^n \sum_{k=0}^{m-1} x^k \, d_p x \, x^{m-1-k}$$

$$= \sum_{j=0}^{n-1} x^j \, d_p x \, x^{n+m-1-j} + \sum_{k=0}^{m-1} x^{n+k} \, d_p x \, x^{m-1-k}$$

$$= \sum_{l=0}^{n+m-1} x^l \, d_p x \, x^{n+m-1-l}$$

$$= d_p x^{n+m}$$

$$= d_p(x^n x^m).$$

The third equality may stump the reader at first. Think about it! Finally, the very definition of Γ_p shows that every element in it is in right standard form. So we have shown that (Γ_p, d_p) is an FODC over the algebra $\mathcal{A} = \mathbb{C}[x]$, But strangely enough, the choice of $p \in \mathcal{A}$ does not seem to play much of a role. At least, so far!

Proposition 2.2. *Suppose that $p, q \in \mathcal{A}$. Then Γ_p is isomorphic to Γ_q as FODCs if and only if $p = q$.*

Proof. The condition $p = q$ is clearly sufficient for the existence of an isomorphism. Thus, let's suppose that we have an isomorphism $\phi : \Gamma_p \to \Gamma_q$ of FODCs. So $d_p x$ (resp., $d_q x$) generates Γ_p (resp., Γ_q) as a free right $\mathcal{A}$-module. Since ϕ intertwines the differentials, we have $\phi(d_p x) = d_q x$; that is, ϕ maps a generator to a generator. But ϕ is also a right $\mathcal{A}$-module morphism, which implies that $\phi(d_p x \, f) = \phi(d_p x) f = (d_q x) f$ for all $f \in \mathcal{A}$. Now ϕ is also a left $\mathcal{A}$-module morphism, so that $\phi(x \, d_p x) = x \, \phi(d_p x)$. But the left side here is

$$\phi(x \, d_p x) = \phi(d_p x \, p) = (d_q x) p,$$

while the right side is

$$x \, \phi(d_p x) = x \, d_q x = (d_q x) q.$$

So we have the equality $(d_q x) p = (d_q x) q$ in Γ_q, which is a free right $\mathcal{A}$-module with generator $d_q x$. It then follows that $p = q$, as desired. ∎

We continue to discuss the FODC (Γ_p, d_p). Since Γ_p is a free right $\mathcal{A}$-module with generator $d_p x$, we can write $d_p f$ for any $f \in \mathcal{A}$ as $d_p f = d_p x (\partial_p f)$ for a *unique* element $\partial_p f \in \mathcal{A}$. The mapping $\partial_p : \mathcal{A} \to \mathcal{A}$ is clearly linear. So it suffices to analyze the action of ∂_p on the basis elements of $\mathcal{A}$, namely, the monomials x^n. One can readily check that $\partial_p 1 = 0$ and $\partial_p x = 1$. Neither of these two expressions depends on p. Also, it seems that this is some sort of partial derivative. Now, in general, we have

$$d_p x^n = \sum_{j=0}^{n-1} x^j \, d_p x \, x^{n-1-j} = \sum_{j=0}^{n-1} d_p x \, p^j x^{n-1-j} = d_p x \sum_{j=0}^{n-1} p^j x^{n-1-j},$$

where the second equality follows from the definition of the left $\mathcal{A}$-module structure. So the result is that

$$\partial_p x^n = \sum_{j=0}^{n-1} p^j x^{n-1-j}.$$

For example,

$$\partial_p x^2 = x + p. \tag{2.2}$$

So $\partial_p x^n$ depends on p for $n \geq 2$. Moreover, by comparing (2.2) with

$$\partial_p x \, x + x \, \partial_p x = 2x,$$

we see that Leibniz's rule holds for ∂_p if and only if $p = x$. Also, (2.2) shows that ∂_p can raise the degree of a polynomial and that ∂_p acting on a monomial is not necessarily a monomial.

Exercise 2.3. *Prove that ∂_p satisfies the following generalization of Leibniz's rule:*

$$\partial_p(fg) = \partial_p(f)g + f(p)\partial_p g,$$

where $f(p)$ is the result of substituting p into f in place of x or, in other words, the composition $f \circ p$ of polynomials. This exercise can be thought of as saying that the usual Leibniz rule becomes deformed as p moves away from x or, equivalently, that the generalized Leibniz rule moves toward the usual Leibniz rule as p approaches x.

Here are two well-known special cases of the operator ∂_q. For the first case, we take $p = qx$ for some $q \in \mathbb{C}$ with $q \neq 1$. Thus, for every integer $n \geq 0$, we have

$$\partial_p x^n = \sum_{j=0}^{n-1} p^j x^{n-1-j} = \sum_{j=0}^{n-1} (qx)^j x^{n-1-j}$$

$$= \sum_{j=0}^{n-1} q^j x^{n-1}$$

$$= \left(\frac{1-q^n}{1-q}\right) x^{n-1}$$

$$= [n]_q x^{n-1},$$

where we have introduced standard notation, $[n]_q := (1 - q^n)/(1 - q)$, the *q-deformation* of the integer n. Note that $\lim_{q \to 1} [n]_q = n$, in accordance with the comment above that the generalized Leibniz rule tends to the usual one as $p \to x$.

Exercise 2.4. *Show that for the choice $p = qx$ as above, we have $\partial_p = D_q$, where for all $f \in \mathcal{A} = \mathbb{C}[x]$, we define*

$$(D_q f)(x) := \frac{f(qx) - f(x)}{qx - x},$$

the well-known q-derivative operator *or* q-deformation of the derivative. *(The reader should also wonder about whether the polynomial in the denominator actually divides the polynomial in the numerator.) What happens if we allow q to be any polynomial in $\mathbb{C}[x]$, including the case $q \equiv 1$?*

For the second case, we take $p = x + c$, where $c \in \mathbb{C}$ with $c \neq 0$. In this case, we simply have

$$\partial_p x^n = \sum_{j=0}^{n-1} p^j x^{n-1-j} = \sum_{j=0}^{n-1} (x+c)^j x^{n-1-j}$$

and that seems to be the end of the matter. But no!

Exercise 2.5. *Show that for this choice of p, we have that $\partial_p = \Delta_c$, where*

$$(\Delta_c f)(x) := \frac{f(x+c) - f(x)}{c}$$

is the discrete difference operator *associated to c. What happens if we allow c to be any polynomial in $\mathbb{C}[x]$, including the case $c \equiv 0$?*

Here when $c \to 0$ (so that $p = x + c \to x$), the operator Δ_c tends to the standard derivative, which satisfies the standard Leibniz rule.

2.3 Notes

The fundamental idea behind an FODC is ancient. It would not have surprised Leibniz in the slightest. But who first thought about how the area of a rectangle changes when the two sides undergo independent changes? Maybe it was Archimedes or some other ancient scientist who lived even earlier. An FODC is a purely algebraic object with totally rigorous infinitesimals. And as such, it has found applications in algebra. For example, see Eisenbud's text [32]. As far as noncommutative geometry is concerned, it has now become folk wisdom that vector fields on a manifold M, which can be thought of algebraically as derivations of the commutative algebra of C^∞-functions on M (real- or complex-valued functions, as one wishes), are not the appropriate objects to be generalized to the noncommutative setting of quantum groups and associated quantum spaces. However, see [16] for this alternative approach of using derivations in the context of quantum matrix groups and associated objects as well as applications of this approach in [17] and [18]. In classical differential geometry, the differential forms are the objects dual to vector fields, and these forms do generalize nicely to give algebraic, infinitesimal structures in all of noncommutative geometry. This point of view in the noncommutative setting is emphasized in Connes' text [13]. We will see later how this concept comes back to the commutative setting of finite groups in Chapter 13 to give us a new way of introducing infinitesimal structures there.

In Section 2.2, we have seen for the first time a q-deformation. These give many, though by no means all, examples of objects in noncommutative geometry. We will comment more on q-deformations in the notes at the end of Chapter 12.

Chapter 3
First-Order Differential Calculus of a Hopf Algebra

3.1 Definitions

We now assume that $\mathcal{A}$ is a Hopf algebra. The *co-multiplication* in $\mathcal{A}$ is denoted by

$$\phi : \mathcal{A} \to \mathcal{A} \otimes \mathcal{A},$$

and the *co-inverse* (also known as the *antipode*) is denoted by

$$\kappa : \mathcal{A} \to \mathcal{A}.$$

In many texts the notation for ϕ is Δ and the notation for κ is S. In particular, $\mathcal{A}$ is an algebra and so it makes sense to talk about first-order differential calculi (FODCs) over it. However, we wish to have some relation of an FODC on $\mathcal{A}$ with the Hopf algebra structure of $\mathcal{A}$ as well. So we now do this.

Recall that Γ already has left and right actions on it by the algebra $\mathcal{A}$. We now look for a left co-action of $\mathcal{A}$ on Γ, where (Γ, d) is an FODC over $\mathcal{A}$. This new structure does not require as much structure on $\mathcal{A}$ and Γ as we have here; we see this in the following definition.

Definition 3.1. Let Γ be a vector space, and let $\mathcal{A}$ be a co-algebra. Then a *left co-action* of $\mathcal{A}$ on Γ (also called a *left co-representation* of $\mathcal{A}$ on Γ) is a linear map $\Phi_\Gamma : \Gamma \to \mathcal{A} \otimes \Gamma$ for which we have the following two commutative diagrams:

$$
\begin{array}{ccc}
\Gamma & \xrightarrow{\Phi_\Gamma} & \mathcal{A} \otimes \Gamma \\
{\scriptstyle \Phi_\Gamma} \downarrow & & \downarrow {\scriptstyle \phi \otimes id} \\
\mathcal{A} \otimes \Gamma & \xrightarrow{id \otimes \Phi_\Gamma} & \mathcal{A} \otimes \mathcal{A} \otimes \Gamma
\end{array}
\tag{3.1}
$$

© Springer International Publishing Switzerland 2015

S.B. Sontz, *Principal Bundles*, Universitext, DOI 10.1007/978-3-319-15829-7_3

and

$$\begin{array}{ccc} \Gamma & \xrightarrow{\Phi_\Gamma} & \mathcal{A} \otimes \Gamma \\ id_\Gamma \downarrow & & \downarrow \varepsilon \otimes id \\ \Gamma & \xrightarrow{\cong} & \mathbb{C} \otimes \Gamma. \end{array} \qquad (3.2)$$

Here $\phi : \mathcal{A} \to \mathcal{A} \otimes \mathcal{A}$ is the *co-multiplication* and $\varepsilon : \mathcal{A} \to \mathbb{C}$ is the *co-unit* of the co-algebra $\mathcal{A}$. One also says that a vector space Γ together with a given left co-action Φ_Γ is a *left $\mathcal{A}$-co-module*.

As usual, we only consider vector spaces over the complex numbers, but any field could have been used instead for this particular definition.

While a diagram often has the advantage of being easy to remember, sometimes the equivalent formula is more useful for doing calculations. The formula equivalent to (3.1) in Sweedler's notation is quite handy. So we write the left co-action as $\Phi_\Gamma(\omega) = \omega^{(0)} \otimes \omega^{(1)} \in \mathcal{A} \otimes \Gamma$ for all $\omega \in \Gamma$. The co-multiplication in this case is written as $\phi(\omega^{(0)}) = \omega^{(00)} \otimes \omega^{(01)}$. Notice that the double indices on the right side of the last expression have a "mixed" interpretation; namely, one index refers to the co-action and the other refers to the co-multiplication. Then the diagram (3.1) is equivalent to

$$\omega^{(00)} \otimes \omega^{(01)} \otimes \omega^{(1)} = \omega^{(0)} \otimes \omega^{(10)} \otimes \omega^{(11)} \qquad (3.3)$$

for all $\omega \in \Gamma$. This formula is familiar in form, if not in meaning. It looks just like the formula expressing the co-associativity of the co-multiplication in a Hopf algebra (or even in a co-algebra). Coincidences like this are not "accidents" in mathematics. What is happening here is that every Hopf algebra is a left co-module over itself with the co-action being the co-multiplication. One more word of caution: Each of the double indices in $\omega^{(10)}$ and $\omega^{(11)}$ refers to the left co-action, in contrast with the double indices on the left side of (3.3).

Similarly, there is an equivalent formula for the diagram (3.2). It is

$$\omega = \varepsilon(\omega^{(0)})\omega^{(1)} \qquad (3.4)$$

for all $\omega \in \Gamma$. Again, this is reminiscent of a formula for Hopf algebras.

Notice that these two diagrams, (3.1) and (3.2), are the dual diagrams to the diagrams (3.5) and (3.6) ahead, which express that a left action of a group G on a vector space V satisfies $g_1(g_2v) = (g_1g_2)v$ and $ev = v$ for all $v \in V$ and all $g_1, g_2 \in G$. Here $e \in G$ is the identity element of G. We denote the action as a map $\alpha : G \times V \to V$ and write $gv = \alpha(g, v)$. Also, we let $\mu_G : G \times G \to G$ denote the multiplication in G and let $\iota : \{e\} \to G$ denote the inclusion map of the singleton subset. Then these diagrams are

$$G \times G \times V \xrightarrow{id_G \times \alpha} G \times V$$
$$\mu_G \times id_V \downarrow \qquad\qquad \downarrow \alpha \qquad\qquad (3.5)$$
$$G \times V \xrightarrow[\alpha]{} V$$

and

$$\{e\} \times V \xrightarrow{\cong} V$$
$$\iota \times id_V \downarrow \qquad\qquad \downarrow id_V \qquad\qquad (3.6)$$
$$G \times V \xrightarrow{\alpha} V.$$

In the second diagram, the upper horizontal arrow is the bijective map given by $(e, v) \mapsto v$ for all $v \in V$. By letting $\mathcal{F}(X)$ denote the complex vector space of all functions $f : X \to \mathbb{C}$, putting $\mathcal{A} = \mathcal{F}(G)$ and $\Gamma = \mathcal{F}(V)$, and applying the contravariant functor $\mathcal{F}$ to these two diagrams, one gets the diagrams (3.1) and (3.2). This procedure is simply known as "pullback." For example, the left action map $\alpha : G \times V \to V$ pulls back to

$$\Gamma = \mathcal{F}(V) \xrightarrow{\alpha^*} \mathcal{F}(G \times V) \cong \mathcal{F}(G) \otimes \mathcal{F}(V) = \mathcal{A} \otimes \Gamma,$$

which is a left co-action.

As an aside, let's note that the definitions of a right co-action and a right co-module are obtained by replacing $\Phi_\Gamma : \Gamma \to \mathcal{A} \otimes \Gamma$ with $_\Gamma\Phi : \Gamma \to \Gamma \otimes \mathcal{A}$. Here are the details:

Definition 3.2. Let Γ be a vector space, and let $\mathcal{A}$ be a co-algebra. Then a *right co-action* of $\mathcal{A}$ on Γ (also called a *right co-representation* of $\mathcal{A}$ on Γ) is a linear map $_\Gamma\Phi : \Gamma \to \Gamma \otimes \mathcal{A}$ for which we have the following two commutative diagrams:

$$\Gamma \xrightarrow{_\Gamma\Phi} \Gamma \otimes \mathcal{A}$$
$$_\Gamma\Phi \downarrow \qquad\qquad \downarrow id \otimes \phi \qquad\qquad (3.7)$$
$$\Gamma \otimes \mathcal{A} \xrightarrow{_\Gamma\Phi \otimes id} \Gamma \otimes \mathcal{A} \otimes \mathcal{A}$$

and

$$\Gamma \xrightarrow{_\Gamma\Phi} \Gamma \otimes \mathcal{A}$$
$$id_\Gamma \downarrow \qquad\qquad \downarrow id \otimes \varepsilon \qquad\qquad (3.8)$$
$$\Gamma \xrightarrow{\cong} \Gamma \otimes \mathbb{C}.$$

One also says that a vector space Γ together with a given right co-action Φ_Γ is a *right $\mathcal{A}$-co-module*.

Exercise 3.1. *Using Sweedler's notation, write down the formulas that are equivalent to the diagrams (3.7) and (3.8).*

Exercise 3.2. *As with almost all definitions, there are trivial examples. Here is the definition of the trivial left co-action of $\mathcal{A}$ on Γ, where these spaces are as above: $T_\Gamma(v) := 1 \otimes v$ for all $v \in \Gamma$. Prove that T_Γ is indeed a left co-action. Then define the trivial right co-action of $\mathcal{A}$ on Γ.*

There is some other terminology that is used for left and right co-actions, and this has been known to lead to a lot of avoidable confusion. The idea behind this is that $\mathcal{A}$ is a Hopf algebra (or maybe something related to a Hopf algebra) and as such it is considered a generalization of $\mathcal{F}(G) := \{f : G \to \mathbb{C}\}$, the dual of a group G. For example, one often says that the co-multiplicative structure ϕ of $\mathcal{A}$ is the multiplicative structure of G even if there is not a classical group G in the game. Now G is just notation for a Hopf algebra. Similarly, the left (resp., right) co-action of $\mathcal{A}$ on a vector space is also called a left (resp., right) action of G. (On what, you might ask. Don't ask!) This can lead to confusion. But it gets worse. Often one simply says "left action" with no further qualifications. And then what is meant is "left co-action"! So you have been duly warned. Watch out!

Now the Γ under consideration here is more than a vector space; it is also an $\mathcal{A}$-bimodule. We will show that $\mathcal{A} \otimes \Gamma$ can also be provided with an $\mathcal{A}$-bimodule structure. As the first step, we provide $\mathcal{A} \otimes \Gamma$ with an $(\mathcal{A} \otimes \mathcal{A})$-bimodule structure. This is given by the linear maps

$$(\mathcal{A} \otimes \mathcal{A}) \otimes (\mathcal{A} \otimes \Gamma) \to \mathcal{A} \otimes \Gamma \qquad (a \otimes b) \otimes (c \otimes \omega) \mapsto ac \otimes b\omega \qquad (3.9)$$

and

$$(\mathcal{A} \otimes \Gamma) \otimes (\mathcal{A} \otimes \mathcal{A}) \to \mathcal{A} \otimes \Gamma \qquad (c \otimes \omega) \otimes (a \otimes b) \mapsto ca \otimes \omega b \qquad (3.10)$$

for $a, b, c \in \mathcal{A}$ and $\omega \in \Gamma$.

Exercise 3.3. *Show that (3.9) makes $\mathcal{A} \otimes \Gamma$ a left $(\mathcal{A} \otimes \mathcal{A})$-module, while (3.10) makes $\mathcal{A} \otimes \Gamma$ a right $(\mathcal{A} \otimes \mathcal{A})$-module. Moreover, show that these two module structures together make $\mathcal{A} \otimes \Gamma$ an $(\mathcal{A} \otimes \mathcal{A})$-bimodule.*
Similarly, show how to make $\Gamma \otimes \mathcal{A}$ into an $(\mathcal{A} \otimes \mathcal{A})$-bimodule.

Since $\mathcal{A}$ is a Hopf algebra, we use the co-multiplication $\phi : \mathcal{A} \to \mathcal{A} \otimes \mathcal{A}$ (the "group structure") to pull back the $(\mathcal{A} \otimes \mathcal{A})$-bimodule structure to make $\mathcal{A} \otimes \Gamma$ into an $\mathcal{A}$-bimodule. Similarly, we make $\Gamma \otimes \mathcal{A}$ into an $\mathcal{A}$-bimodule.

Exercise 3.4. *Show that $\phi : \mathcal{A} \to \mathcal{A} \otimes \mathcal{A}$ makes the Hopf algebra $\mathcal{A}$ into a left $\mathcal{A}$-co-module as well as into a right $\mathcal{A}$-co-module.*

3.2 Left Covariance

We consider an FODC (Γ, d) over $\mathcal{A}$. The motivation for the definition of left covariance of (Γ, d) is that, besides having a left co-action Φ_Γ, we would also like to have these two properties:

1. Φ_Γ is a left $\mathcal{A}$-module morphism; that is,

$$\Phi_\Gamma(a\omega) = \phi(a)\Phi_\Gamma(\omega)$$

for all $a \in \mathcal{A}$ and all $\omega \in \Gamma$. This is equivalent to the commutativity of this diagram:

$$
\begin{array}{ccc}
\mathcal{A} \otimes \Gamma & \overset{id_\mathcal{A} \otimes \Phi_\Gamma}{\longrightarrow} & \mathcal{A} \otimes \mathcal{A} \otimes \Gamma \\
{\scriptstyle l.a.}\downarrow & & \downarrow{\scriptstyle l.a.} \\
\Gamma & \overset{\Phi_\Gamma}{\longrightarrow} & \mathcal{A} \otimes \Gamma.
\end{array}
\tag{3.11}
$$

Here the vertical arrows are the appropriate left actions of $\mathcal{A}$, neither of which has been given its own notation. For example, the arrow on the right is the left action of $\mathcal{A}$ (not of $\mathcal{A} \otimes \mathcal{A}$) on $\mathcal{A} \otimes \Gamma$, as just defined above.

2. $d : \mathcal{A} \to \Gamma$ is a left $\mathcal{A}$-co-module morphism, where the co-multiplication ϕ gives the left $\mathcal{A}$-co-module structure on $\mathcal{A}$ and Φ_Γ gives the left $\mathcal{A}$-co-module structure on Γ. This is equivalent to the commutativity of this diagram:

$$
\begin{array}{ccc}
\mathcal{A} & \overset{d}{\longrightarrow} & \Gamma \\
{\scriptstyle \phi}\downarrow & & \downarrow{\scriptstyle \Phi_\Gamma} \\
\mathcal{A} \otimes \mathcal{A} & \overset{id \otimes d}{\longrightarrow} & \mathcal{A} \otimes \Gamma.
\end{array}
\tag{3.12}
$$

This is equivalent to the identity

$$\Phi_\Gamma(da) = (id \otimes d)\phi(a) = a^{(1)} \otimes da^{(2)}.$$

Suppose that Φ_Γ satisfies both conditions 1 and 2 above and is also a left $\mathcal{A}$-co-module structure on Γ. Then we take an arbitrary element $\omega \in \Gamma$ and will evaluate $\Phi_\Gamma(\omega)$. First, we write $\omega = \sum_k a_k db_k$ (standard form) for some $a_k, b_k \in \mathcal{A}$. Then we have

$$\Phi_\Gamma(\omega) = \Phi_\Gamma(\sum_k a_k db_k)$$

$$= \sum_k \Phi_\Gamma(a_k db_k)$$

$$= \sum_k \phi(a_k)\Phi_\Gamma(db_k)$$

$$= \sum_k \phi(a_k)(id \otimes d)\phi(b_k) \in \mathcal{A} \otimes \Gamma.$$

The last equality follows from condition 2, while the next-to-last equality follows from condition 1. Since ϕ and d are given, this shows that Φ_Γ (if it exists) is unique and is given by the previous formula. Now the representation of ω in standard form is not unique, so this last formula cannot be used in general to define a left co-action Φ_Γ in terms of ϕ and d. But to get the existence of Φ_Γ under these conditions, we will use the property in the next definition.

Definition 3.3. Let (Γ, d) be an FODC over a Hopf algebra $\mathcal{A}$. Let ϕ denote the co-multiplication of $\mathcal{A}$. Suppose that for any elements $a_k, b_k \in \mathcal{A}$ (where $k = 1, \ldots, K$), we have the implication

$$\sum_k a_k db_k = 0 \implies \sum_k \phi(a_k)(id \otimes d)\phi(b_k) = 0 \in \mathcal{A} \otimes \Gamma. \qquad (3.13)$$

Then (Γ, d) is a *left-covariant* FODC.

Theorem 3.1. *Let (Γ, d) be a left-covariant FODC over a Hopf algebra $\mathcal{A}$. Then there exists a unique linear map*

$$\Phi_\Gamma : \Gamma \to \mathcal{A} \otimes \Gamma$$

defined by

$$\Phi_\Gamma(\omega) = \sum_k \phi(a_k)(id \otimes d)\phi(b_k), \qquad (3.14)$$

where $\omega = \sum_k a_k db_k$ is any standard representation of ω, and satisfying the following:

1. *Φ_Γ is an $\mathcal{A}$-bimodule morphism. This means that these equations hold for all $c \in \mathcal{A}$ and all $\omega \in \Gamma$:*

$$\Phi_\Gamma(c\omega) = \phi(c)\Phi_\Gamma(\omega), \qquad (3.15)$$

$$\Phi_\Gamma(\omega c) = \Phi_\Gamma(\omega)\phi(c). \qquad (3.16)$$

   (These conditions can also be given as commutative diagrams.)
2. *Φ_Γ gives Γ the structure of a left $\mathcal{A}$-co-module. This is equivalent to the commutativity of the diagrams (3.1) and (3.2).*
3. *$d : \mathcal{A} \to \Gamma$ is a left $\mathcal{A}$-co-module morphism, where ϕ gives $\mathcal{A}$ a left $\mathcal{A}$-co-module structure. (Here Φ_Γ gives Γ the structure of a left $\mathcal{A}$-co-module, as asserted in*

the previous part of this theorem.) This is equivalent to the commutativity of the diagram (3.12).

Conversely, if $\Phi_\Gamma : \Gamma \to \mathcal{A} \otimes \Gamma$ is a linear map satisfying properties 1–3 above, then Φ_Γ is given by Eq. (3.14) and the implication (3.13) holds.

Proof. We start with the proof of the converse. If $\omega = \sum_k a_k db_k$ is any standard representation of ω, then applying Φ_Γ (whose existence is given) to it and using properties 1–3, one obtains (3.14). Since Φ_Γ is linear, then $\omega = 0 \implies \Phi_\Gamma(\omega) = 0$ holds. But this implication is exactly the same as the implication in (3.13).

We have already shown the existence and uniqueness of Φ_Γ. The rest of the proof is a long, laborious exercise that verifies all the required properties. In other words, nothing unexpected arises in this way of defining Φ_Γ, but this is only known after going through all the following details, which run onto several pages.

To show that Φ_Γ is a left $\mathcal{A}$-module morphism, take an arbitrary element $\omega = \sum_k a_k db_k \in \Gamma$ as before and any $c \in \mathcal{A}$. Then

$$\Phi_\Gamma(c\omega) = \Phi_\Gamma(c \sum_k a_k db_k) = \Phi_\Gamma(\sum_k ca_k db_k)$$

$$= \sum_k \phi(ca_k)(id \otimes d)\phi(b_k) = \sum_k \phi(c)\phi(a_k)(id \otimes d)\phi(b_k)$$

$$= \phi(c) \sum_k \phi(a_k)(id \otimes d)\phi(b_k) = \phi(c)\Phi_\Gamma(\omega),$$

which shows that (3.15) holds. Here we have used that ϕ is a morphism of the algebra $\mathcal{A}$ and that it induces the left $\mathcal{A}$-module structure on $\mathcal{A} \otimes \Gamma$.

Next, we will show that d is a left $\mathcal{A}$-co-module morphism. Using that $da = 1da$ is in standard form and that $\phi(1) = 1 \otimes 1$, we have

$$\Phi_\Gamma(da) = \phi(1)(id \otimes d)\phi(a) = (1 \otimes 1)(id \otimes d)\phi(a) = (id \otimes d)\phi(a)$$

for all $a \in \mathcal{A}$. This shows the commutativity of the diagram (3.12), which exactly says that d is a left $\mathcal{A}$-co-module morphism.

We next prove the commutativity of (3.1), the first condition saying that Γ is a left $\mathcal{A}$-co-module. We take arbitrary elements $a, b \in \mathcal{A}$ and use the fact that the elements $a\,db$ span Γ. First, we compute across and down in the diagram (3.1). Thus, using the definition of Φ_Γ, Sweedler's notation, and so on, we have

$$(\phi \otimes id)\Phi_\Gamma(adb) = (\phi \otimes id)(\phi(a)(id \otimes d)\phi(b))$$

$$= (\phi \otimes id)((a^{(1)} \otimes a^{(2)})(id \otimes d)(b^{(1)} \otimes b^{(2)}))$$

$$= (\phi \otimes id)((a^{(1)} \otimes a^{(2)})(b^{(1)} \otimes db^{(2)}))$$

$$= (\phi \otimes id)((a^{(1)}b^{(1)} \otimes a^{(2)}db^{(2)}))$$

$$= \phi(a^{(1)}b^{(1)}) \otimes a^{(2)}db^{(2)}$$

$$= \phi(a^{(1)})\phi(b^{(1)}) \otimes a^{(2)}db^{(2)}$$

$$= (a^{(11)} \otimes a^{(12)})(b^{(11)} \otimes b^{(12)}) \otimes a^{(2)}db^{(2)}$$

$$= (a^{(11)}b^{(11)} \otimes a^{(12)}b^{(12)}) \otimes a^{(2)}db^{(2)}$$

$$= (a^{(11)} \otimes a^{(12)} \otimes a^{(2)})(b^{(11)} \otimes b^{(12)} \otimes db^{(2)})$$

$$= (a^{(11)} \otimes a^{(12)} \otimes a^{(2)})(id \otimes id \otimes d)(b^{(11)} \otimes b^{(12)} \otimes b^{(2)})$$

$$= (\phi \otimes id)\phi(a)(id \otimes id \otimes d)(\phi \otimes id)\phi(b).$$

This lengthy, though elementary, calculation is typical. It is merely the first of many such that we will be seeing.

Next, we compute down and across in the diagram (3.1) by using the definition of Φ_Γ twice, Sweedler's notation, and so forth, to get

$$(id \otimes \Phi_\Gamma)\Phi_\Gamma(adb) = (id \otimes \Phi_\Gamma)(\phi(a)(id \otimes d)\phi(b))$$

$$= (id \otimes \Phi_\Gamma)((a^{(1)} \otimes a^{(2)})(b^{(1)} \otimes db^{(2)}))$$

$$= (id \otimes \Phi_\Gamma)(a^{(1)}b^{(1)} \otimes a^{(2)}db^{(2)})$$

$$= (a^{(1)}b^{(1)} \otimes \Phi_\Gamma(a^{(2)}db^{(2)}))$$

$$= a^{(1)}b^{(1)} \otimes (\phi(a^{(2)})(id \otimes d)\phi(b^{(2)}))$$

$$= a^{(1)}b^{(1)} \otimes ((a^{(21)} \otimes a^{(22)})(id \otimes d)(b^{(21)} \otimes b^{(22)}))$$

$$= a^{(1)}b^{(1)} \otimes ((a^{(21)} \otimes a^{(22)})(b^{(21)} \otimes db^{(22)}))$$

$$= a^{(1)}b^{(1)} \otimes a^{(21)}b^{(21)} \otimes a^{(22)}db^{(22)}$$

$$= (a^{(1)} \otimes a^{(21)} \otimes a^{(22)})(b^{(1)} \otimes b^{(21)} \otimes db^{(22)})$$

$$= (a^{(1)} \otimes a^{(21)} \otimes a^{(22)})(id \otimes id \otimes d)(b^{(1)} \otimes b^{(21)} \otimes b^{(22)})$$

$$= (id \otimes \phi)\phi(a)(id \otimes id \otimes d)(id \otimes \phi)\phi(b).$$

Finally, the co-associativity of the co-multiplication [namely, $(id \otimes \phi)\phi = (\phi \otimes id)\phi$] gives the commutativity of the diagram (3.1).

Now we turn our attention to the commutativity of the diagram (3.2). We continue with the notation $a, b \in \mathcal{A}$. We recall that in a Hopf algebra we have the identity $(\varepsilon \otimes id)\phi(a) = a$ and that ε is multiplicative. So we have

$$(\varepsilon \otimes id)\Phi_\Gamma(a\,db) = (\varepsilon \otimes id)(\phi(a)(id \otimes d)\phi(b))$$

$$= (\varepsilon \otimes id)((a^{(1)} \otimes a^{(2)})(id \otimes d)(b^{(1)} \otimes b^{(2)}))$$

$$= (\varepsilon \otimes id)\big((a^{(1)} \otimes a^{(2)})(b^{(1)} \otimes db^{(2)})\big)$$
$$= (\varepsilon \otimes id)\big((a^{(1)}b^{(1)} \otimes a^{(2)}db^{(2)})\big)$$
$$= \varepsilon(a^{(1)}b^{(1)}) \otimes a^{(2)}db^{(2)} \cong \varepsilon(a^{(1)}b^{(1)})a^{(2)}db^{(2)}$$
$$= \varepsilon(a^{(1)})\varepsilon(b^{(1)})a^{(2)}db^{(2)}$$
$$= \varepsilon(a^{(1)})a^{(2)}d(\varepsilon(b^{(1)})b^{(2)})$$
$$= \big((\varepsilon \otimes id)\phi(a)\big)d\big((\varepsilon \otimes id)\phi(b)\big)$$
$$= a\,db,$$

where we have also used that $\varepsilon : \mathcal{A} \to \mathbb{C}$, that the action of the scalars $\mathbb{C}$ in $\mathcal{A}$ is central, and that d is linear over $\mathbb{C}$. And this proves the commutativity of the diagram (3.2).

Now it only remains to show (3.16), namely, that Φ_Γ is a right $\mathcal{A}$-module morphism. Note that for $a, b, c \in \mathcal{A}$, Leibniz's rule gives us the identity

$$(adb)c = ad(bc) - abdc.$$

We note that this is a recurrent method for writing expressions such as those that appear here on the left side as a linear combination (with coefficients in $\mathcal{A}$ on the left) of terms, each of which is in standard form. Now we have

$$\Phi_\Gamma\big((adb)c\big) = \Phi_\Gamma\big(ad(bc) - abdc\big)$$
$$= \phi(a)(id \otimes d)\phi(bc) - \phi(ab)(id \otimes d)\phi(c)$$
$$= \phi(a)(id \otimes d)\phi(bc) - \phi(a)\phi(b)(id \otimes d)\phi(c)$$
$$= \phi(a)\big((id \otimes d)\phi(bc) - \phi(b)(id \otimes d)\phi(c)\big)$$
$$= \phi(a)\big((id \otimes d)(\phi(b)\phi(c)) - \phi(b)(id \otimes d)\phi(c)\big)$$
$$= \phi(a)\big((id \otimes d)\phi(b)\big)\phi(c)$$
$$= \Phi_\Gamma\big((adb)\big)\phi(c).$$

We used Leibniz's rule here for $id \otimes d$. Since elements of the form adb span Γ, this finishes the proof of (3.16) and of the theorem. ∎

3.3 Right Covariance

Similarly, one can define a right-covariant FODC and prove a similar theorem. First, here is the definition.

Definition 3.4. Let (Γ, d) be an FODC over a Hopf algebra $\mathcal{A}$. Let ϕ denote the co-multiplication of $\mathcal{A}$. Suppose that for any elements $a_k, b_k \in \mathcal{A}$ (where $k = 1, \ldots, K$), we have the implication

$$\sum_k a_k db_k = 0 \implies \sum_k \phi(a_k)(d \otimes id)\phi(b_k) = 0 \in \mathcal{A} \otimes \Gamma. \tag{3.17}$$

Then (Γ, d) is a *right-covariant* FODC.

Here is the corresponding theorem. Since the proof of this theorem follows the same lines as the theorem in the previous section, we state but do not prove it.

Theorem 3.2. *Let (Γ, d) be a right-covariant FODC over a Hopf algebra $\mathcal{A}$. Then there exists a* unique *linear map*

$$_{\Gamma}\Phi : \Gamma \to \Gamma \otimes \mathcal{A}$$

defined by

$$_{\Gamma}\Phi(\omega) = \sum_k \phi(a_k)(d \otimes id)\phi(b_k), \tag{3.18}$$

where $\omega = \sum_k a_k db_k$ is any standard representation of ω, and satisfying the following:

1. *$_{\Gamma}\Phi$ is an $\mathcal{A}$-bimodule morphism. This means that these equations hold for all $c \in \mathcal{A}$ and all $\omega \in \Gamma$:*

$$_{\Gamma}\Phi(c\omega) = \phi(c)\,_{\Gamma}\Phi(\omega), \tag{3.19}$$

$$_{\Gamma}\Phi(\omega c) = \,_{\Gamma}\Phi(\omega)\phi(c). \tag{3.20}$$

(These conditions can also be given as commutative diagrams.)
2. *$_{\Gamma}\Phi$ gives Γ the structure of a right $\mathcal{A}$-co-module. This is equivalent to the commutativity of diagrams similar to (3.1) and (3.2).*
3. *$d : \mathcal{A} \to \Gamma$ is a right $\mathcal{A}$-co-module morphism, where ϕ gives $\mathcal{A}$ a right $\mathcal{A}$-co-module structure. This is equivalent to the commutativity of a diagram similar to (3.12).*

Conversely, if $_{\Gamma}\Phi : \Gamma \to \Gamma \otimes \mathcal{A}$ is a linear map satisfying properties 1–3 above, then $_{\Gamma}\Phi$ is given by Eq. (3.18) and the implication (3.17) holds.

3.4 Bicovariance

Definition 3.5. An FODC is *bicovariant* if it is both left covariant and right covariant.

Remarks 3.1. *In this definition, no compatibility condition between the left covariant and right covariant structures is required. However, note that both of these one-sided structures are determined by the co-multiplication ϕ and the differential d. Actually, we have the next result about the compatibility of these structures.*

Theorem 3.3. *Suppose (Γ, d) is a bicovariant FODC over a Hopf algebra $\mathcal{A}$. Let Φ_Γ (resp., $_\Gamma\Phi$) denote the left (resp., right) co-action maps. Then this diagram commutes:*

$$
\begin{array}{ccc}
\Gamma & \overset{\Phi_\Gamma}{\longrightarrow} & \mathcal{A} \otimes \Gamma \\
{\scriptstyle _\Gamma\Phi}\downarrow & & \downarrow{\scriptstyle id_\mathcal{A} \otimes _\Gamma\Phi} \\
\Gamma \otimes \mathcal{A} & \overset{\Phi_\Gamma \otimes id_\mathcal{A}}{\longrightarrow} & \mathcal{A} \otimes \Gamma \otimes \mathcal{A}.
\end{array}
\qquad (3.21)
$$

This diagram says that the left co-action and the right co-action commute. So in this sense these two co-actions are always compatible.

Remark. Also, the diagram (3.21) is the dual to the diagram that expresses that a left action and a right action commute, namely:

$$
\begin{array}{ccc}
M & \overset{L}{\longleftarrow} & G \times M \\
{\scriptstyle R}\uparrow & & \uparrow{\scriptstyle id_G \times R} \\
M \times G & \overset{L \times id_G}{\longleftarrow} & G \times M \times G,
\end{array}
\qquad (3.22)
$$

where the classical Lie group G acts on the left (resp., right) of the differential manifold M via the map L (resp., R).

Proof. Take $a, b \in \mathcal{A}$. Going across and down in the diagram (3.21), we have

$$
\begin{aligned}
(id \otimes {}_\Gamma\Phi)\Phi_\Gamma(a\,db) &= (id \otimes {}_\Gamma\Phi)(\phi(a)(id \otimes d)\phi(b)) \quad \text{(def}^n \text{ of } \Phi_\Gamma) \\
&= (id \otimes {}_\Gamma\Phi)(a^{(1)}b^{(1)} \otimes a^{(2)}db^{(2)}) \\
&= a^{(1)}b^{(1)} \otimes {}_\Gamma\Phi(a^{(2)}db^{(2)}) \\
&= a^{(1)}b^{(1)} \otimes (\phi(a^{(2)})(d \otimes id)\phi(b^{(2)})) \quad \text{(def}^n \text{ of } {}_\Gamma\Phi) \\
&= a^{(1)}b^{(1)} \otimes ((a^{(21)} \otimes a^{(22)})(db^{(21)} \otimes b^{(22)})) \\
&= a^{(1)}b^{(1)} \otimes a^{(21)}db^{(21)} \otimes a^{(22)}b^{(22)} \\
&= (a^{(1)} \otimes a^{(21)} \otimes a^{(22)})(b^{(1)} \otimes db^{(21)} \otimes b^{(22)}) \\
&= (a^{(1)} \otimes a^{(21)} \otimes a^{(22)})(id \otimes d \otimes id)(b^{(1)} \otimes b^{(21)} \otimes b^{(22)}) \\
&= ((id \otimes \phi)\phi(a)(id \otimes d \otimes id)(id \otimes \phi)\phi(b)).
\end{aligned}
$$

Now, going down and across in the diagram (3.21) gives a very similar calculation:

$$
\begin{aligned}
(\Phi_\Gamma \otimes id)_\Gamma \Phi(adb) &= (\Phi_\Gamma \otimes id)\big(\phi(a)(d \otimes id)\phi(b)\big) \quad (\text{def}^n \text{ of } {}_\Gamma\Phi)\\
&= (\Phi_\Gamma \otimes id)(a^{(1)}db^{(1)} \otimes a^{(2)}b^{(2)})\\
&= \Phi_\Gamma(a^{(1)}db^{(1)}) \otimes a^{(2)}b^{(2)}\\
&= \phi(a^{(1)})(id \otimes d)\phi(b^{(1)}) \otimes a^{(2)}b^{(2)} \quad (\text{def}^n \text{ of } \Phi_\Gamma)\\
&= (a^{(11)} \otimes a^{(12)})(id \otimes d)(b^{(11)} \otimes b^{(12)}) \otimes a^{(2)}b^{(2)}\\
&= (a^{(11)} \otimes a^{(12)})(b^{(11)} \otimes db^{(12)}) \otimes a^{(2)}b^{(2)}\\
&= a^{(11)}b^{(11)} \otimes a^{(12)}db^{(12)} \otimes a^{(2)}b^{(2)}\\
&= (a^{(11)} \otimes a^{(12)} \otimes a^{(2)})(b^{(11)} \otimes db^{(12)} \otimes b^{(2)})\\
&= (a^{(11)} \otimes a^{(12)} \otimes a^{(2)})(id \otimes d \otimes id)(b^{(11)} \otimes b^{(12)} \otimes b^{(2)})\\
&= \big((\phi \otimes id)\phi(a)\big)(id \otimes d \otimes id)\big((\phi \otimes id)\phi(b)\big).
\end{aligned}
$$

The co-associativity of ϕ then shows that the diagram (3.21) commutes. ∎

The notion of an FODC over a Hopf algebra (rather than simply over an algebra) was introduced precisely so that we can consider bicovariant FODCs. This is based on the situation for a Lie group G where the $C^\infty(G)$-bimodule of smooth sections of the cotangent bundle has actions by G on the left and on the right. Here $C^\infty(G)$ is the *real* vector space (and *commutative* algebra with 1) of C^∞-functions $f : G \to \mathbb{R}$. A standard notation for the cotangent bundle of G is $T^*(G)$, and for its associated space of smooth sections, one notation is $\Gamma(T^*(G))$. In this case the exterior derivative in classical differential geometry is denoted $d : C^\infty(G) \to \Gamma(T^*(G))$. Note that in this classical case, the constructions are functors. In particular, the construction of the FODC from $T^*(G)$ is a functor. In the noncommutative case, this is not so; rather, the FODC is *any* object satisfying various properties (or one may say *axioms*).

Exercise 3.5. *This exercise is a way to check if you have mastered Sweedler's notation by now. So go back to Section 2 on FODCs and notice that the notation for the standard form*

$$
\omega = \sum_k a_k db_k
$$

is used repeatedly. Your problem is to invent a Sweedler-like notation for the standard form (no summation symbol, no subindices) and see how the formulas in Section 3.2 can be simplified while remaining true! If you cannot do this exercise at this point in your studies, please come back to it later.

3.5 Examples

In this section we will give examples of Hopf algebras that are neither commutative nor co-commutative.

We let $q \neq 0$ be a complex number. We start from the *free algebra* over $\mathbb{C}$ on the set of four generators $\{a, b, c, d\}$. We denote this algebra by $\mathbb{C}\{a, b, c, d\}$. To be more specific, we call the set $A = \{a, b, c, d\}$ the *alphabet* and consider the set A^* of all *words* of finite length in this alphabet, where a word is any finite sequence $l_1 l_2 \cdots l_k$ of *letters* $l_j \in A$. Here $k \geq 0$ is any integer called the *length* of the word. For the case $k = 0$, we have what is known as the *empty word*. We define a product on A^* as the binary operation of combining the words w_1 and w_2 by juxtaposing them to $w_1 w_2$. Clearly, if w_1 has length l_1 and w_2 has length l_2, then $w_1 w_2$ has length $l_1 + l_2$. Then $\mathbb{C}\{a, b, c, d\}$ is defined to be the complex vector space with basis A^*. So elements in $\mathbb{C}\{a, b, c, d\}$ are finite linear combinations with coefficients in $\mathbb{C}$ of the elements in A^*. We extend the product on A^* bilinearly to obtain a product on $\mathbb{C}\{a, b, c, d\}$. This product is associative and has unit element 1Λ, where $1 \in \mathbb{C}$ and Λ is the empty word. Hereafter we write this unit element simply as 1. While $\mathbb{C}\{a, b, c, d\}$ is an infinite-dimensional vector space (since the basis A^* is an infinite set), as an algebra $\mathbb{C}\{a, b, c, d\}$ is generated by the four elements $\{a, b, c, d\}$ since all possible products of these four elements give us the basis A^*.

We now consider the two-sided ideal J_q in $\mathbb{C}\{a, b, c, d\}$ generated by these seven elements in $\mathbb{C}\{a, b, c, d\}$, where q is a nonzero complex number:

$$qba - ab, \quad qdb - bd,$$
$$qca - ac, \quad qdc - cd,$$
$$bc - cb, \quad ad - da - (q - q^{-1})bc,$$
$$ad - qbc - 1. \tag{3.23}$$

Leaving the motivation of these relations to the side for now, we introduce the quotient algebra

$$SL_q(2) := \mathbb{C}\{a, b, c, d\}/J_q.$$

Clearly, $SL_q(2)$ is a commutative algebra if and only if $q = 1$. We denote the quotients of the elements of $\{a, b, c, d\}$ by the same symbols. This is a standard abuse of notation. Note that the four elements in $\{a, b, c, d\}$ generate $SL_q(2)$ as an algebra. The notation for this quotient algebra is partially justified by the fact that when one puts the *deformation parameter* $q = 1$, this becomes the function algebra of polynomials in the matrix entries of $SL(2)$, the Lie group of 2×2 matrices with complex entries and determinant 1. For this reason $SL_q(2)$ is called a *quantum function algebra*. Of course, in general this is not a function algebra since it is noncommutative. Further, rather convincing motivation for the definition of $SL_q(2)$

is provided in [45] and Exercise 3.7. Be warned, however, that another convention is used in [45], where the parameter q used here is replaced by q^{-1}.

We now provide $SL_q(2)$ with the structures of a co-multiplication ϕ, a co-unit ε, and a co-inverse κ. This will convert $SL_q(2)$ into a Hopf algebra. It suffices to define these maps on the algebra generators $\{a, b, c, d\}$ of $SL_q(2)$. For example, we define the co-unit on these four generators using this matrix notation:

$$\varepsilon\begin{pmatrix} a & b \\ c & d \end{pmatrix} := \begin{pmatrix} 1 & 0 \\ 0 & 1 \end{pmatrix}.$$

This means that $\varepsilon(a) := 1$, $\varepsilon(b) := 0$, and so forth. The co-multiplication is defined by

$$\phi\begin{pmatrix} a & b \\ c & d \end{pmatrix} := \begin{pmatrix} a & b \\ c & d \end{pmatrix} \otimes \begin{pmatrix} a & b \\ c & d \end{pmatrix}.$$

Again, this is a way of writing four defining equations in matrix form. One of these equations is $\phi(b) := a \otimes b + b \otimes d$. Immediately, this shows that ϕ is not co-commutative. Notice that the definitions of ϕ and ε do not involve q.

Finally, we define the co-inverse by

$$\kappa\begin{pmatrix} a & b \\ c & d \end{pmatrix} := \begin{pmatrix} d & -q^{-1}b \\ -qc & a \end{pmatrix}.$$

It is not difficult to show that κ is a bijection. These maps are defined in the first instance on the domain $\mathbb{C}\{a, b, c, d\}$. Then one must show that the maps factor through the quotient $SL_q(2)$. And then one must show that all the defining properties of a Hopf algebra are satisfied. This is all done very carefully in [45]. Of course, the exact formulas in (3.23) are used in those arguments.

Exercise 3.6. *Very carefully read the proof in [45] that this defines a Hopf algebra. Or prove it yourself.*

For $q \neq 1$, we have a family of Hopf algebras that are neither commutative nor co-commutative. Also, notice that if $q^2 \neq 1$, then $\kappa^2 \neq id$ and κ has finite order if and only if q is a root of unity. Recall that a complex number z is a *root of unity* if an integer $k \geq 1$ exists such that $z^k = 1$.

We still need more theory in order to quickly construct FODCs on this Hopf algebra. We will come back to this later.

Exercise 3.7. *We define the* quantum plane *for $q \in \mathbb{C} \setminus \{0\}$ by*

$$QP_q := \mathbb{C}\{x, y\} / \mathcal{I}_q,$$

where $\mathbb{C}\{x, y\}$ is the free algebra generated by two elements and $\mathcal{I}_q$ is the two-sided ideal generated by the element $xy - qyx$ in $\mathbb{C}\{x, y\}$.

We next define two algebra morphisms $\Phi_L : QP_q \to SL_q(2) \otimes QP_q$ *and* $\Phi_R :$
$QP_q \to QP_q \otimes SL_q(2)$, *respectively, by*

$$\Phi_L\begin{pmatrix} x \\ y \end{pmatrix} = \begin{pmatrix} a & b \\ c & d \end{pmatrix} \otimes \begin{pmatrix} x \\ y \end{pmatrix}$$

and

$$\Phi_R(x \ y) = (x \ y) \otimes \begin{pmatrix} a & c \\ b & d \end{pmatrix}$$

on the generators x, y *of* QP_q. *Show that* Φ_L *(resp.,* Φ_R*) is a left (resp., right) co-action of* $SL_q(2)$ *on the quantum plane* QP_q. *Also, show that when* $q = 1$, *these co-actions are induced via pullback of the standard matrix actions of* $SL(2)$ *on the complex plane* $\mathbb{C}^2$.

One says that $SL_q(2)$ *is a* quantum group *that endows the quantum plane* QP_q *with quantum symmetries via these co-actions. Symmetries like these are a generalization of the classical symmetries that arise from the action of a classical group on a classical space. This geometric point of view is emphasized in [56].*

Exercise 3.8. *Prove that the quantum determinant* $\det_q := ad - qbc$ *is a central element in* $M_q(2)$, *the quotient algebra of* $\mathbb{C}\{a, b, c, d\}$ *modulo the two-sided ideal generated by the first six elements in (3.23). Recall that an element in an algebra is said to be* central *if it commutes with every element in the algebra.*

Exercise 3.9. *As a more challenging exercise, prove that the center of* $M_q(2)$ *is the one-dimensional subspace generated by* $\det_q$ *if and only if* q *is not a root of unity. Recall that the* center *of an algebra is the set of all its central elements.*

There is another Hopf algebra $\tilde{U}_q(sl_2)$ that has a dual pairing with $SL_q(2)$. We briefly describe it now. Let q be a complex number with $q^4 \notin \{0, 1\}$. As an algebra, $\tilde{U}_q(sl_2)$ is defined as the quotient of the free algebra on the four generators, denoted by E, F, K, and K^{-1}, modulo the two-sided ideal generated by the relations that give these equalities in the quotient:

$$KK^{-1} = K^{-1}K = 1,$$

$$KEK^{-1} = qE,$$

$$KFK^{-1} = q^{-1}F,$$

$$EF - FE = \frac{K^2 - K^{-2}}{q - q^{-1}}.$$

Clearly, this is not a commutative algebra.

We define a co-multiplication $\tilde{\phi}$ on the generators of $\tilde{U}_q(sl_2)$ by

$$\tilde{\phi}\begin{pmatrix} K & E \\ F & K^{-1} \end{pmatrix} = \begin{pmatrix} K \otimes K & E \otimes K + K^{-1} \otimes E \\ F \otimes K + K^{-1} \otimes F & K^{-1} \otimes K^{-1} \end{pmatrix}.$$

And this co-multiplication is not co-commutative.

We define a co-unit $\tilde{\varepsilon}$ on the generators of $\tilde{U}_q(sl_2)$ by

$$\tilde{\varepsilon}\begin{pmatrix} K & E \\ F & K^{-1} \end{pmatrix} = \begin{pmatrix} 1 & 0 \\ 0 & 1 \end{pmatrix}.$$

Notice that these definitions for $\tilde{\phi}$ and $\tilde{\varepsilon}$ do not contain the deformation parameter q. Also, the definitions on K^{-1} are not independent of those on K since these maps must be algebra morphisms.

Exercise 3.10. *Prove by direct calculations that the maps $\tilde{\phi}$ and $\tilde{\varepsilon}$ determine a bi-algebra structure on $\tilde{U}_q(sl_2)$.*

Given that we now have a bi-algebra, there is at most one co-inverse κ on it. We define this map on the generators by

$$\tilde{\kappa}\begin{pmatrix} K & E \\ F & K^{-1} \end{pmatrix} = \begin{pmatrix} K^{-1} & -qE \\ -q^{-1}F & K \end{pmatrix}.$$

Note that this map does depend on q.

Exercise 3.11. *Prove that this determines a map $\tilde{\kappa}$ on the bi-algebra $\tilde{U}_q(sl_2)$ that thereby turns it into a Hopf algebra.*

We now define a duality pairing $\tilde{U}_q(sl_2) \times SL_q(2) \to \mathbb{C}$, which we denote by $\langle \cdot, \cdot \rangle$. (See Appendix B for more on duality pairings of Hopf algebras.) Since the pairing map is determined by its values on the algebra generators, we only specify those pairs of generators having nonzero pairing. They are

$$\langle K, a \rangle = q^{-1/2},$$
$$\langle K, d \rangle = q^{1/2},$$
$$\langle E, c \rangle = \langle F, b \rangle = 1.$$

The proof that this defines a unique pairing and that the pairing is nondegenerate for q not equal to a root of unity can be found in [48]. The more adventuresome reader might like to have a go at it.

The pairings with K^{-1} are determined by the above conditions. For example,

$$\langle K^{-1}, a \rangle = \langle \tilde{\kappa}(K), a \rangle = \langle K, \kappa(a) \rangle = \langle K, d \rangle = q^{1/2}.$$

Exercise 3.12. *Prove that*

$$\langle K^{-1}, b \rangle = 0,$$
$$\langle K^{-1}, c \rangle = 0,$$
$$\langle K^{-1}, d \rangle = q^{-1/2}.$$

3.6 Notes

Covariance, especially bicovariance, is a central concept in this theory. We are seeing it here in the setting of FODCs, but this is just the first of many structures we will be considering for which covariance is an essential part of the picture. Later we will introduce *-operations, and then *-bicovariance will come on the center of the stage. This concept is so central, in fact, that I think there is a principle of the "covariance imperative" much like the "categorical imperative." The latter imperative is also to be seen here. In a quite simplistic formulation, this says that all topics in mathematics should be formulated in terms of the structures of category theory: objects, morphisms, functors, natural transformations, and so forth. This is a moral injunction (due to the word "should") and a universal statement (due to the word "all"). Clearly, this is a caricature of what the categorical imperative is (or should be!) about, but it serves a purpose nonetheless. The covariance imperative could never aspire to such universality although it surely does have a decidedly moral overtone. It requires that every structure associated to a quantum group *should* be covariant in an appropriate sense.

In some way, this is how the Erlangen program is expressed in the setting of non-commutative geometry. How this "covariance imperative" is actually implemented in specific contexts is another issue altogether. Viewed this way, quantum groups are made to be used to study their associated covariant noncommutative structures, much as groups are made to be used to study their associated covariant commutative structures. But the "covariance imperative" as an idea in noncommutative geometry may be too clever by half. After all, one point (sic!) of noncommutative geometry is to find new geometric structures in the noncommutative setting, and these new structures need not necessarily closely imitate their classical counterparts. Nor do they even need classical counterparts! The remarks in these paragraphs have a scientific character, or perhaps merely a philosophy of science character.

The examples in Section 3.5 are a particular case of what are known as q-deformations. For $SL_q(2)$, see [81] and [85]. The Hopf algebra $\tilde{U}_q(sl_2)$ is related to the more commonly used Hopf algebra $U_q(sl_2)$. See [51] and [74]. One can realize $U_q(sl_2)$ as a Hopf subalgebra of $\tilde{U}_q(sl_2)$. For this see [48], which, along with [45], is where I learned these examples.

There are also now q-deformations of many other mathematical structures, including the function algebras of the classical Lie groups. See [50, 67] and

[79]. In the sense of duality pairings of Hopf algebras, the universal enveloping algebras of semisimple Lie algebras also have dual q-deformations, which are called *Jimbo–Drinfeld* algebras. See also [15, 43–45], and [53]. The q-deformations give us important examples of Hopf algebras that are neither commutative nor cocommutative. However, not all such Hopf algebras arise this way. There is a lot of research about q-deformations and their relations with noncommutative geometry, including quantum groups. More references are [11, 42, 48, 64, 73], as well as many papers in the section on quantum algebra at arxiv.org.

In this paragraph I dare to make remarks of a historical character. The basic idea here goes back to Abel, Galois, and Lie among others in the 19th century who found that certain problems lead to a more systematic study of the idea of symmetry. This led to the birth of the group. The application of these ideas in a noncommutative setting led to the idea of the quantum group, I believe, more than the other way around. Certainly, the Leningrad school under L. Faddeev came to its ideas about quantum groups from the study of very specific problems in mathematical physics. Apparently, Woronowicz also had motivation from physics for his approach to quantum groups and was surely trying to find a generalization of a Lie group in the noncommutative setting. See his comments in the introduction of [84] for more on this. Much of the research on specific quantum groups has been focused on their corepresentation theory. And these are vector spaces with a covariant structure coming from the quantum group. So who gets the credit for covariance? I leave that as an exercise for the reader. It is one of the few exercises in this book that I have not solved myself.

Chapter 4
Adjoint Co-action

4.1 Right Adjoint Co-action

Now we begin to present material that will be needed for the theorem on bicovariant first-order differential calculi (FODCs). First, we note that the adjoint action of a Lie group G on itself has a counterpart for a Hopf algebra $\mathcal{A}$. We define ad : $\mathcal{A} \to \mathcal{A} \otimes \mathcal{A}$ by

$$\mathrm{ad}(a) := a^{(2)} \otimes \kappa(a^{(1)})a^{(3)}, \tag{4.1}$$

where we are using the *single-digit Sweedler's notation*

$$(\phi \otimes id)\phi(a) = (id \otimes \phi)\phi(a) = a^{(1)} \otimes a^{(2)} \otimes a^{(3)}$$

for the twofold co-product of a. This definition can be motivated by considering the case of a Lie group or that of a finite group. For the latter, see Chapter 13, on finite classical groups. However, this definition already shows that the generalization of a well-known classical structure to the noncommutative theory can be a rather nontrivial task.

Exercise 4.1. *Prove these identities:*

- $\mathrm{ad} = (id \otimes m)(id \otimes \kappa \otimes id)(\sigma_A \otimes id)(id \otimes \phi)\phi$.
- $\mathrm{ad} = (id \otimes m)(\sigma_A \otimes id)(\kappa \otimes id \otimes id)(id \otimes \phi)\phi$.

Theorem 4.1. *The map* ad *is a right co-action of $\mathcal{A}$ on itself. This means that the following two diagrams commute:*

$$
\begin{array}{ccc}
\mathcal{A} & \xrightarrow{\mathrm{ad}} & \mathcal{A} \otimes \mathcal{A} \\
{\scriptstyle \mathrm{ad}}\downarrow & & \downarrow{\scriptstyle id \otimes \phi} \\
\mathcal{A} \otimes \mathcal{A} & \xrightarrow{\mathrm{ad} \otimes id} & \mathcal{A} \otimes \mathcal{A} \otimes \mathcal{A}
\end{array}
\tag{4.2}
$$

© Springer International Publishing Switzerland 2015
S.B. Sontz, *Principal Bundles*, Universitext, DOI 10.1007/978-3-319-15829-7_4

and

$$A \xrightarrow{\;ad\;} A \otimes A$$
$$id_A \downarrow \qquad\qquad \downarrow id \otimes \varepsilon \qquad\qquad (4.3)$$
$$A \xrightarrow{\;\cong\;} A \otimes \mathbb{C}.$$

These diagrams correspond to the diagrams (3.7) and (3.8) in the definition of a right co-action.

Proof. We start with $\mathrm{ad}(a) = a^{(2)} \otimes \kappa(a^{(1)})a^{(3)}$. Going across and down in (4.2), using [B.10], namely, $\phi\kappa = \sigma_A(\kappa \otimes \kappa)\phi$ [see also (A.2) in [86]; cf. (3.5) in Chap. III of [45]], we get

$$(id \otimes \phi)\mathrm{ad}(a) = (id \otimes \phi)\Big(a^{(2)} \otimes \kappa(a^{(1)})a^{(3)}\Big)$$

$$= a^{(2)} \otimes \phi(\kappa(a^{(1)})a^{(3)})$$

$$= a^{(2)} \otimes \phi\big(\kappa(a^{(1)})\big)\,\phi(a^{(3)})$$

$$= a^{(2)} \otimes \sigma_A(\kappa \otimes \kappa)\phi(a^{(1)})\,(a^{(31)} \otimes a^{(32)})$$

$$= a^{(2)} \otimes \sigma_A(\kappa \otimes \kappa)(a^{(11)} \otimes a^{(12)})\,(a^{(31)} \otimes a^{(32)})$$

$$= a^{(2)} \otimes \sigma_A(\kappa(a^{(11)}) \otimes \kappa(a^{(12)}))\,(a^{(31)} \otimes a^{(32)})$$

$$= a^{(2)} \otimes (\kappa(a^{(12)}) \otimes \kappa(a^{(11)}))\,(a^{(31)} \otimes a^{(32)})$$

$$= a^{(2)} \otimes \kappa(a^{(12)})\,a^{(31)} \otimes \kappa(a^{(11)})\,a^{(32)}.$$

Now, going down and across in (4.2), we have that

$$(ad \otimes id)\mathrm{ad}(a) = (ad \otimes id)\Big(a^{(2)} \otimes \kappa(a^{(1)})a^{(3)}\Big)$$

$$= ad(a^{(2)}) \otimes \kappa(a^{(1)})a^{(3)}$$

$$= a^{(22)} \otimes \kappa(a^{(21)})a^{(23)} \otimes \kappa(a^{(1)})a^{(3)}.$$

Using the co-associativity of the co-multiplication, Sweedler's notation simplifies in the two previous formulas. In the first formula, we use

$$a^{(2)} \otimes a^{(12)} \otimes a^{(31)} \otimes a^{(11)} \otimes a^{(32)} = a^{(3)} \otimes a^{(2)} \otimes a^{(4)} \otimes a^{(1)} \otimes a^{(5)},$$

which uses the single-digit Sweedler's notation for the fourfold co-product, while in the second formula we use

$$a^{(22)} \otimes a^{(21)} \otimes a^{(23)} \otimes a^{(1)} \otimes a^{(3)} = a^{(3)} \otimes a^{(2)} \otimes a^{(4)} \otimes a^{(1)} \otimes a^{(5)},$$

which is also Sweedler's notation for the same fourfold co-product. We leave the precise proof of these two identities to the reader. And that finishes the proof of the commutativity of (4.2).

For (4.3), using $\varepsilon\kappa = \varepsilon$ (see B.11), we consider

$$
\begin{aligned}
(id \otimes \varepsilon)(ad(a)) &= (id \otimes \varepsilon)(a^{(2)} \otimes \kappa(a^{(1)})a^{(3)}) \\
&= a^{(2)} \otimes \varepsilon(\kappa(a^{(1)})a^{(3)}) \\
&= a^{(2)} \otimes \varepsilon(\kappa(a^{(1)}))\varepsilon(a^{(3)}) \\
&= a^{(2)} \otimes \varepsilon(a^{(1)})\varepsilon(a^{(3)}) \\
&= a^{(12)} \otimes \varepsilon(a^{(11)})\varepsilon(a^{(2)}) \\
&= \varepsilon(a^{(11)})a^{(12)} \otimes \varepsilon(a^{(2)}) \\
&= a^{(1)} \otimes \varepsilon(a^{(2)}) \\
&= \varepsilon(a^{(2)})a^{(1)} \otimes 1 \\
&= a \otimes 1 \cong a,
\end{aligned}
$$

where we also used $(\varepsilon \otimes id)\phi = id$ and $(id \otimes \varepsilon)\phi = id$. ∎

This result is also proved in Chapter 13, which discusses finite classical groups, in the special case where we take $\mathcal{A} = \mathcal{F}(G)$, the algebra of all functions on a finite group G. The moral coming from that proof is that the special case shown there does not shed much light on how to prove the general case here.

Definition 4.1. *A vector subspace V of $\mathcal{A}$ is* ad-invariant *if*

$$
ad(V) \subset V \otimes \mathcal{A}.
$$

If V is ad-invariant, then the restriction of ad to V (usually also denoted as ad : $V \to V \otimes \mathcal{A}$) is itself a right co-action of $\mathcal{A}$ on the subspace V.

4.2 Left Adjoint Co-action

While we will only be focusing on the right adjoint in the remainder of this book, we note here that there is also a corresponding left adjoint map $ad_L : \mathcal{A} \to \mathcal{A} \otimes \mathcal{A}$ defined by

$$
ad_L(a) := a^{(1)}\kappa(a^{(3)}) \otimes a^{(2)}
$$

for all $a \in \mathcal{A}$.

Exercise 4.2. *In analogy with the proof for the case shown above for the right adjoint co-action, show that this is a left co-action of $\mathcal{A}$ on itself.*

Exercise 4.3. *Show that for $\mathcal{A} = \mathcal{F}(G)$, where G is a group, ad_L is the pullback of the standard left adjoint action of G on itself. (It might help a bit to read first about the right adjoint action of G in Chapter 13.)*

4.3 Notes

This rather brief chapter presents the adjoint representations in the noncommutative setting. Their appearance in the classical theory surely dates back to Lie in the smooth setting, if not a bit earlier in the 19th century in an algebraic setting. These representations also appeared in a quite natural way early in the theory of quantum groups as it was developed at the end of the 20th century. However, as usual, I leave the details to the historians.

Chapter 5
Covariant Bimodules

5.1 Definitions

The study of a Hopf algebra $\mathcal{A}$ is continued here and, in particular, its possible relations with $\mathcal{A}$-bimodules Γ. However, these $\mathcal{A}$-bimodules are not necessarily the Γ in a first-order differential calculus (FODC) (Γ, d) over $\mathcal{A}$, though they could be. Hence, results of this section can and will be applied later to that particular case.

Rather, one should think of an $\mathcal{A}$-bimodule as a concept corresponding to a (smooth) vector bundle in classical differential geometry or, strictly speaking, to the space of smooth sections of a smooth vector bundle. That motivates the continued use of the symbol Γ for these objects. Also, Swan's theorem tells us that the distinction between a vector bundle E and the vector space $\Gamma(E)$ of its sections, though real enough, is not of fundamental importance; these can be thought of in some sense as the same object.

A basic motivating example (as well as a particular case) for this material is a Lie group G and the space of sections of its tangent bundle, which are known of course as *vector fields*. Since there is a naturally defined left action of G on itself, we have an induced left action (given by taking the derivative) of G on its tangent bundle. (Recall the definitions given earlier.) This in turn induces a left action of G on the space of vector fields $\chi(G)$ of G. Typically, $\chi(G)$ has infinite dimension. Those vector fields that are invariant (i.e., do not change) under this action are known as the *left invariant vector fields* on G, which also form a vector space $\chi_L(G)$. However, there is a vector space isomorphism $\chi_L(G) \cong T_e(G)$, where $T_e(G)$ is the tangent space at the identity element $e \in G$. So the subspace $\chi_L(G)$ of $\chi(G)$ is quite small in general. For example, if G is finite dimensional (which is often included as part of the definition of a Lie group), then $\chi_L(G)$ has the same finite dimension as G. But $\chi(G)$ is also a module over the algebra $C^\infty(G)$ of infinitely differentiable (real-valued) functions on G. Actually, in our current terminology $\chi(G)$ is a bimodule over the algebra $C^\infty(G)$. But $\chi_L(G)$ is not (in general) a $C^\infty(G)$-submodule of $\chi(G)$. The relation between $\chi_L(G)$ and $\chi(G)$ is that $\chi(G)$ is a free $C^\infty(G)$-module whose generators can be taken to be any vector space basis of $\chi_L(G)$ or, equivalently, any vector space basis of $T_e(G)$. The considerations of this paragraph apply equally well to any of the spaces of sections of the bundles associated with the tangent bundle (such as the cotangent bundle, tensor bundles, exterior power bundles of these, etc.), all of which have an induced left action of G. Furthermore, there is a theory similar to all this, where we start with the canonical right action of G on G.

For our discussion, the base space of the bundle can be any quantum space equipped with an action of the quantum group, that is, a co-action (left or right) of the Hopf algebra $\mathcal{A}$. Intuitively, this co-action in turn "induces" a co-action in the space of sections of any bundle, and the latter co-action is the important one for now. But rigorously we simply assume that we have an "associated" co-action in a bimodule. And the base space disappears entirely from the discussion! Keeping all this motivation in mind, here are two important definitions.

Definition 5.1. Let Γ be a bimodule over the Hopf algebra $\mathcal{A}$. Suppose that $\Phi_\Gamma :$ $\Gamma \to \mathcal{A} \otimes \Gamma$ is linear. Then we say that (Γ, Φ_Γ) is a *left covariant bimodule* over $\mathcal{A}$ provided that

(i) Φ_Γ is a morphism of $\mathcal{A}$-bimodules and
(ii) Φ_Γ is a left co-action (of $\mathcal{A}$ on Γ).

In part (i) of this definition, the $\mathcal{A}$-bimodule structure on $\mathcal{A} \otimes \Gamma$ is the pullback via the co-multiplication map $\phi : \mathcal{A} \to \mathcal{A} \otimes \mathcal{A}$ of the $\mathcal{A} \otimes \mathcal{A}$-bimodule structure on $\mathcal{A} \otimes \Gamma$.

Definition 5.2. Let Γ be a bimodule over the Hopf algebra $\mathcal{A}$. Suppose that $_\Gamma\Phi :$ $\Gamma \to \Gamma \otimes \mathcal{A}$ is linear. Then we say that $(\Gamma, {}_\Gamma\Phi)$ is a *right covariant bimodule* over $\mathcal{A}$ provided that

(i) $_\Gamma\Phi$ is a morphism of $\mathcal{A}$-bimodules and
(ii) $_\Gamma\Phi$ is a right co-action (of $\mathcal{A}$ on Γ).

Similarly, we also use ϕ here to give $\Gamma \otimes \mathcal{A}$ an $\mathcal{A}$-bimodule structure.

As an aid to memory, we note that the codomain of Φ_Γ, which has the subscript Γ on the right, is $\mathcal{A} \otimes \Gamma$, which also has Γ on the right. But Φ_Γ is called a *left* co-action since the "group" $\mathcal{A}$ appears on the left in the codomain. Another possible notation would be $_\mathcal{A}\Phi_\Gamma$ instead of Φ_Γ; this would clearly indicate that $\mathcal{A}$ appears on the left in the co-domain. But this would be way too many subindices! Similar (and in some sense dual) remarks hold for the right co-action $_\Gamma\Phi$.

Definition 5.3. Let Γ be a bimodule over the Hopf algebra $\mathcal{A}$. Suppose that (Γ, Φ_Γ) is a left covariant bimodule over $\mathcal{A}$ and that $(\Gamma, {}_\Gamma\Phi)$ is a right covariant bimodule over $\mathcal{A}$. Then we say that $(\Gamma, \Phi_\Gamma, {}_\Gamma\Phi)$ is a *bicovariant bimodule* over $\mathcal{A}$ provided that these two co-actions commute; that is, the following diagram is commutative:

$$
\begin{array}{ccc}
\Gamma & \xrightarrow{\Phi_\Gamma} & \mathcal{A} \otimes \Gamma \\
{}_\Gamma\Phi \downarrow & & \downarrow id \otimes {}_\Gamma\Phi \\
\Gamma \otimes \mathcal{A} & \xrightarrow{\Phi_\Gamma \otimes id} & \mathcal{A} \otimes \Gamma \otimes \mathcal{A}.
\end{array}
\tag{5.1}
$$

Remark. Note that this is the diagram that appears in (3.21), where it is a consequence of a theorem. Here it is an essential part of a definition.

Again, a classical and motivating example is $T(G) \to G$, the tangent bundle over a Lie group G. In this case, $\Gamma(T(G))$ is a bimodule over the commutative algebra $C^\infty(G)$. The bimodule structure satisfies $a\rho = \rho a$ for $\rho \in \Gamma(T(G))$ and $a \in C^\infty(G)$. Note that the above definitions apply in the particular case when $\mathcal{A}$ is commutative even though $a\rho$ and ρa may not always be equal.

Definition 5.4. Suppose that (Γ, Φ_Γ) is a left covariant bimodule over $\mathcal{A}$. Then we say that $\omega \in \Gamma$ is *left invariant* if $\Phi_\Gamma(\omega) = 1 \otimes \omega$. We introduce the notation

$$_{\mathrm{inv}}\Gamma := \{\omega \in \Gamma \mid \Phi_\Gamma(\omega) = 1 \otimes \omega\},$$

the set of left invariant elements in Γ. It is clearly a vector subspace of Γ.

Remark. We do not necessarily have that Γ is finite dimensional over $\mathbb{C}$. Consequently, $_{\mathrm{inv}}\Gamma$ also need not be finite dimensional. But we always have Hamel bases.

Some motivation for this definition of a left invariant element is in order. Suppose that $a : G \times M \to M$ is a left action on M. If we have $f : M \to \mathbb{R}$, then the usual definition is that f is *left invariant* provided that $f(g \cdot x) = f(x)$ for all $g \in G$ and all $x \in M$. By pullback we have the induced map

$$a^* : \mathcal{F}(M) \to \mathcal{F}(G \times M) \cong \mathcal{F}(G) \otimes \mathcal{F}(M).$$

When we view it this way, we have that

$$(a^* f)(g, x) = f(a(g, x)) = f(g \cdot x) = f(x) = 1 f(x) = (1 \times f)(g, x),$$

where in this last expression 1 denotes the constant function on G (whose constant value is 1). So $a^* f = 1 \times f \in \mathcal{F}(G \times M)$. But then $1 \times f$ corresponds to $1 \otimes f \in \mathcal{F}(G) \otimes \mathcal{F}(M)$ under the above canonical isomorphism. Modulo this isomorphism we get $a^* f = 1 \otimes f$. Conversely, $a^* f = 1 \otimes f$ implies that f is left invariant.

5.2 Left Covariance

We now start our analysis of left covariant bimodules over $\mathcal{A}$ with the following result.

Proposition 5.1. *Let* (Γ, Φ_Γ) *be a left covariant bimodule over* $\mathcal{A}$. *Then there exists a unique projection* $P : \Gamma \to {}_{\mathrm{inv}}\Gamma$ *such that*

$$P(b\omega) = \varepsilon(b) P(\omega) \tag{5.2}$$

for all $b \in \mathcal{A}$ *and all* $\omega \in \Gamma$.

Moreover, for all $\omega \in \Gamma$, we have that P also satisfies

$$\omega = \omega^{(0)} P(\omega^{(1)}),\tag{5.3}$$

where

$$\Phi_\Gamma(\omega) = \omega^{(0)} \otimes \omega^{(1)}$$

is Sweedler's notation for a co-action.

Remark. $_{\text{inv}}\Gamma$ is a complex vector space. So $P(\omega) \in {}_{\text{inv}}\Gamma$ and $\varepsilon(b) \in \mathbb{C}$ imply that the right side of (5.2) is in $_{\text{inv}}\Gamma$.

Proof. We will exclusively use Sweedler's notation throughout this proof. So $\Phi_\Gamma(\omega) = \omega^{(0)} \otimes \omega^{(1)} \in \mathcal{A} \otimes \Gamma$. We then define

$$P(\omega) := \kappa(\omega^{(0)})\omega^{(1)}.$$

Clearly, P is linear over $\mathbb{C}$.

Next, we prove that Ran $P \subset {}_{\text{inv}}\Gamma$. Take any $\omega \in \Gamma$. We first note that on one hand we have

$$(\phi \otimes id)\Phi_\Gamma(\omega) = (\phi \otimes id)(\omega^{(0)} \otimes \omega^{(1)}) = \omega^{(01)} \otimes \omega^{(02)} \otimes \omega^{(1)},$$

while on the other hand we get

$$(id \otimes \Phi_\Gamma)\Phi_\Gamma(\omega) = (id \otimes \Phi_\Gamma)(\omega^{(0)} \otimes \omega^{(1)}) = \omega^{(0)} \otimes \omega^{(10)} \otimes \omega^{(11)}.$$

But the fact that Φ_Γ is a left co-action implies that these last two expressions are equal. Thus,

$$\omega^{(01)} \otimes \omega^{(02)} \otimes \omega^{(1)} = \omega^{(0)} \otimes \omega^{(10)} \otimes \omega^{(11)},\tag{5.4}$$

which we will use in the following calculation. Now we calculate the object of interest, namely, the left co-action of $P(\omega)$:

$$\begin{aligned}
\Phi_\Gamma\left(P(\omega)\right) &= \Phi_\Gamma\left(\kappa(\omega^{(0)})\omega^{(1)}\right)\\
&= \phi(\kappa(\omega^{(0)}))\,\Phi_\Gamma(\omega^{(1)})\\
&= \phi(\kappa(\omega^{(0)}))(\omega^{(10)} \otimes \omega^{(11)})\\
&= \phi(\kappa(\omega^{(01)}))(\omega^{(02)} \otimes \omega^{(1)})\\
&= \phi(\kappa(\omega^{(01)}))(\omega^{(02)} \otimes 1)(1 \otimes \omega^{(1)})\\
&= (1 \otimes \kappa(\omega^{0}))(1 \otimes \omega^{(1)})
\end{aligned}$$

$$= 1 \otimes (\kappa(\omega^0)\omega^{(1)})$$

$$= 1 \otimes P(\omega),$$

which shows that $P(\omega) \in {}_{inv}\Gamma$ for all $\omega \in \Gamma$, as desired. Here we have also used the identity

$$\phi(\kappa(c^{(1)}))(c^{(2)} \otimes 1) = 1 \otimes \kappa(c)$$

for all $c \in \mathcal{A}$ [see (B.14)].

Now we start with $\omega \in {}_{inv}\Gamma$, which simply means that $\Phi_\Gamma(\omega) = 1 \otimes \omega$. Applying the definition of P using this formula gives

$$P(\omega) = \kappa(1)\omega = 1\omega = \omega;$$

that is, P acts as the identity on ${}_{inv}\Gamma$. And so P is a projection onto ${}_{inv}\Gamma$.

Next, we prove (5.2). To calculate $P(b\omega)$, we first consider

$$\Phi_\Gamma(b\omega) = \phi(b)\Phi_\Gamma(\omega) = (b^{(1)} \otimes b^{(2)})(\omega^{(0)} \otimes \omega^{(1)}) = b^{(1)}\omega^{(0)} \otimes b^{(2)}\omega^{(1)},$$

which implies by the definition of P that

$$P(b\omega) = \kappa(b^{(1)}\omega^{(0)}) b^{(2)}\omega^{(1)} = \kappa(\omega^{(0)})\kappa(b^{(1)}) b^{(2)}\omega^{(1)}$$

$$= \kappa(\omega^{(0)})\varepsilon(b)\omega^{(1)} = \varepsilon(b)\kappa(\omega^{(0)})\omega^{(1)}$$

$$= \varepsilon(b)P(\omega),$$

where we used (B.12) in the third equality. And so (5.2) is proved.

Next, we will show (5.3). Note that the second equality in the following is a consequence of (5.4).

$$\omega^{(0)} P(\omega^{(1)}) = \omega^{(0)}\kappa(\omega^{(10)})\omega^{(11)} = \omega^{(01)}\kappa(\omega^{(02)})\omega^{(1)}$$

$$= \varepsilon(\omega^{(0)})\omega^{(1)} = (\varepsilon \otimes id)\Phi_\Gamma(\omega)$$

$$= \omega.$$

The last equality holds because Φ_Γ is a left co-action.

Finally, to show the uniqueness, suppose that $P' : \Gamma \to {}_{inv}\Gamma$ is surjective, is a projection (i.e., acts as the identity on ${}_{inv}\Gamma$), and satisfies $P'(b\omega) = \varepsilon(b)P'(b)$. Applying P' to (5.3) for any $\omega \in \Gamma$, we get

$$P'(\omega) = P'(\omega^{(0)} P(\omega^{(1)})) = \varepsilon(\omega^{(0)}) P'(P(\omega^{(1)}))$$
$$= \varepsilon(\omega^{(0)}) P(\omega^{(1)}) = P\left(\varepsilon(\omega^{(0)})\omega^{(1)}\right)$$
$$= P((\varepsilon \otimes id)\Phi_\Gamma(\omega)) = P(\omega).$$

Again, the last equality holds because Φ_Γ is a left co-action. Since $\omega \in \Gamma$ is an arbitrary element in (5.3), we conclude that $P' = P$. ∎

We continue with the main structure theorem for left covariant bimodules over $\mathcal{A}$.

Theorem 5.1. *Let (Γ, Φ_Γ) be a left covariant bimodule over $\mathcal{A}$. Suppose that $\{\omega_i\}_{i \in I}$ is a vector space basis of $_{\mathrm{inv}}\Gamma$. Then*

1. *Γ is a free left $\mathcal{A}$-module with basis $\{\omega_i\}_{i \in I}$.*
2. *Γ is a free right $\mathcal{A}$-module with basis $\{\omega_i\}_{i \in I}$.*
3. *There exist linear functionals $f_{ij} \in \mathcal{A}'$, the dual space of $\mathcal{A}$, uniquely determined by the equations*

$$\omega_i b = \sum_{j \in I} (f_{ij} * b)\, \omega_j \tag{5.5}$$

for all $i \in I$ and all $b \in \mathcal{A}$. These linear functionals also satisfy

$$a\omega_i = \sum_{j \in I} \omega_j \left((f_{ij} \circ \kappa^{-1}) * a\right). \tag{5.6}$$

[The convolution product $*$ used here is defined by $g * b := (id \otimes g)\phi(b)$ for $b \in \mathcal{A}$ and $g \in \mathcal{A}'$.] Moreover, these functionals also satisfy

$$f_{ij}(ab) = \sum_{k \in I} f_{ik}(a) f_{kj}(b) \tag{5.7}$$

$$f_{ij}(1) = \delta_{ij} \quad \text{(the Kronecker delta).} \tag{5.8}$$

Writing f for the matrix (f_{ij}), we can express these last two equations as $f(ab) = f(a)f(b)$ and $f(1) = I$, the identity matrix. Hence, f is an algebra morphism *mapping the algebra $\mathcal{A}$ to an algebra of matrices with complex entries. In other words, f is a representation of $\mathcal{A}$. It is called the* structure representation *(SR) of the left covariant bimodule with respect to the basis $\{\omega_i\}_{i \in I}$.*
4. *Let $\{\omega_i'\}_{i \in I}$ be another Hamel basis of the vector space basis of $_{\mathrm{inv}}\Gamma$. (We can take the index set to be I here since the cardinality of a basis of a vector space is independent of its choice.) Let $f_{ij}' \in \mathcal{A}'$ be the corresponding linear functionals as in the previous part of this theorem. Let $C \in \mathrm{End}(_{\mathrm{inv}}\Gamma)$ be the unique invertible linear map of $_{\mathrm{inv}}\Gamma$ to itself such that $\omega_i' = C\omega_i$ for each $i \in I$. Then*

$$f = C^{-1} f' C. \tag{5.9}$$

Remarks. This theorem gives the structure of left covariant bimodules over $\mathcal{A}$. In particular, by the first two parts of this theorem for any $\omega \in \Gamma$, we have that

$$\omega = \sum_{i \in I} a_i \omega_i = \sum_{i \in I} \omega_i b_i$$

for unique elements $a_i, b_i \in \mathcal{A}$. Then the third part shows how these two representations of ω are related. Moreover, using this notation, we have

$$\omega b = \sum_{ij} a_i (f_{ij} * b) \omega_j, \qquad (5.10)$$

$$\Phi_\Gamma(\omega) = \sum_i \phi(a_i)(1 \otimes \omega_i). \qquad (5.11)$$

Free $\mathcal{A}$-modules (free both on the left and on the right) correspond to trivial bundles in the classical case. This is a consequence of the covariance. Nontrivial classical bundles have spaces of sections that are projective, but not free, modules.

We can also consider (5.5) and its consequence (5.10) as *commutation relations*. This is because in classical geometry not only is the algebra commutative, but also the bimodule structure satisfies $a\omega = \omega a$ for all $a \in \mathcal{A}$ and $\omega \in \Gamma$. (I'm speaking totally off the record and parenthetically, but one might be tempted to say that the product $a\omega$ is commutative in the classical case.) But in the quantum case, the formulas (5.5) and (5.10) tell us how to pass an element b acting on the right into something (determined by b) on the left. So these formulas reflect the lack of "commutativity" in the quantum case. Similarly, (5.6) can be thought of as a commutation relation.

For the construction in the reverse direction, suppose that we are given a representation f of the algebra $\mathcal{A}$ by $|I| \times |I|$ matrices with complex entries, where $|I|$ denotes the cardinality of the set I. Then taking arbitrary abstract objects ω_i for each $i \in I$, we consider the free left $\mathcal{A}$-module, call it Γ_f, with basis $\{\omega_i \mid i \in I\}$. Then for any element $\omega = \sum_{i \in I} a_i \omega_i \in \Gamma_f$, where the coefficients $a_i \in \mathcal{A}$ are unique, we define a right action on ω by Eq. (5.10) using the SR f, and we define the left co-action on ω by Eq. (5.11). Then Γ_f is a left covariant bimodule over $\mathcal{A}$. Moreover, using the basis ω_i of Γ_f, we find that its associated SR is exactly f. In this sense, the previous theorem gives a description of all left covariant bimodules over $\mathcal{A}$ in terms of representations. However, two different representations can give rise to isomorphic left covariant bimodules over $\mathcal{A}$. This is basically what part 4 is all about.

By the previous comments, SRs are not needed in classical differential geometry. However, it does apply to that situation if one lets $f_{ij} = \delta_{ij}\varepsilon$, since this choice yields the identity $a\omega = \omega a$. Of course, the SR is a nontrivial structure in the quantum situation. Notice that the quantum situation includes when $\mathcal{A}$ is a commutative algebra but $a\omega = \omega a$ is *not* an identity.

Exercise 5.1. *Write out all the details to show that Γ_f as constructed just above is a left covariant bimodule over $\mathcal{A}$. (This amounts to proving that two diagrams commute.)*

Now we proceed to the proof of Theorem 5.1.

Proof. For part 1, we note that (5.3) says that any $\omega \in \Gamma$ can be written as a linear combination of elements in ${}_{inv}\Gamma$ with coefficients in $\mathcal{A}$ acting on the left. This in turn implies that ω can be written as a linear combination of elements in any vector space basis of ${}_{inv}\Gamma$, say ω_i, with coefficients in $\mathcal{A}$ acting on the left. So this gives the existence of the desired expansion in part 1. To show that it is unique (that is, the left $\mathcal{A}$-module is free), suppose that $\omega = \sum_i a_i \omega_i$ is one such expansion of ω in the basis ω_i of ${}_{inv}\Gamma$. It follows that

$$\Phi_\Gamma(\omega) = \Phi_\Gamma(\sum_i a_i\omega_i) = \sum_i \phi(a_i)\Phi_\Gamma(\omega_i) = \sum_i \phi(a_i)(1 \otimes \omega_i).$$

Then we obtain

$$(id \otimes P)\Phi_\Gamma(\omega) = (id \otimes P)\Big(\sum_i \phi(a_i)(1 \otimes \omega_i)\Big)$$

$$= \sum_i (id \otimes P)\left((a_i^{(1)} \otimes a_i^{(2)}\omega_i)\right)$$

$$= \sum_i a_i^{(1)} \otimes P(a_i^{(2)}\omega_i)$$

$$= \sum_i a_i^{(1)} \otimes \varepsilon(a_i^{(2)})P(\omega_i)$$

$$= \sum_i a_i^{(1)}\varepsilon(a_i^{(2)}) \otimes \omega_i$$

$$= \sum_i a_i \otimes \omega_i \in \mathcal{A} \otimes ({}_{inv}\Gamma).$$

But the fact that the ω_is are a basis over $\mathbb{C}$ of ${}_{inv}\Gamma$ implies that the elements $a_i \in \mathcal{A}$ in the last expression are unique. This finishes the proof of part 1.

For part 3, we start by considering $\omega_j b$ for any $b \in \mathcal{A}$ and any basis element ω_j. So $j \in I$. By part 1, we can write

$$\omega_j b = \sum_{i \in I} F_{ji}(b)\omega_i, \tag{5.12}$$

where the coefficients $F_{ji}(b) \in \mathcal{A}$ are unique. It is straightforward to show that $F_{ji}: \mathcal{A} \to \mathcal{A}$ is linear over the field $\mathbb{C}$ for all $i, j \in I$. Now we consider

$$\sum_i F_{ji}(ab)\omega_i = \omega_j ab$$

$$= \Big(\sum_k F_{jk}(a)\omega_k\Big)b$$

$$= \sum_k F_{jk}(a)(\omega_k b)$$

$$= \sum_{ki} F_{jk}(a)F_{ki}(b)\omega_i,$$

where $a, b \in \mathcal{A}$ and $i, j, k \in I$. By the uniqueness of the coefficients in $\mathcal{A}$ (by part 1 again), we deduce that

$$F_{ji}(ab) = \sum_k F_{jk}(a)F_{ki}(b) \qquad (5.13)$$

for all $a, b \in \mathcal{A}$ and $i, j \in I$. Now we define

$$f_{ji} := \varepsilon F_{ji}.$$

So $f_{ji} : \mathcal{A} \to \mathbb{C}$ is linear, that is, $f_{ji} \in \mathcal{A}'$. Now we use (5.13) to get

$$f_{ji}(ab) = \varepsilon F_{ji}(ab)$$

$$= \varepsilon\Big(\sum_k F_{jk}(a)F_{ki}(b)\Big)$$

$$= \sum_k \varepsilon(F_{jk}(a))\varepsilon(F_{ki}(b))$$

$$= \sum_k f_{jk}(a)f_{ki}(b),$$

which shows (5.7). Next, by taking $b = 1 \in \mathcal{A}$ in (5.12), we obtain

$$\omega_j = \sum_i F_{ji}(1)\omega_i.$$

Then by the uniqueness result of part 1, it follows that $F_{ji}(1) = \delta_{ji}1 \in \mathcal{A}$. Applying ε to this last equation gives $f_{ji}(1) = \delta_{ji}1 = \delta_{ji} \in \mathbb{C}$. And so (5.8) has been shown.

Now we will prove (5.5). This comes down to nothing more than finding a formula for the coefficients $F_{ji}(b)$ on the right side of (5.12). We recall that the definition of the convolution product (in the case of interest here; see [84] for this and other cases) is

$$f * a := (id \otimes f)\phi(a)$$

for $f \in \mathcal{A}'$ and $a \in \mathcal{A}$. So $f * a \in \mathcal{A} \otimes \mathbb{C} \cong \mathcal{A}$. First, we apply Φ_Γ to (5.12), thereby getting for all $j \in I$ that

$$\Phi_\Gamma(\omega_j b) = \sum_i \Phi_\Gamma(F_{ji}(b)\omega_i)$$

$$= \sum_i \phi(F_{ji}(b))\Phi_\Gamma(\omega_i)$$

$$= \sum_i \phi(F_{ji}(b))(1 \otimes \omega_i).$$

But we also have for all $j \in I$ that

$$\Phi_\Gamma(\omega_j b) = \Phi_\Gamma(\omega_j)\phi(b)$$

$$= (1 \otimes \omega_j)\phi(b)$$

$$= (1 \otimes \omega_j)(b^{(1)} \otimes b^{(2)})$$

$$= b^{(1)} \otimes \omega_j b^{(2)}$$

$$= b^{(1)} \otimes \sum_i F_{ji}(b^{(2)})\omega_i$$

$$= \sum_i b^{(1)} \otimes F_{ji}(b^{(2)})\omega_i$$

$$= \sum_i \left((id \otimes F_{ji})(b^{(1)} \otimes b^{(2)})\right)(1 \otimes \omega_i)$$

$$= \sum_i \left((id \otimes F_{ji})\phi(b)\right)(1 \otimes \omega_i).$$

Comparing these last two equations, we find for all $i, j \in I$ that

$$\phi(F_{ji}(b)) = (id \otimes F_{ji})\phi(b).$$

Acting on this last equation with $id \otimes \varepsilon$, we get

$$F_{ji}(b) = (id \otimes \varepsilon)(\phi(F_{ji}(b)))$$

$$= (id \otimes \varepsilon)\left((id \otimes F_{ji})\phi(b)\right)$$

$$= (id \otimes \varepsilon F_{ji})\phi(b)$$

$$= (id \otimes f_{ji})\phi(b)$$

$$= f_{ji} * b.$$

Substituting this back into (5.12), we find that

$$\omega_j b = \sum_i F_{ji}(b)\omega_i = \sum_i (f_{ji} * b)\,\omega_i,$$

which is exactly the formula (5.5) we wished to prove.

To prove the uniqueness of the functionals f_{ij}, we first note that by its definition $F_{ij}(b)$ is unique for all $b \in \mathcal{A}$. So if g_{ij} is another family also satisfying the analog of (5.5), then it follows that $(f_{ij} * b) = (g_{ij} * b)$ for all $b \in \mathcal{A}$. But $\varepsilon(f * a) = f(a)$ for $f \in \mathcal{A}'$ and $a \in \mathcal{A}$ is an identity of Hopf algebras [see (B.16)]. Thus, applying ε to the last equation and using this identity gives

$$f_{ij}(b) = \varepsilon(f_{ij} * b) = \varepsilon(g_{ij} * b) = g_{ij}(b)$$

for all $b \in \mathcal{A}$, which in turn implies $f_{ij} = g_{ij}$.

Next, we will prove (5.6). We will use the identity

$$(f * g) * a = f * (g * a),$$

where $f, g \in \mathcal{A}'$ and $a \in \mathcal{A}$. [Exercise! A bit of help: $f * g := (f \otimes g)\phi \in \mathcal{A}'$ for $f, g \in \mathcal{A}'$.]

Manipulating the right side of (5.6) using (5.5) and the exercise, we see that

$$\sum_{j \in I} \omega_j \left((f_{ij} \circ \kappa^{-1}) * a \right) = \sum_{jk} f_{jk} * \left((f_{ij} \circ \kappa^{-1}) * a \right) \omega_k$$

$$= \sum_k \left(\sum_j \left(f_{jk} * (f_{ij} \circ \kappa^{-1}) \right) * a \right) \omega_k.$$

So (5.6) holds if

$$\sum_j \left(f_{jk} * (f_{ij} \circ \kappa^{-1}) \right) * a = \delta_{ki}\, a.$$

But

$$\delta_{ki}\, a = \delta_{ki}(id \otimes \varepsilon)\phi(a) = (id \otimes \delta_{ki}\varepsilon)\phi(a) = \delta_{ki}\varepsilon * a.$$

So (5.6) holds if

$$\sum_j \left(f_{jk} * (f_{ij} \circ \kappa^{-1}) \right) * a = \delta_{ki}\varepsilon * a$$

for all $a \in \mathcal{A}$. So, in order to prove (5.6), it suffices to show that

$$\sum_j f_{jk} * (f_{ij} \circ \kappa^{-1}) = \delta_{ki} \varepsilon, \qquad (5.14)$$

which we now will do. Note that this formula (5.14) is an equation between elements in $\mathcal{A}'$.

Since κ is surjective, we will show that the two sides of (5.14) evaluate to equal complex numbers when acting on all elements $\kappa(a)$, where $a \in \mathcal{A}$. Starting from the left side of (5.14), we calculate

$$\sum_j \left(f_{jk} * (f_{ij} \circ \kappa^{-1}) \right) (\kappa(a)) = \sum_j \left(f_{jk} \otimes (f_{ij} \circ \kappa^{-1}) \right) \phi(\kappa(a))$$

$$= \sum_j \left(f_{jk} \otimes (f_{ij} \circ \kappa^{-1}) \right) \sigma_{\mathcal{A}} (\kappa \otimes \kappa) \phi(a)$$

$$= \sum_j \left((f_{ij} \circ \kappa^{-1}) \otimes f_{jk} \right) (\kappa \otimes \kappa) \phi(a)$$

$$= \sum_j \left(f_{ij} \otimes f_{jk} \right) (\kappa^{-1} \otimes id)(\kappa \otimes \kappa) \phi(a)$$

$$= \sum_j \left(f_{ij} \otimes f_{jk} \right) (id \otimes \kappa) \phi(a).$$

But an equivalent way of writing (5.7) is $\sum_j f_{ij} \otimes f_{jk} = f_{ik} m_{\mathcal{A}}$, so that

$$\sum_j \left(f_{jk} * (f_{ij} \circ \kappa^{-1}) \right) (\kappa(a)) = f_{ik} \left(m_{\mathcal{A}} (id \otimes \kappa) \phi(a) \right)$$

$$= f_{ik} (\varepsilon(a) 1)$$

$$= \varepsilon(a) f_{ik} (1)$$

$$= \delta_{ik} \varepsilon(\kappa(a)),$$

which by our remarks implies (5.14). Here we have used the identities $m_{\mathcal{A}}(id \otimes \kappa) \phi(a) = \varepsilon(a) 1$ and $\varepsilon(\kappa(a)) = \varepsilon(a)$ [see (B.13) and (B.11)] as well as (5.8).

In much the same way one can show that

$$\sum_j (f_{ji} \circ \kappa^{-1}) * f_{kj} = \delta_{ik} \varepsilon. \qquad (5.15)$$

Note that we are *not* saying that the convolution product is commutative. And it need not be. Furthermore, (5.15) is not an immediate consequence of commutativity.

Exercise 5.2. *Prove (5.15).*

For part 2, we first note that there exists *some* expansion of an arbitrary element $\omega \in \Gamma$ as $\omega = \sum_j \omega_j c_j$ with $c_j \in \mathcal{A}$ because of part 1 and (5.6).

But is this expansion unique? Suppose we have a relation $\sum_k \omega_k b_k = 0$, where only finitely many of the coefficients b_k are nonzero. (Recall that $k \in I$ and that the index set I could have infinite cardinality.) By (5.5), we then have

$$0 = \sum_k \omega_k b_k = \sum_k \sum_j (f_{kj} * b_k) \omega_j = \sum_j \left(\sum_k (f_{kj} * b_k) \right) \omega_j.$$

Now, by part 1, we conclude that $\sum_k f_{kj} * b_k = 0$ for all $j \in I$. So, on the one hand,

$$\sum_j (f_{ji} \circ \kappa^{-1}) * \sum_k f_{kj} * b_k = 0,$$

while on the other hand, using (5.15),

$$\sum_j (f_{ji} \circ \kappa^{-1}) * \sum_k f_{kj} * b_k = \sum_k \sum_j (f_{ji} \circ \kappa^{-1}) * (f_{kj} * b_k)$$

$$= \sum_k \sum_j \left((f_{ji} \circ \kappa^{-1}) * f_{kj} \right) * b_k$$

$$= \sum_k \delta_{ik} \varepsilon * b_k$$

$$= \varepsilon * b_i$$

$$= (id \otimes \varepsilon) \phi(b_i)$$

$$= b_i.$$

So $b_i = 0$ for all $i \in I$, as desired. ∎

5.3 Right Covariance

There is an analogous structure theory for right covariant bimodules over $\mathcal{A}$. We now write out that theory without giving proofs since everything is analogous to the theory given above for left covariant bimodules over $\mathcal{A}$.

Definition 5.5. Suppose that $(\Gamma, {}_\Gamma\Phi)$ is a right covariant bimodule over $\mathcal{A}$. Then we say that $\omega \in \Gamma$ is *right invariant* if ${}_\Gamma\Phi(\omega) = \omega \otimes 1$. We introduce the notation

$$\Gamma_{\mathrm{inv}} := \{\omega \in \Gamma \mid {}_\Gamma\Phi(\omega) = \omega \otimes 1\},$$

the set of right invariant elements in Γ. It is clearly a vector subspace of Γ.

The motivation for this definition is similar (more precisely: dual) to the motivation for left invariant elements. The next theorem gives a complete description of right covariant bimodules.

Theorem 5.2. *Let* $(\Gamma, {}_\Gamma\Phi)$ *be a right covariant bimodule over* $\mathcal{A}$. *Suppose that* $\{\eta_i\}_{i\in J}$ *is a vector space basis of* Γ_{inv}. *Then*

1. Γ *is a free left* $\mathcal{A}$-*module with basis* $\{\eta_i\}_{i\in J}$.
2. Γ *is a free right* $\mathcal{A}$-*module with basis* $\{\eta_i\}_{i\in J}$.
3. *There exist linear functionals* $g_{ij} \in \mathcal{A}'$, *the dual space of* $\mathcal{A}$, *uniquely determined by the equations*

$$\eta_i b = \sum_{j\in J} (b * g_{ij})\, \eta_j \qquad (5.16)$$

for all $i \in J$ *and all* $b \in \mathcal{A}$. *[This is yet another convolution product, which is defined by* $a * f := (f \otimes id)\phi(a)$ *for* $a \in \mathcal{A}$ *and* $f \in \mathcal{A}'$.*] These linear functionals also satisfy*

$$a\eta_i = \sum_{j\in J} \eta_j \left(a * (g_{ij} \circ \kappa^{-1})\right). \qquad (5.17)$$

Moreover, they also satisfy

$$g_{ij}(ab) = \sum_{k\in J} g_{ik}(a)g_{kj}(b), \qquad (5.18)$$

$$g_{ij}(1) = \delta_{ij}. \quad \text{(the Kronecker delta)} \qquad (5.19)$$

Writing g *for the matrix* (g_{ij}), *we can express these last two equations as* $g(ab) = g(a)g(b)$ *and* $g(1) = I$, *the identity matrix. So* g *is called the* structure representation (SR) *of the right covariant bimodule with respect to the basis* $\{\eta_i\}_{i\in I}$.

Remark. The index set J and the SR g here do not necessarily have anything to do with the index set I and the SR f in Theorem 5.1. Remarks hold here for right covariant bimodules over $\mathcal{A}$ that are analogous to the case of left covariant bimodules over $\mathcal{A}$.

5.4 Bicovariance

Now for bicovariant bimodules, we do have something new that requires proof. We claim that this result gives a complete description of bicovariant bimodules; the exact statement follows the proof of the theorem.

Theorem 5.3. *Let $(\Gamma, \Phi_\Gamma, {}_\Gamma\Phi)$ be a bicovariant bimodule over $\mathcal{A}$. Suppose that $\{\omega_i\}_{i \in I}$ is a vector space basis of $_{\text{inv}}\Gamma$, the left invariant elements in Γ. Then*

1. The right co-action on the left invariant basis elements is given by

$$_\Gamma\Phi(\omega_i) = \sum_{j \in I} \omega_j \otimes R_{ji} \qquad (5.20)$$

for all $i \in I$, where the elements $R_{ji} \in \mathcal{A}$ for $i, j \in I$ satisfy

$$\phi(R_{ji}) = \sum_{k \in I} R_{jk} \otimes R_{ki}, \qquad (5.21)$$

$$\varepsilon(R_{ji}) = \delta_{ji}. \qquad (5.22)$$

2. There exists a vector space basis $\{\eta_i\}_{i \in J}$ with the same index set $J = I$ as above for the vector space Γ_{inv} of right invariant elements, so that for all $i \in I$, we have

$$\omega_i = \sum_{j \in I} \eta_j R_{ji}. \qquad (5.23)$$

In particular, dim Γ_{inv} = dim $_{\text{inv}}\Gamma$.

3. With the choice of basis $\{\eta_i\}_{i \in I}$ as in part 2, the linear functionals f_{ij} in Theorem 5.1 and the linear functionals g_{ij} in Theorem 5.2 are the same; that is, $f_{ij} = g_{ij}$ for all $i, j \in I = J$. In short, for the SRs with respect to the appropriately chosen bases, we have $f = g$.

4. For all $j, k \in I$ and all $a \in \mathcal{A}$, we have

$$\sum_i R_{ij}(a * f_{ik}) = \sum_i (f_{ji} * a) R_{ki}. \qquad (5.24)$$

5. Dual to part 1, we also have that the left co-action on the right invariant basis elements of part 2 is given by

$$\Phi_\Gamma(\eta_j) = \sum_i \kappa(R_{ij}) \otimes \eta_i. \qquad (5.25)$$

Remark. The representation f and the matrix $R = (R_{ij})$ (satisfying the appropriate conditions) suffice to define a bicovariant bimodule over $\mathcal{A}$.

Proof. We first prove part 1. Take any element ω of $_{\text{inv}}\Gamma$. So $\Phi_\Gamma(\omega) = 1 \otimes \omega$. The compatibility of the left and right co-actions gives the first equality in this calculation:

$$(\Phi_\Gamma \otimes id)_\Gamma\Phi(\omega) = (id \otimes {}_\Gamma\Phi)\Phi_\Gamma(\omega) = (id \otimes {}_\Gamma\Phi)(1 \otimes \omega) = 1 \otimes {}_\Gamma\Phi(\omega). \qquad (5.26)$$

Claim. This implies that $_\Gamma\Phi(\omega) \in (_\text{inv}\Gamma) \otimes \mathcal{A}$. To show this claim, let $\{a_j\}$ be a basis of $\mathcal{A}$. Then we can expand $_\Gamma\Phi(\omega)$ as

$$_\Gamma\Phi(\omega) = \sum_j \eta_j \otimes a_j$$

for some $\eta_j \in \Gamma$. Applying $(\Phi_\Gamma \otimes id)$ to this and using the compatibility condition (5.26), we obtain

$$\sum \Phi_\Gamma(\eta_j) \otimes a_j = (\Phi_\Gamma \otimes id)(\sum \eta_j \otimes a_j)$$

$$= (\Phi_\Gamma \otimes id)\,_\Gamma\Phi(\omega)$$

$$= 1 \otimes\,_\Gamma\Phi(\omega)$$

$$= 1 \otimes \sum \eta_j \otimes a_j$$

$$= \sum 1 \otimes \eta_j \otimes a_j.$$

Since $\{a_j\}$ is a basis, we conclude that $\Phi_\Gamma(\eta_j) = 1 \otimes \eta_j$ for every j. This in turn simply says that $\eta_j \in\,_\text{inv}\Gamma$ for every j. So it follows that

$$_\Gamma\Phi(\omega) = \sum_j \eta_j \otimes a_j \in\,_\text{inv}\Gamma \otimes \mathcal{A},$$

as claimed.

Since ω_i is a basis over $\mathbb{C}$ of the vector space $_\text{inv}\Gamma$ and the tensor products here are over $\mathbb{C}$, we can write $_\Gamma\Phi(\omega_i)$ in the form (5.20). By definition, the right co-action $_\Gamma\Phi$ satisfies

$$(id \otimes \phi)\,_\Gamma\Phi = (_\Gamma\Phi \otimes id)\,_\Gamma\Phi.$$

Acting on (5.20) with $id \otimes \phi$ and applying the previous formula, we obtain

$$\sum_j \omega_j \otimes \phi(R_{ji}) = (id \otimes \phi) \sum_j \omega_j \otimes R_{ji}$$

$$= (id \otimes \phi)\,_\Gamma\Phi(\omega_i)$$

$$= (_\Gamma\Phi \otimes id)\,_\Gamma\Phi(\omega_i)$$

$$= (_\Gamma\Phi \otimes id) \sum_k \omega_k \otimes R_{ki}$$

$$= \sum_k {}_\Gamma \Phi(\omega_k) \otimes R_{ki}$$

$$= \sum_{jk} \omega_j \otimes R_{jk} \otimes R_{ki}.$$

This then implies that

$$\phi(R_{ji}) = \sum_k R_{jk} \otimes R_{ki},$$

which is exactly (5.21). Next, we apply $id \otimes \varepsilon$ to (5.20), getting

$$\omega_i = (id \otimes \varepsilon) {}_\Gamma \Phi(\omega_i) = \sum_j \omega_j \otimes \varepsilon(R_{ji}) = \sum_j \varepsilon(R_{ji})\omega_j,$$

where the first equality is another defining property of a right co-action. But the ω_ks form a vector space basis of ${}_{\text{inv}}\Gamma$, and so it follows that $\varepsilon(R_{ji}) = \delta_{ji}$, which shows (5.22). This concludes the proof of part 1.

For part 2, we begin with these two identities [see (B.12) and (B.13)]:

$$m_{\mathcal{A}}(\kappa \otimes id)\phi(a) = \varepsilon(a)1 \quad \text{and} \quad m_{\mathcal{A}}(id \otimes \kappa)\phi(a) = \varepsilon(a)1.$$

Using the first of these together with (5.21), we have

$$\delta_{ji}1 = \varepsilon(R_{ji})1$$

$$= m_{\mathcal{A}}(\kappa \otimes id)\phi(R_{ji})$$

$$= \sum_k m_{\mathcal{A}}(\kappa \otimes id)(R_{jk} \otimes R_{ki})$$

$$= \sum_k m_{\mathcal{A}}(\kappa(R_{jk}) \otimes R_{ki})$$

$$= \sum_k \kappa(R_{jk}) R_{ki}.$$

Similarly, using the second identity, one can show that

$$\sum_k R_{jk} \kappa(R_{ki}) = \delta_{ji}1. \tag{5.27}$$

Now we define what will turn out to be the basis for the vector space Γ_{inv} of right invariant elements. Specifically, for each $j \in I$, we define

$$\eta_j := \sum_i \omega_i \kappa(R_{ij}). \tag{5.28}$$

Note that we are using the same index set I for the $\eta's$ as we already had for the basis $\{\omega_i\}_{i \in I}$ of $_{\text{inv}}\Gamma$. We now continue by proving properties of the $\eta's$. To verify (5.23), we calculate

$$\sum_j \eta_j R_{ji} = \sum_j \sum_k \omega_k \kappa(R_{kj}) R_{ji}$$

$$= \sum_k \omega_k \sum_j \kappa(R_{kj}) R_{ji}$$

$$= \sum_k \omega_k \delta_{ki} 1$$

$$= \omega_i,$$

which is exactly (5.23).

Next, to show that η_j is right invariant, we calculate its right co-action:

$$_\Gamma\Phi(\eta_j) = {}_\Gamma\Phi\left(\sum_i \omega_i \kappa(R_{ij})\right)$$

$$= \sum_i {}_\Gamma\Phi(\omega_i \kappa(R_{ij}))$$

$$= \sum_i {}_\Gamma\Phi(\omega_i)\phi(\kappa(R_{ij})).$$

We have a formula for the first factor in this sum, namely, (5.20). To proceed, we examine the second factor here on its own:

$$\phi(\kappa(R_{ij})) = \sigma_\mathcal{A}(\kappa \otimes \kappa)\phi(R_{ij}) = \sigma_\mathcal{A}(\kappa \otimes \kappa)\sum_l R_{il} \otimes R_{lj}$$

$$= \sigma_\mathcal{A}\left(\sum_l \kappa(R_{il}) \otimes \kappa(R_{lj})\right) = \sum_l \kappa(R_{lj}) \otimes \kappa(R_{il}).$$

Putting this back into the previous calculation, we continue as follows:

$$_\Gamma\Phi(\eta_j) = \sum_i {}_\Gamma\Phi(\omega_i)\phi(\kappa(R_{ij}))$$

$$= \sum_{il} {}_\Gamma\Phi(\omega_i)(\kappa(R_{lj}) \otimes \kappa(R_{il}))$$

$$= \sum_{ilm} (\omega_m \otimes R_{mi})(\kappa(R_{lj}) \otimes \kappa(R_{il}))$$

$$= \sum_{ilm} \omega_m \kappa(R_{lj}) \otimes R_{mi}\kappa(R_{il})$$

$$= \sum_{lm} \omega_m \kappa(R_{lj}) \otimes \sum_i R_{mi}\kappa(R_{il})$$

$$= \sum_{lm} \omega_m \kappa(R_{lj}) \otimes \delta_{ml} 1$$

$$= \sum_l \omega_l \kappa(R_{lj}) \otimes 1$$

$$= \eta_j \otimes 1.$$

This says exactly that η_j is right invariant; that is, $\eta_j \in \Gamma_{inv}$ for every $j \in I$.

Next, we will show that any $\eta \in \Gamma_{inv}$ is a unique linear combination of the elements η_j with coefficients in $\mathbb{C}$, namely, that the η_j are a vector space basis of Γ_{inv}. We can write η as

$$\eta = \sum_k \eta_k a_k \tag{5.29}$$

for some elements $a_k \in \mathcal{A}$, since η, being an element in Γ, can first be written as $\eta = \sum_i \omega_i b_i$ for $b_i \in \mathcal{A}$ by Theorem 5.1, part 2, and then (5.23) shows we can write each ω_i as a linear combination of the η'_ks with coefficients in $\mathcal{A}$ acting on the right. However, we want to show that η is a linear combination of the η'_ks with *complex* coefficients, and in (5.29) the coefficients are only known to be in $\mathcal{A}$. So we still have work to do!

Substituting the definition of η_k into (5.29), we get

$$\eta = \sum_l \omega_l \left(\sum_k \kappa(R_{lk}) a_k \right). \tag{5.30}$$

Suppose now that $\eta = 0$. Then since ω_l is a basis for the free right $\mathcal{A}$-module Γ, we see that

$$\sum_k \kappa(R_{lk}) a_k = 0$$

for all $l \in I$. The idea now is that the matrix $\kappa(R_{lk})$ is invertible and so the a'_ks must all be equal to zero. Explicitly, by multiplying this equation by R_{jl} and then summing on $l \in I$, we obtain

$$\sum_k \sum_l R_{jl} \kappa(R_{lk}) a_k = 0.$$

But (5.27) says that $\sum_l R_{jl} \kappa(R_{lk}) = \delta_{jk}$, so that

$$a_j = \sum_k \delta_{jk} a_k = 0$$

for all $j \in I$. Thus, $\eta = 0$ has a unique representation in the form (5.29), and consequently *any* $\eta \in \Gamma$ has a unique representation in that form. We have not yet used the hypothesis that $\eta \in \Gamma_{\text{inv}}$. We now use this hypothesis as well as (5.29) and $\eta_k \in \Gamma_{\text{inv}}$ in this calculation:

$$\eta \otimes 1 = {}_{\Gamma}\Phi(\eta) = \sum_k {}_{\Gamma}\Phi(\eta_k a_k) = \sum_k {}_{\Gamma}\Phi(\eta_k)\phi(a_k)$$

$$= \sum_k (\eta_k \otimes 1)\phi(a_k).$$

On the other hand, by (5.29),

$$\eta \otimes 1 = \sum_k \eta_k a_k \otimes 1 = \sum_k (\eta_k \otimes 1)(a_k \otimes 1).$$

Comparing these last two equations, we conclude for all $k \in I$ that

$$\phi(a_k) = a_k \otimes 1.$$

This places a strong condition on the elements a_k. In fact, by applying $\varepsilon \otimes 1$ to the last equation, we obtain

$$a_k = (\varepsilon \otimes 1)\phi(a_k) = (\varepsilon \otimes 1)(a_k \otimes 1) = \varepsilon(a_k)1 \in \mathbb{C}1 \cong \mathbb{C}.$$

So the formula (5.29) actually expresses an arbitrary $\eta \in \Gamma_{\text{inv}}$ as a linear combination of the $\eta_k's$ with unique *complex* coefficients. In other words, $\{\eta_k \mid k \in I\}$ is a vector space basis of Γ_{inv}, as we claimed. This completes the proof of part 2.

For the proof of part 3, we start from Eqs. (5.17) and (5.28):

$$a\eta_i = \sum_{j \in I} \eta_j \left(a * (g_{ij} \circ \kappa^{-1})\right)$$

$$= \sum_{j,k \in I} \omega_k \kappa(R_{kj})\left(a * (g_{ij} \circ \kappa^{-1})\right),$$

where $a \in \mathcal{A}$ is an arbitrary element. Note that we used $J = I$ here. But by using (5.28) first and then (5.6), we get

$$a\eta_i = a \sum_{j \in I} \omega_j \kappa(R_{ji})$$

$$= \sum_{j \in I} a\omega_j \kappa(R_{ji})$$

$$= \sum_{j,k \in I} \omega_k \big((f_{jk} \circ \kappa^{-1}) * a \big) \kappa(R_{ji}).$$

Combining these last two formulas gives us for every $i \in I$ that

$$\sum_{j,k \in I} \omega_k \big((f_{jk} \circ \kappa^{-1}) * a \big) \kappa(R_{ji}) = \sum_{j,k \in I} \omega_k \kappa(R_{kj}) \big(a * (g_{ij} \circ \kappa^{-1}) \big).$$

Since the ω_k's are a basis for the free left $\mathcal{A}$-module Γ, we immediately conclude that

$$\sum_{j \in I} \big((f_{jk} \circ \kappa^{-1}) * a \big) \kappa(R_{ji}) = \sum_{j \in I} \kappa(R_{kj}) \big(a * (g_{ij} \circ \kappa^{-1}) \big) \qquad (5.31)$$

for all $i, k \in I$. Applying ε to the left side of Eq. (5.31) yields

$$\varepsilon \Big(\sum_{j \in I} \big((f_{jk} \circ \kappa^{-1}) * a \big) \kappa(R_{ji}) \Big) = \sum_{j \in I} \varepsilon \Big(\big((f_{jk} \circ \kappa^{-1}) * a \big) \kappa(R_{ji}) \Big)$$

$$= \sum_{j \in I} \varepsilon \Big(\big((f_{jk} \circ \kappa^{-1}) * a \big) \Big) \varepsilon(\kappa(R_{ji}))$$

$$= \sum_{j \in I} (f_{jk} \circ \kappa^{-1})(a) \varepsilon(R_{ji})$$

$$= \sum_{j \in I} f_{jk}(\kappa^{-1}(a)) \delta_{ji}$$

$$= f_{ik}(\kappa^{-1}(a))$$

for all $i, k \in I$. Here we have used that ε is multiplicative, $\varepsilon \circ \kappa = \varepsilon$, $\varepsilon(f * a) = f(a)$, and $\varepsilon(R_{ji}) = \delta_{ji}$. Applying ε to the right side of Eq. (5.31) gives us (again for all $i, k \in I$) that

$$\varepsilon \Big(\sum_{j \in I} \kappa(R_{kj}) \big(a * (g_{ij} \circ \kappa^{-1}) \big) \Big) = g_{ik}(\kappa^{-1}(a))$$

by a very similar argument, which we leave to the reader. The upshot is that

$$f_{ik}(\kappa^{-1}(a)) = g_{ik}(\kappa^{-1}(a))$$

for all $a \in \mathcal{A}$. Using our standing hypothesis that κ is a bijection, we conclude that $f_{ik} = g_{ik}$ for all $i, k \in I$. And so part 3 is proved.

To prove part 4, we apply κ^{-1} to both sides of (5.31), getting

$$\sum_{j \in I} R_{ji} \kappa^{-1} \big((f_{jk} \circ \kappa^{-1}) * a \big) = \sum_{j \in I} \kappa^{-1} \big(a * (f_{ij} \circ \kappa^{-1}) \big) R_{kj}. \qquad (5.32)$$

Here we used $f_{ij} = g_{ij}$, which we just established, on the right side. We now apply identities to both sides of this equation. Starting with just the second factor on the left side of (5.32), we have

$$\kappa^{-1}\big((f_{jk} \circ \kappa^{-1}) * a\big) = \kappa^{-1}\big((id \otimes f_{jk} \circ \kappa^{-1})\phi(a)\big)$$
$$= \kappa^{-1}\big((id \otimes f_{jk})(id \otimes \kappa^{-1})\phi(a)\big)$$
$$= (id \otimes f_{jk})(\kappa^{-1} \otimes \kappa^{-1})\phi(a)$$
$$= (id \otimes f_{jk})\sigma_{\mathcal{A}}^2(\kappa^{-1} \otimes \kappa^{-1})\phi(a)$$
$$= (id \otimes f_{jk})\sigma_{\mathcal{A}}\phi(\kappa^{-1}(a))$$
$$= (f_{jk} \otimes id)\phi(\kappa^{-1}(a))$$
$$= (\kappa^{-1}(a)) * f_{jk}.$$

The third equality is a little trickier than most, so give it a good look! Then the left side of (5.32) becomes

$$\sum_{j\in I} R_{ji}\,\kappa^{-1}\big((f_{jk} \circ \kappa^{-1}) * a\big) = \sum_{j\in I} R_{ji}\Big((\kappa^{-1}(a) * f_{jk})\Big).$$

We leave it as an exercise for the reader to show by a similar argument that the right side of (5.32) simplifies to

$$\sum_{j\in I} \kappa^{-1}\big(a * (f_{ij} \circ \kappa^{-1})\big) R_{kj} = \sum_{j\in I} \Big(f_{ij} * (\kappa^{-1}(a))\Big) R_{kj}.$$

Equating these two results gives

$$\sum_{j\in I} R_{ji}\Big((\kappa^{-1}(a) * f_{jk})\Big) = \sum_{j\in I} \Big(f_{ij} * (\kappa^{-1}(a))\Big) R_{kj}$$

for all $a \in \mathcal{A}$.

Finally, substituting in $\kappa(a)$ in place of a, we obtain that

$$\sum_{j\in I} R_{ji}(a * f_{jk}) = \sum_{j\in I} (f_{ij} * a) R_{kj},$$

which is Eq. (5.24) in part 4.

Next, to show part 5, we evaluate the left co-action on the right invariant basis elements η_j of part 2. So we calculate

$$\Phi_\Gamma(\eta_j) = \Phi_\Gamma\left(\sum_i \omega_i \kappa(R_{ij})\right)$$

$$= \sum_i \Phi_\Gamma(\omega_i \kappa(R_{ij}))$$

$$= \sum_i \Phi_\Gamma(\omega_i) \phi(\kappa(R_{ij}))$$

$$= \sum_i (1 \otimes \omega_i) \phi(\kappa(R_{ij}))$$

$$= \sum_{il} (1 \otimes \omega_i)(\kappa(R_{lj}) \otimes \kappa(R_{il}))$$

$$= \sum_{il} (\kappa(R_{lj}) \otimes \omega_i \kappa(R_{il}))$$

$$= \sum_l \left(\kappa(R_{lj}) \otimes \sum_i \omega_i \kappa(R_{il})\right)$$

$$= \sum_l \kappa(R_{lj}) \otimes \eta_l.$$

This is the formula (5.25) in part 5. And that is the last thing to prove in this theorem. ∎

5.5 Notes

In this topic, covariant bimodules, we see the previously introduced idea of covariance introduced in a more general setting. This is a prelude to the next chapter as well as to the rest of the book. The material in this chapter comes from Woronowicz's paper [86].

Chapter 6
Covariant First-Order Differential Calculi

We already have a structure theorem of first-order differential calculi (FODCs). We are now going to continue to the next stage in that analysis, which will culminate in structure theorems for covariant FODCs (left and right) and for bicovariant FODCs.

6.1 Universal FODC

Recall that $\mathcal{A}^2 = \ker m_{\mathcal{A}} \subset \mathcal{A} \otimes \mathcal{A}$ is an $\mathcal{A}$-bimodule, where $m_{\mathcal{A}} : \mathcal{A} \otimes \mathcal{A} \to \mathcal{A}$ is the multiplication map of $\mathcal{A}$. We will use the notation $U := \mathcal{A}^2$ to keep in mind that (U, D) is the universal FODC over $\mathcal{A}$ (see Theorem 2.1).

To continue this discussion, we will suppose that $\mathcal{A}$ is a Hopf algebra with co-multiplication ϕ. We next define a linear map $\Phi_L : (\mathcal{A} \otimes \mathcal{A}) \to \mathcal{A} \otimes (\mathcal{A} \otimes \mathcal{A})$ and another linear map $_R\Phi : (\mathcal{A} \otimes \mathcal{A}) \to (\mathcal{A} \otimes \mathcal{A}) \otimes \mathcal{A}$ of $\mathcal{A}$ on the common domain $(\mathcal{A} \otimes \mathcal{A})$ by

$$\Phi_L := (m_{\mathcal{A}} \otimes id \otimes id)(id \otimes \sigma_{\mathcal{A}} \otimes id)(\phi \otimes \phi), \tag{6.1}$$

$$_R\Phi := (id \otimes id \otimes m_{\mathcal{A}})(id \otimes \sigma_{\mathcal{A}} \otimes id)(\phi \otimes \phi), \tag{6.2}$$

where $\sigma_{\mathcal{A}} : \mathcal{A} \otimes \mathcal{A} \to \mathcal{A} \otimes \mathcal{A}$ is the twist map $\sigma_{\mathcal{A}}(a \otimes b) := b \otimes a$ and $m_{\mathcal{A}} : \mathcal{A} \otimes \mathcal{A} \to \mathcal{A} \otimes \mathcal{A}$ is the multiplication map $m_{\mathcal{A}}(a \otimes b) := ab$. These will turn out to be co-actions, as we will prove momentarily.

Also, by using Sweedler's notation $\phi(a) = a^{(1)} \otimes a^{(2)}$ and $\phi(b) = b^{(1)} \otimes b^{(2)}$, we have these formulas:

$$\Phi_L(a \otimes b) = a^{(1)}b^{(1)} \otimes a^{(2)} \otimes b^{(2)},$$

$$_R\Phi(a \otimes b) = a^{(1)} \otimes b^{(1)} \otimes a^{(2)}b^{(2)}.$$

The following is a key result.

© Springer International Publishing Switzerland 2015
S.B. Sontz, *Principal Bundles*, Universitext, DOI 10.1007/978-3-319-15829-7_6

Theorem 6.1. *Let A be a Hopf algebra.*

1. *$((A \otimes A), \Phi_L, {}_R\Phi)$ is a bicovariant bimodule over A.*
2. *U is an invariant A-subbimodule of $(A \otimes A)$. We denote the restrictions of Φ_L and ${}_R\Phi$ to U by Φ_U and ${}_U\Phi$, respectively. So $(U, \Phi_U, {}_U\Phi)$ is a bicovariant bimodule over A.*
3. *The universal FODC (U, D) is bicovariant with respect to the co-actions Φ_U and ${}_U\Phi$. (Recall that the differential $D : A \to U$ is defined by $Da = a \otimes 1 - 1 \otimes a$.)*

Proof. We start by proving that Φ_L is a left co-action of A on $(A \otimes A)$. The identity we must show is $(id_A \otimes \Phi_L)\Phi_L = (\phi \otimes id_{(A \otimes A)})\Phi_L$. (We leave it to the reader to write down the corresponding diagram.) We take an element $a \otimes b \in (A \otimes A)$ and evaluate both sides of the claimed identity. First, for the right side, we get

$$
\begin{aligned}
(\phi \otimes id_{(A \otimes A)})\Phi_L(a \otimes b) &= (\phi \otimes id_{(A \otimes A)})(a^{(1)}b^{(1)} \otimes a^{(2)} \otimes b^{(2)}) \\
&= \phi(a^{(1)}b^{(1)}) \otimes a^{(2)} \otimes b^{(2)} \\
&= \phi(a^{(1)})\phi(b^{(1)}) \otimes a^{(2)} \otimes b^{(2)} \\
&= ((a^{(11)} \otimes a^{(12)})(b^{(11)} \otimes b^{(12)})) \otimes a^{(2)} \otimes b^{(2)} \\
&= a^{(11)}b^{(11)} \otimes a^{(12)}b^{(12)} \otimes a^{(2)} \otimes b^{(2)} \\
&= (a^{(11)} \otimes a^{(12)} \otimes a^{(2)} \otimes 1)(b^{(11)} \otimes b^{(12)} \otimes 1 \otimes b^{(2)}).
\end{aligned}
$$

Then, for the left side, we have

$$
\begin{aligned}
(id_A \otimes \Phi_L)\Phi_L(a \otimes b) &= (id_A \otimes \Phi_L)(a^{(1)}b^{(1)} \otimes a^{(2)} \otimes b^{(2)}) \\
&= a^{(1)}b^{(1)} \otimes \Phi_L(a^{(2)} \otimes b^{(2)}) \\
&= a^{(1)}b^{(1)} \otimes a^{(21)}b^{(21)} \otimes a^{(22)} \otimes b^{(22)} \\
&= (a^{(1)} \otimes a^{(21)} \otimes a^{(22)} \otimes 1)(b^{(1)} \otimes b^{(21)} \otimes 1 \otimes b^{(22)}).
\end{aligned}
$$

The co-associativity of ϕ shows that these two expressions are equal.

The proof that ${}_R\Phi$ is a right co-action is quite similar and will be left to the reader. The compatibility of Φ_L and ${}_R\Phi$ is the equation

$$
(id_A \otimes {}_R\Phi)\Phi_L = (\Phi_L \otimes id_A){}_R\Phi,
$$

which again the reader is encouraged to express as a diagram. This result boils down to an application of the co-associativity of the co-multiplication again. We leave the proof to the reader.

The map Φ_L must be shown to be an A-bimodule morphism as well. Recall that $\Phi_L : (A \otimes A) \to A \otimes (A \otimes A)$. The parentheses are important! We take $a, b, c \in A$. Then to show that Φ_L is a right A-module morphism, we calculate

$$
\Phi_L((a \otimes b)c) = \Phi_L(a \otimes bc) = a^{(1)}(bc)^{(1)} \otimes a^{(2)} \otimes (bc)^{(2)}
$$

as well as

$$\Phi_L(a \otimes b)\phi(c) = \left(a^{(1)}b^{(1)} \otimes (a^{(2)} \otimes b^{(2)})\right)\left(c^{(1)} \otimes c^{(2)}\right)$$

$$= a^{(1)}b^{(1)}c^{(1)} \otimes (a^{(2)} \otimes b^{(2)})c^{(2)}$$

$$= a^{(1)}b^{(1)}c^{(1)} \otimes (a^{(2)} \otimes b^{(2)}c^{(2)})$$

$$= a^{(1)}b^{(1)}c^{(1)} \otimes a^{(2)} \otimes b^{(2)}c^{(2)}.$$

Notice the importance of the parentheses in aiding us to put the factors of the action of c pulled back by ϕ on the correct factors in the triple product. Now the fact that ϕ is a multiplicative morphism tells us that $\phi(bc) = \phi(b)\phi(c)$, which translates immediately into $(bc)^{(1)} \otimes (bc)^{(2)} = b^{(1)}c^{(1)} \otimes b^{(2)}c^{(2)}$. This should be enough information for the reader to see that

$$\Phi_L((a \otimes b)c) = \Phi_L(a \otimes b)\phi(c),$$

and so Φ_L is indeed a right $\mathcal{A}$-module morphism. Similarly, one proves that Φ_L is also a left $\mathcal{A}$-module morphism.

Finally, to finish with the details for the first part, one proves (in the usual similar way) that $_R\Phi$ is also an $\mathcal{A}$-bimodule morphism.

We now discuss the second part of this theorem. First, we show that U is an $\mathcal{A}$-subbimodule of $(\mathcal{A} \otimes \mathcal{A})$. [Recall that $(\mathcal{A} \otimes \mathcal{A})$ is an $\mathcal{A}$-bimodule. See Appendix A.] So let $u = \sum_k a_k \otimes b_k \in U$ be arbitrary. Thus, $\sum_k a_k b_k = 0$. Take any $c \in \mathcal{A}$. Then $cu = \sum_k ca_k \otimes b_k$ satisfies $\sum_k ca_k b_k = c \sum_k a_k b_k = 0$ so that $cu \in U$ follows. Therefore, U is a left $\mathcal{A}$-submodule of $(\mathcal{A} \otimes \mathcal{A})$. And similarly, U is also a right $\mathcal{A}$-submodule of $(\mathcal{A} \otimes \mathcal{A})$.

To show $\Phi_U(u) \in \mathcal{A} \otimes U$ for all $u \in U$, we will prove $(id \otimes m_{\mathcal{A}})\Phi_U(u) = 0$. But

$$(id \otimes m_{\mathcal{A}})\Phi_U = (id \otimes m_{\mathcal{A}})(m_{\mathcal{A}} \otimes id \otimes id)(id \otimes \sigma_{\mathcal{A}} \otimes id)(\phi \otimes \phi)$$

$$= m_{\mathcal{A} \otimes \mathcal{A}}(\phi \otimes \phi)$$

$$= \phi m_{\mathcal{A}},$$

where $m_{\mathcal{A} \otimes \mathcal{A}}$ is the multiplication in the algebra $\mathcal{A} \otimes \mathcal{A}$ and the last equality holds because ϕ is a multiplicative morphism. Evaluating both sides on $u \in U = \ker m_{\mathcal{A}}$, we get $(id \otimes m_{\mathcal{A}})\Phi_U(u) = \phi m_{\mathcal{A}}(u) = 0$, as desired. Similarly, one shows that $_U\Phi(u) \in U \otimes \mathcal{A}$ for all $u \in U$. Writing this as diagrams (arrow notation), we have

$$\Phi_U : U \to \mathcal{A} \otimes U \quad \text{and} \quad _U\Phi : U \to U \otimes \mathcal{A}.$$

Then it follows immediately that restriction gives us that $(U, \Phi_U, {}_U\Phi)$ is a bicovariant bimodule over $\mathcal{A}$. Hence, Φ_U (resp., $_U\Phi$) is a left (resp., right) co-action of $\mathcal{A}$ on U.

For the third part of the theorem, one has to prove that the co-actions Φ_U and $_U\Phi$ come from the FODC structure. Explicitly, according to the conditions (3.13) and (3.17), one has to show for all $a_k, b_k \in \mathcal{A}$ that

$$\sum_k a_k Db_k = 0 \implies \sum_k \phi(a_k)(id_\mathcal{A} \otimes D)\phi(b_k) = 0,$$

$$\sum_k a_k Db_k = 0 \implies \sum_k \phi(a_k)(D \otimes id_\mathcal{A})\phi(b_k) = 0.$$

And then one must show that the co-actions corresponding to these two conditions are Φ_U and $_U\Phi$, respectively. This finishes our discussion of the third part. ∎

Exercise 6.1. *Using this outline, prove the third part of the previous theorem.*

So in our quest to understand bicovariant FODCs, we have produced one example of such an object, namely, the universal FODC. Since by Theorem 2.1 *all* FODCs are quotients of the universal FODC (by subbimodules), that quest boils down to finding which of these quotients also are bicovariant. So we wish to identify all of those $\mathcal{A}$-subbimodules $\mathcal{N}$ of U for which the associated FODC $(\Gamma_\mathcal{N}, d_\mathcal{N})$ is bicovariant. Actually, we will give the necessary and sufficient condition first for left covariance and next for right covariance. This comes down to understanding how subobjects and quotients work in this context. Of course, these are standard considerations from the perspective of category theory. The only subtle point is that covariance is more than the existence of a co-action (or co-actions in the case of bicovariance) but also includes a compatibility condition between the co-action(s) and the FODC structure. These conditions are given in (3.13) and (3.17), respectively. It is exactly these conditions that allow us to define the left co-action and the right co-action, respectively.

First, we present a detail whose importance is discussed ahead.

Lemma 6.1. *Let (Γ, d) and (Γ', d') be isomorphic FODCs over $\mathcal{A}$. Suppose that $\psi : \Gamma \to \Gamma'$ is any isomorphism of FODCs. Then*

- *(Γ, d) is left covariant if and only if (Γ', d') is left covariant. In this case, ψ is an isomorphism of left co-modules over $\mathcal{A}$.*
- *(Γ, d) is right covariant if and only if (Γ', d') is right covariant. In this case, ψ is an isomorphism of right co-modules over $\mathcal{A}$.*
- *(Γ, d) is bicovariant if and only if (Γ', d') is bicovariant. In this case, ψ is an isomorphism of left and right co-modules over $\mathcal{A}$.*

The proof consists of noting that the conditions (3.13) and (3.17) are preserved under isomorphisms of FODCs. This lemma is used in conjugation with the following key result. For example, given a left covariant FODC, we know it is isomorphic, but only as an FODC, to some FODC $(\Gamma_\mathcal{N}, d_\mathcal{N})$. Then this lemma says that this is also an isomorphism of left covariant FODCs. In other words, $(\Gamma_\mathcal{N}, d_\mathcal{N})$ is left covariant and thus satisfies the first condition in the following theorem. We then get a classification of all left covariant FODCs.

Theorem 6.2. *Let* $\mathcal{N}$ *be an* $\mathcal{A}$-*subbimodule of* U. *Put* $\Gamma_{\mathcal{N}} = U/\mathcal{N}$ *and let* $\pi_{\mathcal{N}}$:
$U \to \Gamma_{\mathcal{N}}$ *be the quotient map. Set* $d_{\mathcal{N}} = \pi_{\mathcal{N}} \circ D$. *So by Theorem 2.1,* $(\Gamma_{\mathcal{N}}, d_{\mathcal{N}})$
is an FODC over $\mathcal{A}$ *and, moreover, any FODC over* $\mathcal{A}$ *is isomorphic as an FODC*
to $(\Gamma_{\mathcal{N}}, d_{\mathcal{N}})$ *for some* $\mathcal{N}$ *that is an* $\mathcal{A}$-*submodule of* U. *Then we have the following*
important equivalences:

- $(\Gamma_{\mathcal{N}}, d_{\mathcal{N}})$ *is a left covariant FODC over* $\mathcal{A}$ *if and only if* $\mathcal{N}$ *is left invariant [i.e.,*
  $\Phi_U(\mathcal{N}) \subset \mathcal{A} \otimes \mathcal{N}]$.
- $(\Gamma_{\mathcal{N}}, d_{\mathcal{N}})$ *is a right covariant FODC over* $\mathcal{A}$ *if and only if* $\mathcal{N}$ *is right invariant*
  [i.e., ${}_U\Phi(\mathcal{N}) \subset \mathcal{N} \otimes \mathcal{A}]$.
- $(\Gamma_{\mathcal{N}}, d_{\mathcal{N}})$ *is a bicovariant FODC over* $\mathcal{A}$ *if and only if it is both a left and a right*
  covariant FODC over $\mathcal{A}$.

Remark. An essential ingredient here is that we are taking quotients of U, which
is in and of itself a bicovariant bimodule over $\mathcal{A}$. In order to analyze a general
FODC (Γ, d), we use Lemma 6.1 and Theorem 2.1 to reduce to the case $(\Gamma_{\mathcal{N}}, d_{\mathcal{N}})$
considered in this theorem.

Proof. We start with the first part. Suppose that $(\Gamma_{\mathcal{N}}, d_{\mathcal{N}})$ is a left covariant FODC
over $\mathcal{A}$. Then by Theorem 3.1, there is a left co-action $\Phi_{\mathcal{N}}$ of $\mathcal{A}$ on $\Gamma_{\mathcal{N}}$, namely,
$\Phi_{\mathcal{N}} : \Gamma_{\mathcal{N}} \to \mathcal{A} \otimes \Gamma_{\mathcal{N}}$. Moreover, Theorem 3.1 gives us an explicit formula for $\Phi_{\mathcal{N}}$.
Then the reader can use this to prove that the following diagram commutes:

$$
\begin{array}{ccc}
U & \xrightarrow{\Phi_U} & \mathcal{A} \otimes U \\
{\scriptstyle \pi_{\mathcal{N}}} \downarrow & & \downarrow {\scriptstyle id_{\mathcal{A}} \otimes \pi_{\mathcal{N}}} \\
\Gamma_{\mathcal{N}} & \xrightarrow{\Phi_{\mathcal{N}}} & \mathcal{A} \otimes \Gamma_{\mathcal{N}}.
\end{array}
\qquad (6.3)
$$

Putting this diagram into words, one says that the left co-action $\Phi_{\mathcal{N}}$ is the quotient
of the left co-action Φ_U.

Now pick $\alpha \in \mathcal{N} \subset U$ and chase it through this diagram. Going down gives zero.
So the down–across chase gives zero. Next, chasing across and down, we see that
$\Phi_U(\alpha) \in \ker(id_{\mathcal{A}} \otimes \pi_{\mathcal{N}}) = \mathcal{A} \otimes \mathcal{N}$. In short, we have shown that $\Phi_U(\mathcal{N}) \subset \mathcal{A} \otimes \mathcal{N}$.

Conversely, we now suppose that $\Phi_U(\mathcal{N}) \subset \mathcal{A} \otimes \mathcal{N}$. To show that $(\Gamma_{\mathcal{N}}, d_{\mathcal{N}})$
is a left covariant FODC, we consider elements $a_k, b_k \in \mathcal{A}$ that satisfy
$\sum_k a_k d_{\mathcal{N}} b_k = 0$. The tricky bit is that we do not yet have the map $\Phi_{\mathcal{N}}$, and
consequently neither do we have the commutative diagram (6.3). Now using that
$\pi_{\mathcal{N}}$ is a left $\mathcal{A}$-morphism, we do have

$$
\pi_{\mathcal{N}}\left(\sum_k a_k Db_k\right) = \sum_k a_k \pi_{\mathcal{N}} Db_k = \sum_k a_k d_{\mathcal{N}} b_k = 0.
$$

And this implies that $\sum_k a_k Db_k \in \mathcal{N}$. So by the hypothesis, we see that

$$
\Phi_U\left(\sum_k a_k Db_k\right) \in \mathcal{A} \otimes \mathcal{N}.
$$

Motivated by the diagram (6.3), we observe that this implies

$$(id_{\mathcal{A}} \otimes \pi_{\mathcal{N}})\big(\Phi_U\big(\sum_k a_k Db_k\big)\big) = 0.$$

But we can evaluate the left side to get

$$(id_{\mathcal{A}} \otimes \pi_{\mathcal{N}})\big(\Phi_U\big(\sum_k a_k Db_k\big)\big) = (id_{\mathcal{A}} \otimes \pi_{\mathcal{N}})\big(\sum_k \phi(a_k)(id_{\mathcal{A}} \otimes D)\phi(b_k)\big)$$

$$= \sum_k \phi(a_k)(id_{\mathcal{A}} \otimes \pi_{\mathcal{N}} D)\phi(b_k)$$

$$= \sum_k \phi(a_k)(id_{\mathcal{A}} \otimes d_{\mathcal{N}})\phi(b_k).$$

In summary, we have shown that $\sum_k \phi(a_k)(id_{\mathcal{A}} \otimes d_{\mathcal{N}})\phi(b_k) = 0$ follows from the assumption that $\sum_k a_k d_{\mathcal{N}} b_k = 0$. And so we have proved that $(\Gamma_{\mathcal{N}}, d_{\mathcal{N}})$ is a left covariant FODC over $\mathcal{A}$. This concludes the proof of the first part.

For the second part, we introduce the notation $_{\mathcal{N}}\Phi$ for the right co-action of the right covariant FODC. But, as usual, we leave the very similar proof to the reader.

For the third part, we need only concern ourselves with the extra detail having to do with the commutativity of the left co-action and the right co-action. And this is only for the proof of the converse implication. But if $(\Gamma_{\mathcal{N}}, d_{\mathcal{N}})$ is both left and right covariant, then the left co-action $\Phi_{\mathcal{N}}$ is the quotient of the left co-action Φ_U in U, while the right co-action $_{\mathcal{N}}\Phi$ is the quotient of the right co-action $_U\Phi$ in U. However, we have already seen that the co-actions Φ_U and $_U\Phi$ in U commute. And this readily implies the result that $(\Gamma_{\mathcal{N}}, d_{\mathcal{N}})$ is bicovariant. ∎

6.2 Isomorphic Universal FODCs

It is a general principle that universal objects in a category are unique up to isomorphism. Nonetheless, as we shall see in this section, one isomorphic copy sometimes is more useful than its clones.

To study this topic further, we will use the operator

$$r : \mathcal{A} \otimes \mathcal{A} \to \mathcal{A} \otimes \mathcal{A}$$

defined for $a, b \in \mathcal{A}$ by

$$r(a \otimes b) := (ab^{(1)}) \otimes b^{(2)} \tag{6.4}$$

in Sweedler's notation. Another way of writing this is

$$r(a \otimes b) = (a \otimes 1)\phi(b). \tag{6.5}$$

But maybe the best way is as the composition $r = (m_A \otimes id)(id \otimes \phi)$, namely,

$$\mathcal{A} \otimes \mathcal{A} \xrightarrow{id \otimes \phi} \mathcal{A} \otimes \mathcal{A} \otimes \mathcal{A} \xrightarrow{m_A \otimes id} \mathcal{A} \otimes \mathcal{A}, \qquad (6.6)$$

as one can readily verify. As r is the composition of two linear maps, it follows immediately that r is also linear.

We will also use the operator

$$s : \mathcal{A} \otimes \mathcal{A} \to \mathcal{A} \otimes \mathcal{A}$$

defined for $a, b \in \mathcal{A}$ by $s(a \otimes b) := b^{(1)} \otimes (ab^{(2)})$ in Sweedler's notation. Another way of writing this is $s(a \otimes b) = (1 \otimes a)\phi(b)$. And again this is a composition of standard maps, that is, $s = (id \otimes m_A)(\sigma_A \otimes id)(id \otimes \phi)$, as the reader can check. And since s is the composition of linear maps, we also have that s is linear.

Next, we state and prove an important basic property of r, namely, that it is a bijection. Before doing this, we first define the linear map, which will turn out to be its inverse. This is $p : \mathcal{A} \otimes \mathcal{A} \to \mathcal{A} \otimes \mathcal{A}$, which is defined by

$$p : (a \otimes b) \mapsto (a \otimes 1)(\kappa \otimes id)\phi(b).$$

Another way to write p is as the composition $p = (m_A \otimes id)(id \otimes \kappa \otimes id)(id \otimes \phi)$, that is,

$$\mathcal{A} \otimes \mathcal{A} \xrightarrow{id \otimes \phi} \mathcal{A} \otimes \mathcal{A} \otimes \mathcal{A} \xrightarrow{id \otimes \kappa \otimes id} \mathcal{A} \otimes \mathcal{A} \otimes \mathcal{A} \xrightarrow{m_A \otimes id} \mathcal{A} \otimes \mathcal{A}, \qquad (6.7)$$

as the reader should check.

To remove a bit of the mystery surrounding the definitions of r and p, we will now see what they are for the special case $\mathcal{A} = \mathcal{F}(G) = \{f : G \to \mathbb{C}\}$, where G is a finite group. Recall that δ_g is the characteristic function on G whose support is the singleton $\{g\} \subset G$ for all $g \in G$. Also, recall that the set $\{\delta_g \mid g \in G\}$ is a basis of $\mathcal{F}(G)$. Then for all $g, h \in G$, we calculate

$$r(\delta_g \otimes \delta_h) = (m_A \otimes id)(id \otimes \phi)(\delta_g \otimes \delta_h)$$

$$= (m_A \otimes id)\Big(\delta_g \otimes \sum_k \delta_k \otimes \delta_{k^{-1}h}\Big)$$

$$= (m_A \otimes id) \sum_k (\delta_g \otimes \delta_k \otimes \delta_{k^{-1}h})$$

$$= \sum_k (\delta_g \delta_k \otimes \delta_{k^{-1}h})$$

$$= \delta_g \otimes \delta_{g^{-1}h}.$$

Defining $R(x, y) := (x, xy)$ for $x, y \in G$, we can compute that $R^* = r$.

On the other hand,

$$p(\delta_g \otimes \delta_h) = (m_A \otimes id)(id \otimes \kappa \otimes id)(id \otimes \phi)(\delta_g \otimes \delta_h)$$

$$= (m_A \otimes id)(id \otimes \kappa \otimes id) \sum_k (\delta_g \otimes \delta_k \otimes \delta_{k^{-1}h})$$

$$= (m_A \otimes id) \sum_k (\delta_g \otimes \delta_{k^{-1}} \otimes \delta_{k^{-1}h})$$

$$= \sum_k (\delta_g \delta_{k^{-1}} \otimes \delta_{k^{-1}h})$$

$$= \delta_g \otimes \delta_{gh}.$$

So defining $P(x, y) := (x, x^{-1}y)$ for $x, y \in G$, we can compute that $P^* = p$. So in this case the maps r and p are rather transparent objects, and it is clear that each is the inverse of the other. Moreover, they are pullbacks of maps that are essentially transition functions in the second argument. This is discussed in more detail in [80] in the chapter on multiplicative unitaries. While we will not discuss this example further, the reader might keep it in mind when trying to understand the rest of this section.

Theorem 6.3. *The linear map r is a bijection whose inverse r^{-1} is given by the map p. So r is an isomorphism of vector spaces.*

Proof. This proof is complicated enough to qualify being called nontrivial even though no new ideas are needed for it. We are simply going to show that $rp = pr = id_{A \otimes A}$. So this is an elementary result with a long proof.

We first note that $r((a \otimes 1)\beta) = (a \otimes 1)r(\beta)$ and $p((a \otimes 1)\beta) = (a \otimes 1)p(\beta)$ for all $a \in A$ and all $\beta \in A \otimes A$, as the reader can check. So it suffices to show that rp and pr act as the identity on elements of the form $1 \otimes b$. This makes things a bit more transparent since $r(1 \otimes b) = b^{(1)} \otimes b^{(2)} = \phi(b)$, while $p(1 \otimes b) = \kappa(b^{(1)}) \otimes b^{(2)} = (\kappa \otimes id)\phi(b)$. Let's recall the basic properties of the co-inverse κ. One is that the following diagram is commutative:

$$A \xrightarrow{\phi} A \otimes A \xrightarrow{\kappa \otimes id} A \otimes A \xrightarrow{m} A$$

with ε going down from A to $\mathbb{C}$ on the left, η going up to A on the right, and id from $\mathbb{C}$ to $\mathbb{C}$ along the bottom.

The other property is that the corresponding diagram with $id \otimes \kappa$ instead of $\kappa \otimes id$ is also commutative. Here the map η is the unit of the algebra A, namely, the map defined by $\eta(1) := 1$. On the other hand, $\varepsilon : A \to \mathbb{C}$ is the co-unit of the co-algebra A. We also recall another identity for Hopf algebras: $(\eta\varepsilon \otimes id)\phi(a) = 1 \otimes a$ for all $a \in A$ [see (B.15)].

We now consider the composition pr acting on $1 \otimes b$. First, we have

$$r : 1 \otimes b \mapsto \phi(b) = b^{(1)} \otimes b^{(2)}$$

and then

$$p : b^{(1)} \otimes b^{(2)} \mapsto (b^{(1)} \otimes 1)p(1 \otimes b^{(2)})$$
$$= (b^{(1)} \otimes 1)(\kappa(b^{(21)}) \otimes b^{(22)})$$
$$= b^{(1)}\kappa(b^{(21)}) \otimes b^{(22)}.$$

So it remains to show $b^{(1)}\kappa(b^{(21)}) \otimes b^{(22)} = 1 \otimes b$. To get a grip on this, we note that $b^{(1)} \otimes b^{(21)} \otimes b^{(22)} = (id_{\mathcal{A}} \otimes \phi)\phi(b) = (\phi \otimes id_{\mathcal{A}})\phi(b)$ by the co-associativity of ϕ. So, using the properties just cited, we have

$$pr(1 \otimes b) = (m_{\mathcal{A}} \otimes id_{\mathcal{A}})(id_{\mathcal{A}} \otimes \kappa \otimes id_{\mathcal{A}})(\phi \otimes id_{\mathcal{A}})\phi(b)$$
$$= ([m_{\mathcal{A}}(id_{\mathcal{A}} \otimes \kappa)\phi] \otimes id_{\mathcal{A}})\phi(b)$$
$$= (\eta\varepsilon \otimes id_{\mathcal{A}})\phi(b)$$
$$= 1 \otimes b,$$

where $m_{\mathcal{A}} : \mathcal{A} \otimes \mathcal{A} \to \mathcal{A}$ is the multiplication map of $\mathcal{A}$.

If we are considering the case when $\mathcal{A}$ is finite dimensional (and then $\mathcal{A} \otimes \mathcal{A}$ is also finite dimensional), we can immediately conclude that $rp = id_{\mathcal{A} \otimes \mathcal{A}}$ as well. Otherwise, see the next exercise. ∎

Exercise 6.2. *Show by a similar elementary (though possibly lengthy) calculation that rp is the identity on $\mathcal{A} \otimes \mathcal{A}$.*

Because of this, from now on we will write r^{-1} instead of p.

Exercise 6.3. *Let $t : \mathcal{A} \otimes \mathcal{A} \to \mathcal{A} \otimes \mathcal{A}$ be defined by*

$$t(a \otimes b) = (b \otimes 1)\sigma_{\mathcal{A}}(id \otimes \kappa^{-1})\phi(a),$$

where $\sigma_{\mathcal{A}} : \mathcal{A} \otimes \mathcal{A} \to \mathcal{A} \otimes \mathcal{A}$ is the interchange map $a \otimes b \mapsto b \otimes a$. Show that t is the inverse to s. (This again might be lengthy, yet still elementary.)

Exercise 6.4. *Let V be any complex vector space. Put $\Gamma := \mathcal{A} \otimes V$. Define*

$$\Phi : \Gamma = \mathcal{A} \otimes V \to \mathcal{A} \otimes \Gamma = \mathcal{A} \otimes \mathcal{A} \otimes V$$

by $\Phi = \phi \otimes id_V$. Show that (Γ, Φ) is a left $\mathcal{A}$-co-module. This left $\mathcal{A}$-co-module is called the free left $\mathcal{A}$-co-module over V.

Write out and prove some analogous statements, and then define what a free right *$\mathcal{A}$-co-module over V is.*

Proposition 6.1. *Let* $\Gamma = \mathcal{A} \otimes V$ *be a free left co-module with left co-action* $\Phi = \phi \otimes id_V$, *as above. Then the set of left invariant elements of* Γ *is given by* $_{\text{inv}}\Gamma = 1 \otimes V$.

Proof. First, consider any $1 \otimes v \in 1 \otimes V$. Then $(\phi \otimes id)(1 \otimes v) = \phi(1) \otimes v = 1 \otimes 1 \otimes v$. So $1 \otimes v$ is left invariant.

Conversely, suppose that $\alpha \in \Gamma = \mathcal{A} \otimes V$ is left invariant. We must show that $\alpha \in 1 \otimes V$. If $\alpha = 0$, then $\alpha = 0 = 1 \otimes 0 \in 1 \otimes V$, and we are done. So we suppose that $\alpha \neq 0$ from now on. But then we can write α in the form

$$\alpha = \sum_{k \in I} a_k \otimes v_k,$$

where I is a finite, nonempty index set and the set of vectors $\{ v_k \mid k \in I \}$ is linearly independent in V. Since α is left invariant, we have

$$\Phi(\alpha) = (\phi \otimes id)\Big(\sum_{k \in I} a_k \otimes v_k \Big) = 1 \otimes \sum_{k \in I} a_k \otimes v_k = \sum_{k \in I} 1 \otimes a_k \otimes v_k.$$

But we also have

$$(\phi \otimes id)\Big(\sum_{k \in I} a_k \otimes v_k \Big) = \sum_{k \in I} \phi(a_k) \otimes v_k.$$

Applying $id \otimes \varepsilon \otimes id$ to both sides of

$$\sum_{k \in I} 1 \otimes a_k \otimes v_k = \sum_{k \in I} \phi(a_k) \otimes v_k,$$

we obtain

$$\sum_{k \in I} \varepsilon(a_k) 1 \otimes v_k = \sum_{k \in I} (id \otimes \varepsilon)\phi(a_k) \otimes v_k = \sum_{k \in I} a_k \otimes v_k.$$

Since the v_ks are linearly independent, we conclude that $a_k = \varepsilon(a_k) 1$ for all $k \in I$. But then we substitute this back into the formula for α to get

$$\alpha = \sum_{k \in I} a_k \otimes v_k$$

$$= \sum_{k \in I} \varepsilon(a_k) 1 \otimes v_k$$

$$= \sum_{k \in I} 1 \otimes \varepsilon(a_k) v_k$$

$$= 1 \otimes \sum_{k \in I} \varepsilon(a_k) v_k.$$

And so $\alpha \in 1 \otimes V$, as desired. Note that we have used $\varepsilon(a_k) \in \mathbb{C}$ here. ∎

We note that for the free right co-module $\Gamma = V \otimes \mathcal{A}$ with its right co-action $id \otimes \phi$, we have that the right invariant elements are analogously given by $\Gamma_{\mathrm{inv}} = V \otimes 1$.

Exercise 6.5. *Prove* $\mathrm{ad}(a) = s\!\left(r^{-1}(1 \otimes a)\right)$ *for all* $a \in \mathcal{A}$, *where* ad *is the adjoint, a right co-action of* $\mathcal{A}$ *on itself [see (4.1)].*

Proposition 6.2. *The restriction of* r *to* U *(also denoted by* r*) satisfies*

$$r : U \to \mathcal{A} \otimes \ker \varepsilon. \qquad (6.8)$$

Moreover, this is a bijection so that r *is a vector space isomorphism between* U *and* $\mathcal{A} \otimes \ker \varepsilon$. *Similarly,*

$$s : U \to \ker \varepsilon \otimes \mathcal{A}, \qquad (6.9)$$

which is also a bijection. So s *is a vector space isomorphism between* U *and* $\mathcal{A} \otimes \ker \varepsilon$.

Proof. Equation (B.23) [see also (A.15) in [86]] is $m_{\mathcal{A}} r^{-1} = id \otimes \varepsilon$, where we are using the result of the last theorem, namely, that r is a bijection. Next, for $u \in U = \ker m_{\mathcal{A}}$, we have that

$$((id \otimes \varepsilon)r)(u) = m_{\mathcal{A}} r^{-1} r(u) = m_{\mathcal{A}} u = 0.$$

This says $r(U) \subset \mathcal{A} \otimes \ker \varepsilon$. Also, (6.8) is injective, being the restriction of a bijection.

It remains to show that (6.8) is surjective. But we already know that $r : \mathcal{A} \otimes \mathcal{A} \to \mathcal{A} \otimes \mathcal{A}$ is surjective. Consider elements $a \in \mathcal{A}$ and $b \in \ker \varepsilon$. Then it suffices to show that $r^{-1}(a \otimes b) \in U$. But

$$m_{\mathcal{A}} r^{-1}(a \otimes b) = (id \otimes \varepsilon)(a \otimes b) = a \otimes \varepsilon(b) = 0,$$

which exactly means that $r^{-1}(a \otimes b) \in \ker m_{\mathcal{A}} = U$.

The result for s follows just as easily from Eq. (B.24). ∎

The domain and co-domain spaces of r and s have more structure than just that of vector spaces. It is important to establish how r and s relate to these other structures. First off, all these spaces are $\mathcal{A}$-bimodules. We already have above that U is a $\mathcal{A}$-subbimodule of $\mathcal{A} \otimes \mathcal{A}$. But what about the other two spaces?

Definition 6.1. For $a \otimes b \in \mathcal{A} \otimes \ker \varepsilon$ and $c \in \mathcal{A}$, we define $c \cdot (a \otimes b) := ca \otimes b$ and $(a \otimes b) \cdot c := (a \otimes b)\phi(c)$.

Also, for $a \otimes b \in \ker \varepsilon \otimes \mathcal{A}$ and $c \in \mathcal{A}$, we define $c \cdot (a \otimes b) := a \otimes cb$ and $(a \otimes b) \cdot c := (a \otimes b)\phi(c)$.

Exercise 6.6. *Show that* $\mathcal{A} \otimes \ker \varepsilon$ *and* $\ker \varepsilon \otimes \mathcal{A}$ *are* $\mathcal{A}$*-bimodules with these definitions.*

Now here is the relationship of r and s to these $\mathcal{A}$-bimodules.

Theorem 6.4. *The maps*

$$r : U \to \mathcal{A} \otimes \ker \varepsilon$$

$$s : U \to \ker \varepsilon \otimes \mathcal{A}$$

are isomorphisms of $\mathcal{A}$*-bimodules.*

Proof. Since these are already known to be vector space isomorphisms, it suffices to show that each of them preserves the $\mathcal{A}$-bimodule structures. But the identities (B.19)–(B.22) say precisely that. (Those identities were, of course, the motivation for the previous definition of the $\mathcal{A}$-bimodule structures.) ∎

In the next theorem, we will consider $\mathcal{A} \otimes \ker \varepsilon$ as the free left $\mathcal{A}$-co-module over $\ker \varepsilon$. We also consider $\ker \varepsilon \otimes \mathcal{A}$ as the free right $\mathcal{A}$-co-module over $\ker \varepsilon$. But U is also a left and a right $\mathcal{A}$-co-module via the co-actions Φ_U and $_U\Phi$, respectively. We next examine how r and s relate to these structures.

Theorem 6.5. *The map* $r : U \to \mathcal{A} \otimes \ker \varepsilon$ *is a left* $\mathcal{A}$*-co-module isomorphism. The assertion that* r *is a left* $\mathcal{A}$*-co-module morphism simply means that the following diagram is commutative:*

$$
\begin{array}{ccc}
U & \overset{r}{\longrightarrow} & \mathcal{A} \otimes \ker \varepsilon \\
{\scriptstyle \Phi_U} \downarrow & & \downarrow {\scriptstyle \phi \otimes id} \\
\mathcal{A} \otimes U & \overset{id \otimes r}{\longrightarrow} & \mathcal{A} \otimes \mathcal{A} \otimes \ker \varepsilon.
\end{array}
\qquad (6.10)
$$

Similarly, the map $s : U \to \ker \varepsilon \otimes \mathcal{A}$ *is a right* $\mathcal{A}$*-co-module isomorphism, so that this diagram is commutative:*

$$
\begin{array}{ccc}
U & \overset{s}{\longrightarrow} & \ker \varepsilon \otimes \mathcal{A} \\
{\scriptstyle _U\Phi} \downarrow & & \downarrow {\scriptstyle id \otimes \phi} \\
U \otimes \mathcal{A} & \overset{s \otimes id}{\longrightarrow} & \ker \varepsilon \otimes \mathcal{A} \otimes \mathcal{A}.
\end{array}
\qquad (6.11)
$$

Remarks. We already have proved that r is a bijection. So the assertion that (6.10) is commutative is the new information here about r.

According to a previous exercise, the vertical arrow on the right side of the diagram (6.10) is the left co-action of $\mathcal{A}$ on $\mathcal{A} \otimes \ker \varepsilon$. The upshot is that (U, Φ_U) is (or strictly speaking, is *isomorphic to*) a free left $\mathcal{A}$-co-module. Moreover, we have an explicit isomorphism, namely, r, mapping U to the free left $\mathcal{A}$-co-module $\mathcal{A} \otimes \ker \varepsilon$.

Similar remarks apply to the diagram (6.11).

Proof. The diagram (6.10) is the restriction to U (in the upper left corner) of the following diagram:

$$
\begin{array}{ccc}
\mathcal{A} \otimes \mathcal{A} & \xrightarrow{\;r\;} & \mathcal{A} \otimes \mathcal{A} \\
\Phi_U \downarrow & & \downarrow \phi \otimes id \\
\mathcal{A} \otimes \mathcal{A} \otimes \mathcal{A} & \xrightarrow{id \otimes r} & \mathcal{A} \otimes \mathcal{A} \otimes \mathcal{A}.
\end{array}
\qquad (6.12)
$$

Actually, here we are using the formula

$$
\Phi_U = (m_A \otimes id \otimes id)(id \otimes \sigma_A \otimes id)(\phi \otimes \phi)
$$

to define Φ_U on the larger domain $\mathcal{A} \otimes \mathcal{A}$; that is, for any $a, b \in \mathcal{A}$, we put $\Phi_U(a \otimes b) = a^{(1)}b^{(1)} \otimes a^{(2)} \otimes b^{(2)}$.

So it suffices to show that (6.12) commutes. We take $a \otimes b \in \mathcal{A} \otimes \mathcal{A}$ and chase it through (6.12). First, going across and down, we have

$$
a \otimes b \xrightarrow{\;r\;} ab^{(1)} \otimes b^{(2)} \xrightarrow{\phi \otimes id} \phi(ab^{(1)}) \otimes b^{(2)}
$$
$$
= \phi(a)\phi(b^{(1)}) \otimes b^{(2)}
$$
$$
= (\phi(a) \otimes 1)(\phi(b^{(1)}) \otimes b^{(2)})
$$
$$
= (\phi(a) \otimes 1)((\phi \otimes id)\phi(b)).
$$

Next, going down and across, we have

$$
a \otimes b \xrightarrow{\Phi_U} a^{(1)}b^{(1)} \otimes a^{(2)} \otimes b^{(2)} \xrightarrow{id \otimes r} a^{(1)}b^{(1)} \otimes (a^{(2)} \otimes 1)\phi(b^{(2)})
$$
$$
= (a^{(1)} \otimes a^{(2)} \otimes 1)(b^{(1)} \otimes \phi(b^{(2)}))
$$
$$
= (\phi(a) \otimes 1)((id \otimes \phi)\phi(b)).
$$

By the co-associativity of ϕ, we see that (6.12) commutes.

As is to be expected, we leave the very similar proof of the commutativity of (6.11) to the reader. ∎

Exercise 6.7 (Optional). *Explain what it means to say that this theorem is functorial. And then prove it.*

We next have an important corollary to this result.

Corollary 6.1. *An element $u \in U$ is left invariant with respect to the left co-action Φ_U if and only if $u = r^{-1}(1 \otimes a)$ for some $a \in \ker \varepsilon$.*

Similarly, an element $u \in U$ is right invariant with respect to the right co-action $_U\Phi$ if and only if $u = s^{-1}(b \otimes 1)$ for some $b \in \ker \varepsilon$.

Proof. We know by Proposition 6.1 that all the left invariant elements in the free left co-module $(\mathcal{A} \otimes \ker \varepsilon, \phi \otimes id)$ are given by the elements $1 \otimes a$, where $a \in \ker \varepsilon$. We then use the isomorphism r^{-1} from the theorem to obtain the corresponding left invariant elements in (U, Φ_U). So they are exactly those elements of the form $r^{-1}(1 \otimes a)$ for some $a \in \ker \varepsilon$.

As is now our customary practice, we leave the very similar proof of the second part of this theorem to the reader. ∎

Exercise 6.8. *Define $U_L := \mathcal{A} \otimes \ker \varepsilon$. We recall that $r : U \to U_L$ is an $\mathcal{A}$-bimodule isomorphism as well as a left $\mathcal{A}$-co-module isomorphism.*

Show that the universal differential D for U maps over to the differential $D_L := rD$ for U_L, which is then given by $D_L a = \phi(a) - a \otimes 1$ for all $a \in \mathcal{A}$.

Prove directly that $D_L : \mathcal{A} \to U_L$ is a derivation and that every element in U_L can be written in standard form.

Now convince yourself that Theorem 2.1 translates under the isomorphism r to the following statement:

Theorem 6.6. *Let $\mathcal{N}$ be an $\mathcal{A}$-subbimodule of U_L. Define $\Gamma_{L,\mathcal{N}} := U_L/\mathcal{N}$, the quotient $\mathcal{A}$-bimodule. Let $\pi_{L,\mathcal{N}} : U_L \to \Gamma_{L,\mathcal{N}}$ be the quotient map. Define $d_{L,\mathcal{N}} = \pi_{L,\mathcal{N}} \circ D_L : \mathcal{A} \to \Gamma_{L,\mathcal{N}}$. (We drop the subscript $\mathcal{N}$ when context indicates which $\mathcal{A}$-subbimodule is being used.) Then $(\Gamma_{L,\mathcal{N}}, d_{L,\mathcal{N}})$ is an FODC over $\mathcal{A}$.*

Moreover, any FODC over $\mathcal{A}$ is isomorphic to $(\Gamma_{L,\mathcal{N}}, d_{L,\mathcal{N}})$ for some $\mathcal{A}$-subbimodule $\mathcal{N}$ of U_L.

The model $U_L = \mathcal{A} \otimes \ker \varepsilon$ of the universal FODC is discussed in [48], Chapter 14, where it is denoted as $\Omega^1(\mathcal{A})$.

Exercise 6.9. *This exercise parallels the previous one, except that now the isomorphism of interest is s.*

Define $U_R := \ker \varepsilon \otimes \mathcal{A}$. Recall that $s : U \to U_R$ is an $\mathcal{A}$-bimodule isomorphism as well as a right $\mathcal{A}$-co-module isomorphism.

Show that the universal differential D for U maps over to the differential $D_R := sD$ for U_R, which is then given by $D_R a = \phi(a) - 1 \otimes a$ for all $a \in \mathcal{A}$.

Prove directly that $D_R : \mathcal{A} \to U_R$ is a derivation and that every element in U_R can be written in standard form.

Now convince yourself that Theorem 2.1 translates under the isomorphism s to the following statement:

Theorem 6.7. *Let $\mathcal{N}$ be an $\mathcal{A}$-subbimodule of U_R. Define $\Gamma_{R,\mathcal{N}} := U_R/\mathcal{N}$, the quotient $\mathcal{A}$-bimodule. Let $\pi_{R,\mathcal{N}} : U_R \to \Gamma_{R,\mathcal{N}}$ be the quotient map. Define $d_{R,\mathcal{N}} = \pi_{R,\mathcal{N}} \circ D_R : \mathcal{A} \to \Gamma_{R,\mathcal{N}}$. (We drop the subscript $\mathcal{N}$ when context indicates which $\mathcal{A}$-subbimodule is being used.) Then $(\Gamma_{R,\mathcal{N}}, d_{R,\mathcal{N}})$ is an FODC over $\mathcal{A}$.*

Moreover, any FODC over $\mathcal{A}$ is isomorphic to $(\Gamma_{R,\mathcal{N}}, d_{R,\mathcal{N}})$ for some $\mathcal{A}$-subbimodule $\mathcal{N}$ of U_R.

One moral of this lengthy story is that we have three isomorphic models for the universal FODC: $(U, D), (U_L, D_L)$, and (U_R, D_R). We use the model that is most convenient in a given application. Another moral is that the universal FODC of $\mathcal{A}$ is constructed directly from $\mathcal{A}$ itself in a well-determined way; no other information is required. Or, saying this in more abstract terms, the universal FODC is a functor of $\mathcal{A}$.

6.3 Structure Theorems

We now state the structure theorem concerning left covariant FODCs.

Theorem 6.8. *Let $\mathcal{A}$ be a Hopf algebra. Suppose that $\mathcal{R} \subset \mathrm{Ker}\,\varepsilon$ is a right ideal in $\mathcal{A}$. Define $\mathcal{N}_{\mathcal{R}} := \mathcal{A} \otimes \mathcal{R} \subset \mathcal{A} \otimes \mathrm{Ker}\,\varepsilon = U_L$, which we will often simply denote as $\mathcal{N}$. Then $\mathcal{N}$ is an $\mathcal{A}$-subbimodule of U_L and the FODC $(\Gamma_{L,\mathcal{N}}, d_{L,\mathcal{N}})$ is left covariant.*

Furthermore, any left covariant FODC over $\mathcal{A}$ is isomorphic to $(\Gamma_{L,\mathcal{N}}, d_{L,\mathcal{N}})$, for some $\mathcal{N} = \mathcal{N}_{\mathcal{R}}$, where $\mathcal{R} \subset \mathrm{Ker}\,\varepsilon$ is a right ideal in $\mathcal{A}$.

Recall that

$$\Gamma_{L,\mathcal{N}} = \frac{U_L}{\mathcal{A} \otimes \mathcal{R}} = \frac{\mathcal{A} \otimes \mathrm{ker}\,\varepsilon}{\mathcal{A} \otimes \mathcal{R}} = \mathcal{A} \otimes \left(\frac{\mathrm{ker}\,\varepsilon}{\mathcal{R}}\right).$$

The last equality is really a natural isomorphism that comes from the fact that $\mathcal{A} \otimes \cdot$ is a right exact functor.

Exercise 6.10. *If you do not know what a right exact functor is, then this might be the time to consult a text on homological algebra and see why this concept applies here. You will also have to read there what "natural" means in this context. However, this is not necessary. Instead, if you wish, your exercise is to prove directly using linear algebra that an isomorphism between these vector spaces exists and do not worry for now about what makes this isomorphism "natural."*

Also, as we will see momentarily in the proof of Theorem 6.8, the left co-action on $\Gamma_{L,\mathcal{N}}$ under this identification is $\phi \otimes id$, where id is the identity map of $\mathrm{ker}\,\varepsilon/\mathcal{R}$. This means that $\Gamma_{L,\mathcal{N}}$ is a free left co-module over $\mathcal{A}$. And that implies that the set of left invariant elements in $\Gamma_{L,\mathcal{N}}$ is given by

$$_{\mathrm{inv}}\Gamma_{L,\mathcal{N}} = 1 \otimes \left(\frac{\mathrm{ker}\,\varepsilon}{\mathcal{R}}\right) \cong \frac{\mathrm{ker}\,\varepsilon}{\mathcal{R}},$$

where the isomorphism here is natural. Therefore, if (Γ, d) is any left covariant FODC over $\mathcal{A}$ and its associated right ideal is $\mathcal{R}$, then we have $_{\mathrm{inv}}\Gamma \cong \mathrm{ker}\,\varepsilon/\mathcal{R}$, but this isomorphism depends on the choice of the isomorphism $\Gamma \cong \Gamma_{L,\mathcal{N}}$. For those who have learned the term, we note that this last isomorphism is not natural.

We will presently prove this theorem. But first here is the corresponding structure theorem concerning right covariant FODCs, whose proof we leave to the reader.

Theorem 6.9. *Let $\mathcal{A}$ be a Hopf algebra. Suppose that $\mathcal{R} \subset \mathrm{Ker}\,\varepsilon$ is a right ideal in $\mathcal{A}$. Define $\mathcal{M}_{\mathcal{R}} := \mathcal{R} \otimes \mathcal{A}$, which we will often denote as $\mathcal{M}$. Then $\mathcal{M}$ is a subbimodule of U_R and the FODC $(\Gamma_{R,\mathcal{M}}, d_{R,\mathcal{M}})$ is right covariant.*

Moreover, any right covariant FODC over $\mathcal{A}$ is isomorphic to $(\Gamma_{R,\mathcal{M}}, d_{R,\mathcal{M}})$, for some $\mathcal{M} = \mathcal{M}_{\mathcal{R}}$, where $\mathcal{R} \subset \mathrm{Ker}\,\varepsilon$ is a right ideal in U_R.

Proof of Theorem 6.8. To prove the first part of the theorem, we let $\mathcal{R} \subset \mathrm{Ker}\,\varepsilon$ be a right ideal in $\mathcal{A}$. Then we claim that $\mathcal{N} = \mathcal{A} \otimes \mathcal{R}$ is an $\mathcal{A}$-subbimodule of $\mathcal{A} \otimes \mathrm{Ker}\,\varepsilon = U_L$. To see this, take $a, b \in \mathcal{A}$ and $c \in \mathcal{R}$. Then for the left action of a on $b \otimes c \in \mathcal{N}$, we obtain

$$a \cdot (b \otimes c) = (a \otimes 1)(b \otimes c) = ab \otimes c \in \mathcal{A} \otimes \mathcal{R},$$

while for the right action, we have

$$(b \otimes c) \cdot a = (b \otimes c)\phi(a) = (b \otimes c)(a^{(1)} \otimes a^{(2)}) = ba^{(1)} \otimes ca^{(2)} \in \mathcal{A} \otimes \mathcal{R}.$$

We used the hypothesis that $\mathcal{R}$ is a right ideal in $\mathcal{A}$ to conclude that $ca^{(2)} \in \mathcal{R}$. The left co-action on U_L is the map $\phi \otimes id$. Applying this to $\mathcal{N}$, we find that

$$(\phi \otimes id)(\mathcal{N}) = (\phi \otimes id)(\mathcal{A} \otimes \mathcal{R}) = \phi(\mathcal{A}) \otimes \mathcal{R} \subset \mathcal{A} \otimes \mathcal{A} \otimes \mathcal{R} = \mathcal{A} \otimes \mathcal{N},$$

which says that the submodule $\mathcal{N}$ is invariant under the left co-action. Hence, this co-action passes to the quotient, making it into a left covariant FODC. This proves the first part of the theorem.

We recall that (Γ, d) is a left covariant FODC over $\mathcal{A}$ if and only if this condition holds:

$$\sum_k a_k db_k = 0 \implies \sum_k \phi(a_k)(id \otimes d)\phi(b_k) = 0 \qquad (6.13)$$

for all $a_k, b_k \in \mathcal{A}$ (cf. Theorem 3.1). As noted in Lemma 6.1, this condition is invariant under isomorphisms of FODCs. This tells us that if two FODCs are isomorphic as FODCs and one of them is left covariant, then the other is also left covariant. Furthermore, in this case any FODC isomorphism between them is automatically an isomorphism of left co-modules. And *mutatis mutandis* for right covariant FODCs.

To prove the last assertion of the theorem, we assume that we have a left covariant FODC, say (Γ, d). We have to show that it is isomorphic as an FODC to $(\Gamma_{L,\mathcal{N}}, d_{L,\mathcal{N}})$, where $\mathcal{N}$ is as stated in the theorem. But we already know that (Γ, d) is isomorphic to $(\Gamma_{\mathcal{P}}, d_{\mathcal{P}})$ for *some* $\mathcal{A}$-subbimodule $\mathcal{P}$ of U. Using the map r, we obtain that (Γ, d) is isomorphic to some quotient of U_L by some $\mathcal{A}$-subbimodule $\mathcal{N}$ of U_L. [Clearly, $\mathcal{N} = r(\mathcal{P})$.] To simplify the argument, we can assume that (Γ, d) simply *is* a quotient of U_L by some $\mathcal{A}$-subbimodule $\mathcal{N}$ of U_L. But we do not yet have that $\mathcal{N} = \mathcal{N}_{\mathcal{R}}$, where $\mathcal{R} \subset \mathrm{Ker}\,\varepsilon$ is a right ideal in $\mathcal{A}$. However, the left covariance of (Γ, d) implies that

$$(\phi \otimes id)(\mathcal{N}) \subset \mathcal{A} \otimes \mathcal{N}.$$

Now we know that any element ω in the left covariant submodule $\mathcal{N}$ can be written as a linear combination of left invariant elements in $\mathcal{N}$ with coefficients in $\mathcal{A}$. But according to Proposition 6.1, the left invariant elements (in U_L and hence also in $\mathcal{N}$) have the form $1 \otimes a$ for some $a \in \ker \varepsilon$. So we have

$$\omega = \sum_k c_k \cdot (1 \otimes a_k) = \sum_k c_k \otimes a_k \qquad (6.14)$$

for $a_k \in \ker \varepsilon$ and $c_k \in \mathcal{A}$ for all k. Define $\mathcal{R}$ to be the linear span of all the $a_k's$ arising this way. (This will turn out to be the right ideal we are seeking.) By the very definition of $\mathcal{R}$, Eq. (6.14) above shows that $\mathcal{N} = \mathcal{A} \otimes \mathcal{R}$. It remains to show that $\mathcal{R} \subset \operatorname{Ker} \varepsilon$ and that it is a right ideal in $\mathcal{A}$. But the elements of $\mathcal{R}$ are linear combinations of the $a_k's$ and $\varepsilon(a_k) = 0$ for all k. So $\mathcal{R} \subset \operatorname{Ker} \varepsilon$ follows. Next, to show that $\mathcal{R}$ is a right ideal, we take $b \in \mathcal{R}$ and $a \in \mathcal{A}$. Then we have to show $ba \in \mathcal{R}$. But $1 \otimes b \in \mathcal{A} \otimes \mathcal{R} = \mathcal{N}$. Since $\mathcal{N}$ is a bimodule, we have that

$$(1 \otimes b) \cdot a = (1 \otimes b)\phi(a) \in \mathcal{N} = \mathcal{A} \otimes \mathcal{R},$$

which implies that

$$(\varepsilon \otimes id)\big((1 \otimes b)\phi(a)\big) \in \mathbb{C} \otimes \mathcal{R} \cong \mathcal{R}.$$

Finally, we calculate

$$(\varepsilon \otimes id)\big((1 \otimes b)\phi(a)\big) = (\varepsilon \otimes id)\big((1 \otimes b)(a^{(1)} \otimes a^{(2)})\big)$$

$$= (\varepsilon \otimes id)(a^{(1)} \otimes ba^{(2)})$$

$$= \varepsilon(a^{(1)}) \, ba^{(2)}$$

$$= b \, \varepsilon(a^{(1)}) \, a^{(2)}$$

$$= ba.$$

And so we conclude that $ba \in \mathcal{R}$, as desired. ∎

Theorem 6.10. *Suppose we have the setup of Theorem 6.8 as follows in the next paragraph:*

Let $\mathcal{A}$ be a Hopf algebra. Suppose that $\mathcal{R} \subset \operatorname{Ker} \varepsilon$ is a right ideal in $\mathcal{A}$. Define $\mathcal{M} := \mathcal{A} \otimes \mathcal{R} \subset \mathcal{A} \otimes \operatorname{Ker} \varepsilon = U_L$. Then by Theorem 6.8, we already have established that $\mathcal{M}$ is an $\mathcal{A}$-subbimodule of U_L and the FODC $(\Gamma, d) := (\Gamma_{L,\mathcal{M}}, d_{L,\mathcal{M}})$ is left covariant.

Then (Γ, d) is bicovariant if and only if $\mathcal{R}$ is ad-invariant.

Proof. We first suppose that $\mathcal{R}$ is ad-invariant, where $\mathcal{R} \subset \mathrm{Ker}\,\varepsilon$ is a right ideal in $\mathcal{A}$. Then by Theorem 6.8, $(\Gamma, d) := (\Gamma_{L,\mathcal{M}}, d_{L,\mathcal{M}})$ is left covariant. We will work with the universal FODC (U, D) and use the maps r and s to relate it to the other models U_L and U_R, respectively. Moreover, in the proof of Theorem 6.8, we established that $\mathcal{N} = r^{-1}(\mathcal{A} \otimes \mathcal{R})$ is an $\mathcal{A}$-submodule of U and that $\mathcal{N}$ is invariant under the left co-action. Recall that the left co-action, denoted here as Φ_U, of $\mathcal{A}$ on U commutes with the right co-action, denoted $_U\Phi$, on U; that is,

$$(\Phi_U \otimes id)\,_U\Phi = (id \otimes\,_U\Phi)\,\Phi_U. \tag{6.15}$$

Now the left invariant elements in $\mathcal{N}$ are $_{\mathrm{inv}}\mathcal{N} = r^{-1}(1 \otimes \mathcal{R})$. So we take $a \in \mathcal{R}$ and see what happens to the left invariant element $r^{-1}(1 \otimes a)$ under the right action $_U\Phi$. We claim that this is again a left invariant element, but now under the left co-action $\Phi_U \otimes id : U \otimes \mathcal{A} \to \mathcal{A} \otimes U \otimes \mathcal{A}$. (And yes, this *is* a left co-action of $\mathcal{A}$ on $U \otimes \mathcal{A}$.) So we calculate

$$\begin{aligned}
(\Phi_U \otimes id)\,_U\Phi\big(r^{-1}(1 \otimes a)\big) &= (id \otimes\,_U\Phi)\Phi_U\big(r^{-1}(1 \otimes a)\big) \\
&= (id \otimes\,_U\Phi)(id \otimes r^{-1})((\phi \otimes 1)(1 \otimes a)) \\
&= (id \otimes\,_U\Phi)(id \otimes r^{-1})(1 \otimes 1 \otimes a) \\
&= (id \otimes\,_U\Phi)\big(1 \otimes r^{-1}(1 \otimes a)\big) \\
&= 1 \otimes\,_U\Phi\big(r^{-1}(1 \otimes a)\big).
\end{aligned}$$

We used (6.15) in the first equality and the diagram (6.10) in the second equality. So the element $_U\Phi\big(r^{-1}(1 \otimes a)\big) \in U \otimes \mathcal{A}$ is left invariant under the co-action of $\Phi_U \otimes id$, as claimed. Now

$$U \otimes \mathcal{A} = r^{-1}(\mathcal{A} \otimes \mathrm{ker}\,\varepsilon) \otimes \mathcal{A} = (r^{-1} \otimes id)(\mathcal{A} \otimes \mathrm{ker}\,\varepsilon \otimes \mathcal{A}),$$

and so its set of left invariant elements is

$$_{\mathrm{inv}}(U \otimes \mathcal{A}) = (r^{-1} \otimes id)(1 \otimes \mathrm{ker}\,\varepsilon \otimes \mathcal{A}).$$

So

$$_U\Phi\big(r^{-1}(1 \otimes a)\big) = (r^{-1} \otimes id)(1 \otimes q) \tag{6.16}$$

for some $q \in \mathrm{ker}\,\varepsilon \otimes \mathcal{A}$. We claim that $q = \mathrm{ad}(a)$. Given this claim and the hypothesis that $\mathrm{ad}(\mathcal{R}) \subset \mathcal{R} \otimes \mathcal{A}$, we conclude that

$$\begin{aligned}
_U\Phi\big(r^{-1}(1 \otimes a)\big) &= (r^{-1} \otimes id)(1 \otimes q) \\
&= (r^{-1} \otimes id)(1 \otimes \mathrm{ad}(a)) \in (r^{-1} \otimes id)(1 \otimes \mathcal{R} \otimes \mathcal{A}),
\end{aligned}$$

since ad$(a) \in \mathcal{R} \otimes \mathcal{A}$. But

$$(r^{-1} \otimes id)(1 \otimes \mathcal{R} \otimes \mathcal{A}) = {}_{\text{inv}}(\mathcal{A} \otimes \mathcal{R}) \otimes \mathcal{A} = {}_{\text{inv}}(\mathcal{N}) \otimes \mathcal{A}.$$

Since $r^{-1}\big((1 \otimes a)\big) \in {}_{\text{inv}}\mathcal{N}$ was arbitrary, we get ${}_U\Phi({}_{\text{inv}}\mathcal{N}) \subset {}_{\text{inv}}(\mathcal{N}) \otimes \mathcal{A}$. Next, we recall that every element in $\mathcal{N}$ is a linear combination of elements in ${}_{\text{inv}}\mathcal{N}$ with coefficients in $\mathcal{A}$ to conclude that ${}_U\Phi(\mathcal{N}) \subset \mathcal{N} \otimes \mathcal{A}$. This implies that Γ is right covariant. But it is also left covariant. Also, the left and right co-actions are compatible, as noted in (6.15). So Γ is bicovariant.

But we still have to prove the claim that $q = \text{ad}(a)$. We do this by applying $\varepsilon \otimes id \otimes id$ to (6.16). So the left side yields

$$(\varepsilon \otimes id \otimes id)\big({}_U\Phi\big(r^{-1}(1 \otimes a)\big)\big) = s\big((r^{-1}(1 \otimes a))\big) = \text{ad}(a). \qquad (6.17)$$

Here we used the identity $s = (\varepsilon \otimes id \otimes id) \, {}_U\Phi$ (see the next exercise). Now we write $q = \sum_i q_i' \otimes q_i''$ for the next calculation. So for the right side of (6.16), we get

$$(\varepsilon \otimes id \otimes id)\big((r^{-1} \otimes id)(1 \otimes q)\big) = \big((\varepsilon \otimes id)r^{-1} \otimes id\big)(1 \otimes q)\big)$$

$$= \sum_i \big((\varepsilon \otimes id)r^{-1} \otimes id\big)(1 \otimes q_i' \otimes q_i'')$$

$$= \sum_i (\varepsilon \otimes id)r^{-1}(1 \otimes q_i') \otimes q_i''$$

$$= \sum_i \varepsilon(1)q_i' \otimes q_i''$$

$$= \sum_i q_i' \otimes q_i''$$

$$= q, \qquad (6.18)$$

where we used (B.25) in the fourth equality. Comparing (6.17) and (6.18), we see that $q = \text{ad}(a)$.

Conversely, suppose that (Γ, d) is bicovariant. According to Theorem 6.9, there exists a right ideal $\mathcal{S} \subset \ker \varepsilon$ in $\mathcal{A}$ such that $\mathcal{N} = s^{-1}(\mathcal{S} \otimes \mathcal{A})$, where $\mathcal{N} = r^{-1}(\mathcal{A} \otimes \mathcal{R})$ as in the previous part of this proof. So from these two expressions for $\mathcal{N}$, we get

$$r^{-1}(\mathcal{A} \otimes \mathcal{R}) = s^{-1}(\mathcal{S} \otimes \mathcal{A}), \qquad (6.19)$$

which implies that

$$(\varepsilon \otimes id)r^{-1}(\mathcal{A} \otimes \mathcal{R}) = (\varepsilon \otimes id)s^{-1}(\mathcal{S} \otimes \mathcal{A}). \qquad (6.20)$$

Now identity (B.25) is $(\varepsilon \otimes id)r^{-1} = \varepsilon(a)b$, and this implies that

$$(\varepsilon \otimes id)r^{-1}(A \otimes \mathcal{R}) = \mathcal{R}.$$

And (B.26) says $(\varepsilon \otimes id)s^{-1} = \varepsilon(b)a$, which implies $s^{-1}(S \otimes A) = S$. And so (6.20) becomes $\mathcal{R} = S$. Then (6.19) becomes

$$r^{-1}(A \otimes \mathcal{R}) = s^{-1}(\mathcal{R} \otimes A),$$

which is equivalent to

$$sr^{-1}(A \otimes \mathcal{R}) = \mathcal{R} \otimes A.$$

Finally, this implies

$$\mathrm{ad}(\mathcal{R}) = sr^{-1}(1 \otimes \mathcal{R}) \subset \mathcal{R} \otimes A,$$

which simply says by definition that $\mathcal{R}$ is ad-invariant. ∎

Exercise 6.11. *Prove* $s = (\varepsilon \otimes id \otimes id)_U \Phi$.

Exercise 6.12. *Suppose that* (Γ, d) *is bicovariant. Continuing to use the notation of the previous theorem, prove that* $rs^{-1}(\mathcal{R} \otimes 1) \subset A \otimes \mathcal{R}$. *Then find a formula for* $rs^{-1}(a \otimes 1)$ *for* $a \in A$. *This then gives a necessary condition for* (Γ, d) *to be bicovariant. Is this condition also sufficient?*

The previous theorem raises our interest in the right ideals $\mathcal{R} \subset \ker \varepsilon$, which are ad-invariant. Clearly, $\mathcal{R} = 0$, the smallest possibility, is one such (even two-sided) ideal. In that case, the bicovariant FODC is U_L. What about the largest possibility, namely, $\mathcal{R} = \ker \varepsilon$? In that case, the corresponding FODC is $\Gamma = 0$, which is clearly bicovariant. So $\ker \varepsilon$ must be an ad-invariant right ideal in A. Now, it is actually a two-sided ideal since that is a property of the kernel of any multiplicative morphism. But it may not be quite as obvious that it is also ad-invariant, so we now verify this directly in the next proposition.

Proposition 6.3. *The two-sided ideal* $\ker \varepsilon$ *is ad-invariant in* A.

Proof. So we consider an arbitrary element $a \in \ker \varepsilon$. Then by the definition of ad-invariant, we must show that $\mathrm{ad}(a) \in \ker \varepsilon \otimes A$. So using the definition of ad, standard identities, and $\varepsilon(a) = 0$, we calculate

$$
\begin{aligned}
(\varepsilon \otimes id)\mathrm{ad}(a) &= (\varepsilon \otimes id)(a^{(2)} \otimes \kappa(a^{(1)})a^{(3)}) \\
&= \varepsilon(a^{(2)}) \otimes \kappa(a^{(1)})a^{(3)} \\
&\cong \varepsilon(a^{(2)})\kappa(a^{(1)})a^{(3)} \\
&= \kappa(a^{(1)})a^{(2)} \\
&= m_A(\kappa \otimes id)\phi(a)
\end{aligned}
$$

$$= \varepsilon(a)1$$
$$= 0.$$

So $\mathrm{ad}(a) \in \ker(\varepsilon \otimes id) = \ker \varepsilon \otimes \mathcal{A}$. And this shows the result. $\blacksquare$

It turns out that Eq. (6.16) in the proof of the previous theorem deserves to be singled out as a result itself.

Corollary 6.2. *Let $\mathcal{R} \subset \ker \varepsilon$ be an* ad-*invariant right ideal in $\mathcal{A}$. Then the following diagram is commutative:*

$$
\begin{array}{ccccc}
\mathcal{R} & \cong & 1 \otimes \mathcal{R} & \xrightarrow{r^{-1}} & {}_{\mathrm{inv}}U \\
{\scriptstyle ad} \downarrow & & & & \downarrow {\scriptstyle U}\Phi \\
\mathcal{R} \otimes \mathcal{A} & \cong & 1 \otimes \mathcal{R} \otimes \mathcal{A} & \xrightarrow{r^{-1} \otimes id_A} & {}_{\mathrm{inv}}U \otimes \mathcal{A}.
\end{array}
\qquad (6.21)
$$

Notice that the horizontal arrows are injections. A particular case of this diagram is when we take $\mathcal{R} = \ker \varepsilon$, since $\ker \varepsilon$ is an ad-*invariant ideal in $\mathcal{A}$. Then that particular case, together with the general case (6.21), gives us this commutative diagram:*

$$
\begin{array}{ccccc}
(\ker \varepsilon)/\mathcal{R} & \cong & 1 \otimes (\ker \varepsilon)/\mathcal{R} & \xrightarrow{r^{-1}} & {}_{\mathrm{inv}}\Gamma \\
{\scriptstyle ad} \downarrow & & & & \downarrow {\scriptstyle \Gamma}\Phi \\
(\ker \varepsilon)/\mathcal{R} \otimes \mathcal{A} & \cong & 1 \otimes (\ker \varepsilon)/\mathcal{R} \otimes \mathcal{A} & \xrightarrow{r^{-1} \otimes id_A} & {}_{\mathrm{inv}}\Gamma \otimes \mathcal{A}.
\end{array}
\qquad (6.22)
$$

In this diagram the horizontal arrows are isomorphisms. Also, Γ here is the model FODC $\Gamma_{L,\mathcal{N}}$ introduced in Theorem 6.8.

The proof consists of noting that the diagram (6.21) corresponds to (6.16). The vertical arrow on the left in the diagram (6.22) is the right co-action ad, passed to the quotient by $\mathcal{R}$. It is also denoted as ad and is also called the *right adjoint co-action*. (It is a right co-action!) It is, by definition, the unique linear map $(\ker \varepsilon)/\mathcal{R} \to (\ker \varepsilon)/\mathcal{R} \otimes \mathcal{A}$, making this diagram commutative, where $\pi_{\mathcal{R}}$ is the quotient map:

$$
\begin{array}{ccc}
\ker \varepsilon & \xrightarrow{\ ad\ } & \ker \varepsilon \otimes \mathcal{A} \\
{\scriptstyle \pi_{\mathcal{R}}} \downarrow & & \downarrow {\scriptstyle \pi_{\mathcal{R}} \otimes id} \\
(\ker \varepsilon)/\mathcal{R} & \xrightarrow{\ ad\ } & (\ker \varepsilon)/\mathcal{R} \otimes \mathcal{A}.
\end{array}
\qquad (6.23)
$$

This shows that ${}_{\mathrm{inv}}\Gamma = (\ker \varepsilon)/\mathcal{R}$ is ad-invariant.

Exercise 6.13. *Prove* ad $: (\ker \varepsilon)/\mathcal{R} \to (\ker \varepsilon)/\mathcal{R} \otimes \mathcal{A}$ *is a right co-action.*

The diagram (6.22) says that the right co-action ${}_{\Gamma}\Phi$ when restricted to the left invariant elements is "equal" to the right adjoint co-action on $(\ker \varepsilon)/\mathcal{R}$.

6.4   Quantum Germs Map

We now define and discuss the "quantum germs" map $\pi : \mathcal{A} \to \Gamma$, where (Γ, d) is a left covariant FODC. This structure is discussed in [21] and [22]. However, we use a different approach although we end up with the same map.

Definition 6.2. Let (Γ, d) be the left covariant FODC $(\Gamma_{L\mathcal{N}}, d_{L\mathcal{N}})$ as defined in Theorem 6.8. Let $\mathcal{R} \subset \ker \varepsilon$ be the corresponding right ideal in $\mathcal{A}$, that is, $\Gamma = \mathcal{A} \otimes (\ker \varepsilon / \mathcal{R})$. Define the *quantum germs map* $\pi : \mathcal{A} \to \Gamma$ for all $a \in \mathcal{A}$ by

$$\pi(a) := 1 \otimes [a - \varepsilon(a)1]_{\mathcal{R}} = 1 \otimes \pi_{\mathcal{R}}(a - \varepsilon(a)1), \qquad (6.24)$$

where $[\cdot]_{\mathcal{R}}$ means the equivalence class in $\ker \varepsilon / \mathcal{R}$.

Since $a - \varepsilon(a)1 \in \ker \varepsilon$, we see that π is well defined. Also, $a \in \ker \varepsilon$ implies that $\pi(a) = 1 \otimes \pi_{\mathcal{R}}(a)$. Here are some more basic properties of the quantum germs map.

Proposition 6.4. *Let (Γ, d) be the left covariant FODC $(\Gamma, d) \equiv (\Gamma_{L\mathcal{N}}, d_{L\mathcal{N}})$. Then we have the following assertions:*

1. $\pi(a) \in {}_{\mathrm{inv}}\Gamma$ *for all $a \in \mathcal{A}$.*
2. $\pi : \mathcal{A} \to {}_{\mathrm{inv}}\Gamma$ *is a surjection. Actually, the restriction $\pi : \ker \varepsilon \to {}_{\mathrm{inv}}\Gamma$ is already a surjection.*
3. $\pi(1) = 0$.
4. $\ker \pi = \mathcal{R} + \mathbb{C}1$.
5. *For $\theta \in {}_{\mathrm{inv}}\Gamma$ written as $\theta = \pi(a)$ for some $a \in \ker \varepsilon$, we have that $\mathrm{ad}(\theta) = \mathrm{ad}(\pi(a)) = (\pi \otimes id)\mathrm{ad}(a)$.*
   More generally, $\mathrm{ad}(\pi(a)) = (\pi \otimes id)\mathrm{ad}(a)$ for all $a \in \mathcal{A}$.
6. $da = a^{(1)}\pi(a^{(2)})$ *for all $a \in \mathcal{A}$.*
7. $d\kappa(a) = -\pi(a^{(1)})\kappa(a^{(2)})$ *for all $a \in \mathcal{A}$.*
8. *(Maurer–Cartan formula) $d\pi(a) = -\pi(a^{(1)})\pi(a^{(2)})$ for all $a \in \mathcal{A}$.*

Proof. Statement 1 follows immediately from the characterization in Proposition 6.1 of the left invariant elements in $\Gamma = \mathcal{A} \otimes (\ker \varepsilon / \mathcal{R})$.

For statement 2, suppose that $1 \otimes [b]_{\mathcal{R}} \in 1 \otimes (\ker \varepsilon / \mathcal{R}) = {}_{\mathrm{inv}}\Gamma$. Then $b \in \ker \varepsilon \subset \mathcal{A}$; that is, $\varepsilon(b) = 0$. Thus, $\pi(b) = 1 \otimes [b - \varepsilon(b)1]_{\mathcal{R}} = 1 \otimes [b]_{\mathcal{R}}$. And this shows that π is a surjection.

Statement 3 follows by substituting $a = 1$ into the definition of π and then using $\varepsilon(1) = 1$.

For statement 4, first consider $r \in \mathcal{R} \subset \ker \varepsilon$ and $\lambda \in \mathbb{C}$. Then

$$\pi(r + \lambda 1) = \pi(r) + \lambda\pi(1) = \pi(r) = 1 \otimes [r]_{\mathcal{R}} = 0.$$

This shows that $\mathcal{R} + \mathbb{C}1 \subset \ker \pi$. For the opposite inclusion, note that $\pi(a) = 0$ implies $a - \varepsilon(a)1 \in \mathcal{R}$, which in turn implies

$$a \in \mathcal{R} + \varepsilon(a)1 \subset \mathcal{R} + \mathbb{C}1.$$

The first part of statement 5 is simply the diagram (6.23) rewritten as a formula, using the identification of π and $\pi_{\mathcal{R}}$ on $\ker \varepsilon$. For the second part of statement 5, we remark that $\mathcal{A} = \ker \varepsilon \oplus \mathbb{C}1$ since $a = (a - \varepsilon(a)1) + \varepsilon(a)1$. So by linearity, it only remains to prove that $\text{ad}(\pi(a)) = (\pi \otimes id)\text{ad}(a)$ for $a \in \mathbb{C}1$. But in that case, both sides of the desired equality are equal to zero.

We prove statement 6 for (U_L, D_L), since it then passes to the quotient. Recall that $D_L = \phi(a) - a \otimes 1$. So we have

$$
\begin{aligned}
a^{(1)}\pi(a^{(2)}) &= a^{(1)} \otimes \left(1 \otimes (a^{(2)} - \varepsilon(a^{(2)})1)\right) \\
&= a^{(1)} \otimes a^{(2)} - a^{(1)} \otimes \varepsilon(a^{(2)})1 \\
&= \phi(a) - \varepsilon(a^{(2)})a^{(1)} \otimes 1 \\
&= \phi(a) - a \otimes 1 \\
&= D_L a.
\end{aligned}
$$

Statement 8, the Maurer–Cartan formula, is just for the record for the time being. The astute reader will have already noticed that on the left side the differential d has not been defined for elements in Γ. Also, on the right side the product of two elements in Γ has not yet been defined. We will come back to this in Section 10.3. ∎

Exercise 6.14. *Prove statement 7 of the proposition.*
Hint: This may be easier to solve after reading the rest of this section.

Remarks 6.1. *Since $\mathcal{R} \cap \mathbb{C}1 = 0$, we can write $\mathcal{R} + \mathbb{C}1 = \mathcal{R} \oplus \mathbb{C}1$.*

Now we will identify our definition of π with the one given in [22].

Theorem 6.11. *For the left covariant FODC $(\Gamma, d) \equiv (\Gamma_{L\mathcal{N}}, d_{L\mathcal{N}})$, we have*

$$
\pi(a) = \kappa(a^{(1)})da^{(2)} \tag{6.25}
$$

for all $a \in \mathcal{A}$.

Proof. The differential $d = d_{L\mathcal{N}}$ is given by $d = \pi_{L\mathcal{N}}D_L$. Recall that $D_L : \mathcal{A} \to U_L$ is given by the formula

$$
D_L a = \phi(a) - a \otimes 1 = a^{(1)} \otimes a^{(2)} - a \otimes 1.
$$

Then, if we consider it an element in $\mathcal{A} \otimes (\ker \varepsilon / \mathcal{R})$, we have

$$
da = a^{(1)} \otimes [a^{(2)}]_{\mathcal{R}} - a \otimes [1]_{\mathcal{R}}.
$$

Then we calculate

$$
\kappa(a^{(1)})da^{(2)} = \kappa(a^{(1)})\left(a^{(21)} \otimes [a^{(22)}]_{\mathcal{R}} - a^{(2)} \otimes [1]_{\mathcal{R}}\right)
$$

$$= \kappa(a^{(1)})a^{(21)} \otimes [a^{(22)}]_\mathcal{R} - \kappa(a^{(1)})a^{(2)} \otimes [1]_\mathcal{R}$$

$$= \kappa(a^{(1)})a^{(21)} \otimes [a^{(22)}]_\mathcal{R} - \varepsilon(a)1 \otimes [1]_\mathcal{R}$$

$$= \kappa(a^{(11)})a^{(12)} \otimes [a^{(2)}]_\mathcal{R} - \varepsilon(a)1 \otimes [1]_\mathcal{R}$$

$$= \varepsilon(a^{(1)})1 \otimes [a^{(2)}]_\mathcal{R} - \varepsilon(a)1 \otimes [1]_\mathcal{R}$$

$$= 1 \otimes [\varepsilon(a^{(1)})a^{(2)}]_\mathcal{R} - 1 \otimes [\varepsilon(a)1]_\mathcal{R}$$

$$= 1 \otimes [a]_\mathcal{R} - 1 \otimes [\varepsilon(a)1]_\mathcal{R}$$

$$= 1 \otimes [a - \varepsilon(a)1]_\mathcal{R}$$

$$= \pi(a).$$

Here we have used various standard identities of Hopf algebras. For example, the fourth equality is an application of the co-associativity of ϕ. ∎

We now have a definition for the quantum germs map π for the specific left covariant FODCs $(\Gamma_{L,\mathcal{N}}, d_{L,\mathcal{N}})$. We next want to define π for an arbitrary left covariant FODC (Γ, d) over $\mathcal{A}$. One way to do this would be to take some FODC isomorphism $\psi : \Gamma \to \Gamma_{L,\mathcal{N}}$ for an appropriate $\mathcal{N}$ and then define $\tilde{\pi}$ on Γ as $\psi^{-1}\pi\psi$. Then one would have to show that the resulting map $\tilde{\pi}$ does not depend on the particular choice of ψ and gives the same definition as before on $(\Gamma_{L,\mathcal{N}}, d_{L,\mathcal{N}})$. And then one could rename $\tilde{\pi}$ as π.

But we will take a different approach. We can use (6.25) as a definition for any left covariant FODC (Γ, d) over $\mathcal{A}$, since the right side makes sense in such a context. Note that the right side of (6.24) does not make sense in the context of a general left covariant FODC.

Definition 6.3. Let (Γ, d) be a left covariant FODC over $\mathcal{A}$. Then we define the *quantum germs map* $\pi : \mathcal{A} \to \Gamma$ by

$$\pi(a) := \kappa(a^{(1)})da^{(2)} \tag{6.26}$$

for all $a \in \mathcal{A}$.

By the previous remarks, this is an extension of this structure to a larger class of objects. Proposition 6.4 also holds in this broader context. Here it is:

Proposition 6.5. *Let (Γ, d) be any left covariant FODC. Then π satisfies all the properties in Proposition 6.4.*

Proof. The proof simply consists of noting that all of those properties are expressed in terms of properties of left covariant FODCs and that these are preserved under FODC isomorphisms. ∎

We record yet another identity.

Proposition 6.6. *For all $a \in \mathcal{A}$, we have*

$$\pi(a) = \kappa(a^{(1)})\, da^{(2)} = -d\kappa(a^{(1)})\, a^{(2)}.$$

Proof. This first equality is the definition. The second follows from applying the differential d to the identity $\kappa(a^{(1)})a^{(2)} = \varepsilon(a)1$ and then applying Leibniz's rule as well as $d\,1 = 0$. ∎

We now will introduce a right $\mathcal{A}$-structure on the complex vector space $_{\mathrm{inv}}\Gamma$ of left invariant elements of a left covariant FODC (Γ, d). We will do this in a two-step process, much like our definition of the quantum germs map π. So we first consider a special case of a left covariant FODC, namely, $(\Gamma, d) = (\Gamma_{L\mathcal{N}}, d_{L\mathcal{N}})$. Then the left invariant elements are given by

$$_{\mathrm{inv}}\Gamma_{L\mathcal{N}} = 1 \otimes \left(\frac{\ker \varepsilon}{\mathcal{R}} \right).$$

Since $\ker \varepsilon$ is a two-sided ideal in $\mathcal{A}$ and $\mathcal{R}$ is a right ideal in $\mathcal{A}$, then the quotient $\ker \varepsilon / \mathcal{R}$ is a right $\mathcal{A}$-module. Explicitly, the right action of $a \in \mathcal{A}$ on $[b]_{\mathcal{R}} \in \ker \varepsilon / \mathcal{R}$ is given by $[b]_{\mathcal{R}} \cdot a = [ba]_{\mathcal{R}}$, where $\cdot$ denotes the right action.

Transporting this right action to $_{\mathrm{inv}}\Gamma_{L\mathcal{N}}$ in the natural way gives us this definition of a right action of $\mathcal{A}$:

$$\left(1 \otimes [b]_{\mathcal{R}}\right) \circ a := 1 \otimes [ba]_{\mathcal{R}}, \tag{6.27}$$

where $\circ$ denotes the right action, $b \in \ker \varepsilon$, and $a \in \mathcal{A}$. Of course, we again have a definition that does not work for the general case. So we try to write this definition as a formula that has a meaning for any left covariant FODC and then use that formula for the general case.

Proposition 6.7. *Let $\pi : \mathcal{A} \to {}_{\mathrm{inv}}\Gamma_{L\mathcal{N}}$ be the quantum germs map. Then for all $a, b \in \mathcal{A}$, we have*

$$\pi(a) \circ b = \pi(ab - \varepsilon(a)b), \tag{6.28}$$

$$\pi(ab) = \varepsilon(a)\pi(b) + \pi(a) \circ b. \tag{6.29}$$

Remarks 6.2. *Since any element in $_{\mathrm{inv}}\Gamma_{L\mathcal{N}}$ is equal to $\pi(a)$ for some $a \in \mathcal{A}$, (6.28) completely determines the right action $\circ$.*

Proof. Here is the simple computation for (6.28):

$$\pi(a) \circ b = \left(1 \otimes [a - \varepsilon(a)1]_{\mathcal{R}}\right) \circ b$$

$$= 1 \otimes [ab - \varepsilon(a)b]_{\mathcal{R}}$$

$$= \pi(ab - \varepsilon(a)b).$$

If only they were all so simple!

Identity (6.29) follows immediately from (6.28) since

$$\pi(a) \circ b = \pi(ab - \varepsilon(a)b) = \pi(ab) - \varepsilon(a)\pi(b).$$

So (6.29) and (6.28) are essentially the same identity. ∎

Since (6.28) is expressed in terms of structures that make sense for any left invariant FODC, we can now extend the definition of the right action ∘ to that larger context by simply taking (6.28) to be a definition.

Definition 6.4. Let Γ be a left covariant FODC over $\mathcal{A}$. For any $a, b \in \mathcal{A}$, define

$$\pi(a) \circ b := \pi(ab - \varepsilon(a)b), \qquad (6.30)$$

where $\pi : \mathcal{A} \to {}_{\mathrm{inv}}\Gamma$ is the quantum germs map.

Exercise 6.15. *Verify directly using the definition of right action that the expression defined in (6.30) is indeed a right action of $\mathcal{A}$ on ${}_{\mathrm{inv}}\Gamma$.*

Proposition 6.8. *Let Γ be a left covariant FODC over $\mathcal{A}$. For any $\theta \in {}_{\mathrm{inv}}\Gamma$ and $a \in \mathcal{A}$, we have that*

$$\theta \circ a = \kappa(a^{(1)})\theta a^{(2)}.$$

Proof. Write $\theta = \pi(b)$ for some $b \in \ker \varepsilon$. Then we calculate

$$\begin{aligned}
\theta \circ a &= \pi(b) \circ a \\
&= \pi\big((ba) - \varepsilon(b)\pi(a)\big) \\
&= \pi(ba) \\
&= \kappa((ba)^{(1)})d((ba)^{(2)}) \\
&= \kappa(b^{(1)}a^{(1)})d(b^{(2)}a^{(2)}) \\
&= \kappa(a^{(1)})\kappa(b^{(1)})\big((db^{(2)})a^{(2)} + b^{(2)}da^{(2)}\big) \\
&= \kappa(a^{(1)})\kappa(b^{(1)})db^{(2)}a^{(2)} + \kappa(a^{(1)})\kappa(b^{(1)})b^{(2)}da^{(2)} \\
&= \kappa(a^{(1)})\pi(b)a^{(2)} + \kappa(a^{(1)})\varepsilon(b)da^{(2)} \\
&= \kappa(a^{(1)})\theta a^{(2)},
\end{aligned}$$

as desired. ∎

The following computation will be used over and over again. Take an element $\theta \in {}_{\mathrm{inv}}\Gamma$. Then we write $\theta = \pi(a)$ for some $a \in \ker \varepsilon$. First, we note that

$$\theta^{(0)} \otimes \theta^{(1)} = \mathrm{ad}(\theta)$$

$$= \mathrm{ad}(\pi(a))$$

$$= (\pi \otimes id)\mathrm{ad}(a)$$

$$= (\pi \otimes id)\big(a^{(2)} \otimes \kappa(a^{(1)})a^{(3)}\big)$$

$$= \pi(a^{(2)}) \otimes \kappa(a^{(1)})a^{(3)}.$$

This allows us to take $\theta^{(0)} = \pi(a^{(2)})$ and $\theta^{(1)} = \kappa(a^{(1)})a^{(3)}$.

For a left covariant FODC Γ, we immediately obtain from the structure theorem this relation between Γ and its left invariant elements:

$$\Gamma = \mathcal{A} \otimes {}_{\mathrm{inv}}\Gamma. \tag{6.31}$$

Using this relation, we define a map

$$\pi_{\mathrm{inv}} := \varepsilon \otimes id : \Gamma \to {}_{\mathrm{inv}}\Gamma,$$

that is,

$$\pi_{\mathrm{inv}} := \varepsilon \otimes id : \Gamma = \mathcal{A} \otimes {}_{\mathrm{inv}}\Gamma \xrightarrow{\varepsilon \otimes id} \mathbb{C} \otimes {}_{\mathrm{inv}}\Gamma \cong {}_{\mathrm{inv}}\Gamma.$$

Recall that we also defined a map $P : \Gamma \to {}_{\mathrm{inv}}\Gamma$ in Proposition 5.1. This map P can be found in Lemma 2.2 in [86].

Proposition 6.9. *Let Γ be a left covariant FODC over $\mathcal{A}$. Then $P = \pi_{\mathrm{inv}}$.*

Proof. We use the uniqueness result of Proposition 5.1. So we have to show that π_{inv} is a surjective projection that satisfies $\pi_{\mathrm{inv}}(b\omega) = \varepsilon(b)\pi_{\mathrm{inv}}(\omega)$ for all $b \in \mathcal{A}$ and all $\omega \in \Gamma$. First, for any $\eta \in {}_{\mathrm{inv}}\Gamma$, we have that

$$\pi_{\mathrm{inv}}(1 \otimes \eta) = (\varepsilon \otimes id)(1 \otimes \eta) = \varepsilon(1)\eta = \eta.$$

This shows that π_{inv} is surjective and acts as the identity on ${}_{\mathrm{inv}}\Gamma \cong 1 \otimes {}_{\mathrm{inv}}\Gamma$; that is, π_{inv} is a projection onto ${}_{\mathrm{inv}}\Gamma$. Finally, for $b \in \mathcal{A}$ and $\omega = a \otimes \eta$ with $a \in \mathcal{A}$ and $\eta \in {}_{\mathrm{inv}}\Gamma$, we calculate

$$\pi_{\mathrm{inv}}(b\omega) = \pi_{\mathrm{inv}}(b(a \otimes \eta))$$

$$= \pi_{\mathrm{inv}}((ba) \otimes \eta)$$

$$= (\varepsilon \otimes id)((ba) \otimes \eta)$$

$$= \varepsilon(ba) \otimes \eta$$

$$= \varepsilon(b)\varepsilon(a) \otimes \eta$$

$$= \varepsilon(b)(\varepsilon(a) \otimes \eta)$$
$$= \varepsilon(b)(\varepsilon \otimes id)(a \otimes \eta)$$
$$= \varepsilon(b)\pi_{\mathrm{inv}}(\omega).$$

And that finishes the proof. ∎

We showed the following identity in the course of the previous proof.

Corollary 6.3. *Let Γ be a left covariant FODC over $\mathcal{A}$. Then for all $a \in \mathcal{A}$ and all $\omega \in \Gamma$, we have*

$$\pi_{\mathrm{inv}}(a\omega) = \varepsilon(a)\pi_{\mathrm{inv}}(\omega).$$

Note that $\ker \pi_{\mathrm{inv}} = \ker (\varepsilon \otimes id) = \ker (\varepsilon) \otimes_{\mathrm{inv}} \Gamma$ since

$$0 \to \ker (\varepsilon) \to \mathcal{A} \xrightarrow{\varepsilon} \mathbb{C} \to 0$$

is a short, exact sequence, which transforms into a short, exact sequence under application of the functor $\cdot \otimes_{\mathrm{inv}} \Gamma$. This is also a quick way to see that π_{inv} is surjective.

Corollary 6.4. *Let Γ be a left covariant FODC over $\mathcal{A}$. Then for all $\omega \in \Gamma$, we have that*

$$\pi_{\mathrm{inv}}(\omega) = \kappa(\omega^{(0)})\omega^{(1)}, \tag{6.32}$$

where $\Phi_{\Gamma}(\omega) = \omega^{(0)} \otimes \omega^{(1)} \in \mathcal{A} \otimes \Gamma$.

Proof. The right side of (6.32) is nothing other than the formula given for P in Proposition 5.1. So this result follows immediately from $P = \pi_{\mathrm{inv}}$. ∎

Here is a nice relation between the left invariant projection map π_{inv} and the quantum germs map π.

Theorem 6.12. *For all $a \in \mathcal{A}$, we have that*

$$\pi_{\mathrm{inv}}(da) = \pi(a). \tag{6.33}$$

Proof. To prove this identity, take $a \in \mathcal{A}$ and use the covariance of d with respect to Φ_{Γ} to get

$$\Phi_{\Gamma}(da) = (id \otimes d)\phi(a) = (id \otimes d)(a^{(1)} \otimes a^{(2)}) = a^{(1)} \otimes da^{(2)},$$

using Sweedler's notation. Then by the formulas (6.32) and (6.26), we see that

$$\pi_{\mathrm{inv}}(da) = \kappa(a^{(1)})da^{(2)} = \pi(a). \quad \blacksquare$$

So the quantum germs map sends a to the left invariant projection of its differential. Since this is an identity, one can use it to define the quantum germs map as the composition $\pi_{\text{inv}}\, d$ if one wishes. Here it is:

Definition 6.5 (Alternative definition of quantum germs map). The *quantum germs map* $\pi : \mathcal{A} \to {}_{\text{inv}}\Gamma$ is defined by

$$\pi := \pi_{\text{inv}}\, d.$$

The point of emphasizing this alternative definition is that it may give the reader extra insight into the properties of the quantum germs map. It might also help gain a better understanding of Section 5 of [86], where the operator $P\,d$ is used repeatedly but never given its own notation or name. Recall that $P = \pi_{\text{inv}}$. This definition of π is also used in [48], Chapter 14, where it is denoted as ω.

6.5 Quantum Left Invariant Vector Fields

Definition 6.6. Let Γ be a left invariant FODC associated to the right ideal $\mathcal{R} \subset \ker \varepsilon$. Then the *space $\mathcal{X}$ of quantum left invariant vector fields* of Γ is defined as

$$\mathcal{X} := \{X : \mathcal{A} \to \mathbb{C} \mid X \text{ is linear and } \mathcal{R} \oplus \mathbb{C}1_{\mathcal{A}} \subset \ker X\}. \tag{6.34}$$

The elements of $\mathcal{X}$ are called *quantum left invariant vector fields*, which is often abbreviated as *quantum LIVFs*. We also simply say *quantum vector field* when the left invariance is clear by context.

Clearly, $\mathcal{X}$ is a complex vector space. The space $\mathcal{X}$ is also called the *quantum tangent space* in [48], but the elements in that space are unnamed there. This terminology is motivated by the classical case in which the tangent space $T_e(G)$ at the identity element e of a Lie group G is isomorphic as a vector space to the space of left invariant vector fields on G. However, in [86] neither the space nor its objects are given names. It seems better to have names both for the objects and for the space of all such objects.

In this approach to defining quantum vector fields X, there is no apparent attempt to require that X is a derivation or some "quantum" generalization of a derivation. Such an alternative approach would most likely lead to a *unique* space of quantum vector fields. (Compare with [16, 17] and [18].) But in the definition here, we have a structure that depends on the choice of the left invariant FODC Γ. The idea is to construct a space of quantum vector fields that is dual to the space ${}_{\text{inv}}\Gamma$ of left invariant 1-forms. And this definition fulfills that condition, as we shall show later on. In particular, just as the dimension of the vector space ${}_{\text{inv}}\Gamma$ can depend on the choice of the left invariant FODC Γ, so also the dimension of $\mathcal{X}$ will depend on that choice.

Here is some motivation for this definition. Consider the case of a Lie group G of dimension n. We take a chart of local coordinates on G such that the identity $e \in G$ corresponds to $0 \in \mathbb{R}^n$. Then any smooth $f : G \to \mathbb{R}$ has a Taylor expansion that can be written in these local coordinates for $a = (a_1, \ldots, a_n) \in \mathbb{R}^n$ in the coordinate patch as

$$f(a) = f(0) + \langle \omega, X_a \rangle + O(|a|^2) \quad \text{as} \quad a \to 0 \in \mathbb{R}^n, \tag{6.35}$$

where $\omega = \omega_1 dx_1 + \cdots + \omega_n dx_n$ is any 1-form at 0 with

$$\omega_j = \frac{\partial f}{\partial x_j}(0)$$

(e.g., $\omega = df$ will do) and $X_a = a_1 \frac{\partial}{\partial x_1} + \cdots + a_n \frac{\partial}{\partial x_n}$ is a tangent vector at 0. The brackets $\langle \cdot, \cdot \rangle$ denote the usual pairing of 1-forms and tangent vectors; that is,

$$\langle \omega, X_a \rangle = \omega_1 a_1 + \cdots + \omega_n a_n.$$

The co-unit $\varepsilon : C^\infty(G) \to \mathbb{R}$ is given by $\varepsilon(f) := f(0)$ in local coordinates, and so $\ker \varepsilon = \{f \mid f(0) = 0\}$. The Taylor expansion (6.35) for $f \in \ker \varepsilon$ in local coordinates is then

$$f(a) = \langle \omega, X_a \rangle + O(|a|^2) \quad \text{as} \quad a \to 0 \in \mathbb{R}^n. \tag{6.36}$$

We know that the space $_{\mathrm{inv}}\Gamma$ of left invariant 1-forms on G can be identified with the cotangent space at 0 as well as with $\ker \varepsilon / \mathcal{R}$ for some right ideal $\mathcal{R} \subset \ker \varepsilon$. To isolate the 1-form in (6.36), we have to eliminate the remainder term $O(|a|^2)$, which denotes a C^∞-function g that satisfies $|g(a)| \leq C|a|^2$ for all a in some neighborhood of 0. And, in fact, all such functions g arise this way, as one can see by using (6.36) to define f provided that g is given. By a common abuse of notation, we let $O(|a|^2)$ now denote the set of all such functions g. The upshot is that $\mathcal{R} = O(|a|^2)$. Notice that $O(|a|^2)$ is an ideal in $C^\infty(G)$ and $O(|a|^2) \subset \ker \varepsilon$.

So $\ker \varepsilon / \mathcal{R}$ can be identified with the space of all 1-forms ω at 0. Now the tangent vectors at 0 are dual to the 1-forms at 0 and so can be identified as linear functionals $X : \ker \varepsilon \to \mathbb{R}$ that pass to the quotient $\ker \varepsilon / \mathcal{R}$, that is, those such that $\mathcal{R} \subset \ker X$. This condition is part of Definition (6.34). The other condition, namely, $\mathbb{C}1_\mathcal{A} \subset \ker X$ [equivalently, $X(1) = 0$], is indeed a remnant of the classical condition that X is a derivation as well as being a requirement that the domain of X is $\ker \varepsilon \oplus \mathbb{C}1_\mathcal{A} = \mathcal{A}$. Of course, the motivation given here is in terms of vector spaces over the reals $\mathbb{R}$ as is usual in classical differential geometry, while the theory developed here is over the complex numbers $\mathbb{C}$.

While we do not need to identify $\mathcal{R} = O(|a|^2)$ further to motivate the definition of a quantum LIVF, it is enlightening to do so since this ideal can be described neatly in terms of the Hopf algebra structure. First, we note that $(\ker \varepsilon)^2 \subset O(|a|^2)$ since $f, g \in \ker \varepsilon \subset O(|a|)$ implies that

$$fg \in O(|a|)O(|a|) \subset O(|a|^2).$$

In order to show the opposite inclusion $O(|a|^2) \subset (\ker \varepsilon)^2$, we invoke the following lemma from elementary analysis.

Lemma 6.2. *Let $f : B \to \mathbb{C}$ be of class C^∞, where $B \subset \mathbb{R}^n$ is a nonempty open ball centered at 0. Suppose that $f \in O(|a|^2)$. Then for all $x \in B$, we have*

$$f(x) = \sum_{j\,k=1}^{n} x_j x_k \int_0^1 dt\,(1-t) \frac{\partial^2 f}{\partial x_j \partial x_k}(tx).$$

Exercise 6.16. *Prove this lemma. (**Hint:** Use the fact that $f \in O(|a|^2)$ implies that $f(0) = 0$ and $(\text{grad } f)(0) = 0$. Then the rest is all elementary calculus.)*
   Next, show that $\psi_{jk} : B \to \mathbb{C}$ defined for all $x \in B$ by

$$\psi_{jk}(x) := \int_0^1 dt(1-t) \frac{\partial^2 f}{\partial x_j \partial x_k}(tx)$$

is a C^∞-function, given that f is C^∞.
   Finally, show that $\psi_{jk}(0) = \frac{1}{2}(\text{Hess } f)_{jk}(0)$, where Hess f is the Hessian matrix of f.

So given these results, we have written any C^∞-function $f \in O(|a|^2)$ in the form $f(x) = \sum_{jk} x_j x_k \psi_{jk}(x)$. Since the coordinate functions x_j and x_k are in $\ker \varepsilon$ and ψ_{jk} is C^∞, it follows that $f \in (\ker \varepsilon)^2$. Putting this all together gives $\mathcal{R} = O(|a|^2) = (\ker \varepsilon)^2$.

Exercise 6.17 (Optional). *In this discussion we have (quite deliberately!) blurred the distinction between functions defined globally on all of G and functions defined only in some neighborhood of $e \in G$, that is, in some open ball in Euclidean space. The reader is invited to verify that $\mathcal{R} = (\ker \varepsilon)^2$ is really okay in general.*

Quantum right invariant vector fields are defined similarly with respect to a right invariant FODC.

Theorem 6.13. *There is a unique bilinear pairing $\mathcal{X} \times \Gamma \to \mathbb{C}$ denoted by $(\cdot, \cdot)$ that satisfies*

$$(X, a\theta) = \varepsilon(a)(X, \theta), \tag{6.37}$$

$$(X, da) = X(a) \tag{6.38}$$

for $X \in \mathcal{X}$, $a \in \mathcal{A}$ and $\theta \in \Gamma$.
   This is a nondegenerate pairing when restricted to $\mathcal{X} \times_{\text{inv}} \Gamma \to \mathbb{C}$. For all $a \in \mathcal{A}$, it also satisfies

$$(X, \pi(a)) = X(a). \tag{6.39}$$

Proof. Suppose that there exists a bilinear pairing satisfying (6.37) and (6.38). Take any $\theta \in \Gamma$ and write $\theta = \sum_j a_j \, db_j \in \Gamma$ in standard form with $a_j, b_j \in \mathcal{A}$. Then for $X \in \mathcal{X}$, we calculate

$$(X, \theta) = \sum_j (X, a_j \, db_j)$$

$$= \sum_j \varepsilon(a_j)(X, db_j)$$

$$= \sum_j \varepsilon(a_j) X(b_j)$$

$$= X\Big(\sum_j \varepsilon(a_j) b_j\Big).$$

This shows that the pairing is uniquely determined by (6.37) and (6.38).

So, since we have no other choice, we define the pairing for $X \in \mathcal{X}$ by

$$(X, \theta) := X\Big(\sum_j \varepsilon(a_j) b_j\Big),$$

where we write $\theta = \sum_j a_j \, db_j \in \Gamma$ in standard form with $a_j, b_j \in \mathcal{A}$. The bilinearity of the pairing is rather clear. To show that the definition does not depend on the particular choice of standard form, we suppose $\sum_j a_j \, db_j = 0$. Then we must show that $X(\sum_j \varepsilon(a_j) \, db_j) = 0$ for any $X \in \mathcal{X}$. Consider

$$0 = \pi_{\text{inv}}\Big(\sum_j a_j \, db_j\Big)$$

$$= \sum_j \varepsilon(a_j) \, \pi_{\text{inv}}(db_j)$$

$$= \sum_j \varepsilon(a_j) \, \pi(b_j)$$

$$= \pi\Big(\sum_j \varepsilon(a_j) \, b_j\Big).$$

But this implies that $\sum_j \varepsilon(a_j) b_j \in \ker \pi = \mathcal{R} \oplus \mathbb{C} 1_{\mathcal{A}} \subset \ker X$. And so we see that $X(\sum_j \varepsilon(a_j) b_j) = 0$, as desired.

We next verify (6.38). Let $\theta = da$. So we can write $\theta = 1 da$ in standard form. Then $(X, da) = X(\varepsilon(1)a) = X(a)$. The verification of (6.37) is also straightforward and will be left to the reader.

Next, we calculate

$$
\begin{aligned}
(X, \pi(a)) &= (X, \kappa(a^{(1)})da^{(2)}) \\
&= X\big(\varepsilon(\kappa(a^{(1)})a^{(2)})\big) \\
&= X\big(\varepsilon(a^{(1)})a^{(2)}\big) \\
&= X(a).
\end{aligned}
$$

To show nondegeneracy, first suppose that $\theta_0 \neq 0$ for some $\theta_0 \in {}_{\mathrm{inv}}\Gamma$. But by basic linear algebra, there exists a linear map $X_0 : \mathcal{A} \to \mathbb{C}$ such that $\ker X_0 = \mathcal{R} \oplus \mathbb{C}1_{\mathcal{A}}$. So $X_0 \in \mathcal{X}$. But $\theta_0 = \pi(a_0)$ for some $a_0 \in \mathcal{A}$ with $a_0 \notin \ker \pi = \mathcal{R} \oplus \mathbb{C}1_{\mathcal{A}}$. Therefore, $(X_0, \theta_0) = (X_0, \pi(a_0)) = X_0(a_0) \neq 0$.

Next, suppose that $X \neq 0$; that is, $X(a) \neq 0$ for some $a \in \mathcal{A}$. Then we have that $(X, \pi(a)) = X(a) \neq 0$, where $\pi(a) \in {}_{\mathrm{inv}}\Gamma$. And so the degeneracy of the pairing is established. ∎

By the nondegeneracy of this pairing, we have the first equality in the following while the second comes from the isomorphism ${}_{\mathrm{inv}}\Gamma \cong \ker \varepsilon / \mathcal{R}$:

$$
\dim_{\mathbb{C}} \mathcal{X} = \dim_{\mathbb{C}}({}_{\mathrm{inv}}\Gamma) = \dim_{\mathbb{C}} \left(\frac{\ker \varepsilon}{\mathcal{R}} \right).
$$

We say that this cardinal number is the *dimension* of the FODC Γ. Be warned that this is *not* the same as the dimension of Γ as a complex vector space! Moreover, $\dim_{\mathbb{C}}({}_{\mathrm{inv}}\Gamma)$ in general depends on the choice of the FODC Γ and not just on $\mathcal{A}$. We knew this implicitly already because of the isomorphism ${}_{\mathrm{inv}}\Gamma \cong \ker \varepsilon / \mathcal{R}$.

We now pick any basis $\{X_i \mid i \in I\}$ of the complex vector space $\mathcal{X}$, and let $\{\omega_i \mid i \in I\}$ be the dual basis of ${}_{\mathrm{inv}}\Gamma$ with respect to the pairing. This means that $(X_i, \omega_j) = \delta_{ij}$, the Kronecker delta. By Theorem 5.1, there is a family of functionals $\{f_{ij} : \mathcal{A} \to \mathbb{C} \mid i, j \in I\}$ doubly indexed by the same set I and with various properties as indicated there. We will use all of this notation in the following. We also recall the definition of the convolution of a linear functional $g : \mathcal{A} \to \mathbb{C}$, and an element $a \in \mathcal{A}$ is defined to be

$$
g * a := (id \otimes g)\phi(a) = a^{(1)} \otimes g(a^{(2)}) \in \mathcal{A} \otimes \mathbb{C} \cong \mathcal{A}.
$$

We then have $\varepsilon(g * a) = g(a)$ [see (B.16)].

Proposition 6.10. *Let (Γ, d) be a left invariant FODC over $\mathcal{A}$. Then we have for $a, b \in \mathcal{A}$ that*

$$
da = \sum_{i \in I}(X_i * a)\omega_i, \tag{6.40}
$$

$$
X_i(ab) = \varepsilon(a)X_i(b) + \sum_{j \in I} X_j(a) f_{ji}(b) \quad \text{for all } i \in I. \tag{6.41}
$$

Remark. Since $f_{ji} = \delta_{ji}\varepsilon$ in the classical case, we can view (6.41) as a deformation of Leibniz's rule for the quantum LIVF X_i in the chosen basis. And since any nonzero quantum LIVF X can be realized as an element in a basis of $\mathcal{X}$ (but in general not in a unique way), we can view any quantum LIVF as a deformation of a classical derivation of a function at the origin. However, even though (6.41) holds for each element X_i of a basis of $\mathcal{X}$, it does not extend nicely to a formula for arbitrary $X = \sum_i \lambda_i X_i \in \mathcal{X}$ because of the second term.

Proof. Let $a \in \mathcal{A}$. Then by Eq. (6.39), for each $i \in I$ we have

$$\langle X_i, \pi(a) \rangle = X_i(a).$$

Since $\{X_i \mid i \in I\}$ is a basis whose dual basis is $\{\omega_i \mid i \in I\}$, this means that $\pi(a) = \sum_i X_i(a)\omega_i$. Therefore, by part 6 of Proposition 6.4, we obtain

$$da = a^{(1)}\pi(a^{(2)}) = \sum_i a^{(1)} X_i(a^{(2)})\,\omega_i = \sum_i (X_i * a)\omega_i.$$

And this proves (6.40). To prove (6.41), we use (6.40) three times and the identity (5.5) for the f_{ij}s in the following, where $a, b \in \mathcal{A}$:

$$\sum_{i \in I} (X_i * ab)\omega_i = d(ab)$$

$$= (da)b + a(db)$$

$$= \sum_{j \in I}(X_j * a)\omega_j\, b + a\sum_{i \in I}(X_i * b)\omega_i$$

$$= \sum_{i,j \in I}(X_j * a)(f_{ji} * b)\,\omega_i + \sum_{i \in I} a(X_i * b)\omega_i.$$

Since $\{\omega_i \mid i \in I\}$ is a basis, we can equate the coefficients of each ω_i on each side of the last equation to get for each $i \in I$ and all $a, b \in \mathcal{A}$ that

$$X_i * ab = \sum_{j \in I}(X_j * a)(f_{ji} * b) + a(X_i * b). \tag{6.42}$$

Evaluating ε on both sides of this and using (B.16), we get

$$X_i(ab) = \sum_{j \in I} X_j(a) f_{ji}(b) + \varepsilon(a) X_i(b), \tag{6.43}$$

which proves (6.41). ∎

We record an identity proved in the course of the previous proof.

Corollary 6.5. *For all $a \in \mathcal{A}$, we have $\pi(a) = \sum_i X_i(a)\omega_i$.*

We can let a quantum LIVF X act on the algebra $\mathcal{A}$ of "quantum functions" by $a \mapsto X * a \in \mathcal{A}$. In this way we can think of a quantum LIVF X as a quantum generalization of a classical LIVF. Moreover, the formula (6.43) allows us to interpret X_i as a deformed derivation acting on $\mathcal{A}$.

Now we consider the Hopf algebra $\mathcal{A}^o$ introduced in Appendix B, where the definition for the Hopf algebra $\mathcal{A}^o$ and other details are given.

We assume from now on that $\mathcal{X}$ is a finite-dimensional vector space. Then each quantum LIVF $X \in \mathcal{X}$ satisfies $X \in \mathcal{A}^o$. Similarly, the functionals f_{ij} also satisfy $f_{ij} \in \mathcal{A}^o$. Moreover, for all $a, b \in \mathcal{A}$, we have by (6.43) that

$$(\Delta^o X_i)(a \otimes b) = X_i(ab)$$

$$= \sum_{j \in I} X_j(a) f_{ji}(b) + \varepsilon(a) X_i(b)$$

$$= \sum_{j \in I} (X_j \otimes f_{ji})(a \otimes b) + (\varepsilon \otimes X_i)(a \otimes b),$$

which implies that

$$\Delta^o X_i = \sum_{j \in I} (X_j \otimes f_{ji}) + \varepsilon \otimes X_i.$$

Theorem 6.14. *Let $\mathcal{X}$ be a finite-dimensional subspace of $\mathcal{A}'$, the dual space of a given Hopf algebra $\mathcal{A}$. Then $\mathcal{X}$ is the space of quantum LIVFs of a left covariant FODC Γ over $\mathcal{A}$ if and only if every $X \in \mathcal{X}$ satisfies the conditions*

$$(a)\ \mathbb{C}1_{\mathcal{A}} \subset \ker X \quad (\Longleftrightarrow\ X(1_{\mathcal{A}}) = 0),$$

$$(b)\ \Delta^o X - \varepsilon \otimes X \in \mathcal{X} \otimes \mathcal{A}^o.$$

The left covariant FODC over $\mathcal{A}$ that satisfies these two conditions is unique up to isomorphism of FODCs over $\mathcal{A}$.

Remark. Conditions (a) and (b) are linear in X, and so it suffices to verify them on a basis of $\mathcal{X}$.

Proof. By the above, we already know that the space $\mathcal{X}$ of quantum LIVFs satisfies conditions (a) and (b). So suppose now that (a) and (b) hold. Define

$$\mathcal{R} := \{a \in \ker \varepsilon \mid X(a) = 0 \text{ for all } X \in \mathcal{X}\} = \{a \in \ker \varepsilon \mid \mathcal{X}(a) = 0\}.$$

Suppose that $a \in \mathcal{R}$ and $b \in \mathcal{A}$. Then for every $X \in \mathcal{X}$, we have

$$X(ab) = X(ab) - \varepsilon(a)X(b) \in \mathcal{X}(a) \otimes \mathcal{A}^o(b) = 0$$

by condition (b) and the definition of $\mathcal{R}$. This implies that $ab \in \mathcal{R}$, and so $\mathcal{R}$ is a right ideal contained in $\ker \varepsilon$. By the structure theorem, there exists a left covariant FODC Γ over $\mathcal{A}$ corresponding to $\mathcal{R}$. We denote the space of quantum LIVFs associated to Γ by $\mathcal{X}_\Gamma$. Thus,

$$\mathcal{X}_\Gamma := \{ X \in \mathcal{A}' \mid \mathcal{R} \oplus \mathbb{C}1_\mathcal{A} \subset \ker X \}.$$

To finish this proof, it suffices to show that $\mathcal{X}_\Gamma = \mathcal{X}$. First, we suppose that $X \in \mathcal{X}$. By condition (a), we have $\mathbb{C}1_\mathcal{A} \subset \ker X$. And by the definition of $\mathcal{R}$, we have that $\mathcal{R} \subset \ker X$. It follows that $X \in \mathcal{X}_\Gamma$, which then itself implies that $\mathcal{X} \subset \mathcal{X}_\Gamma$.

To show the equality of these vector spaces, let $k = \dim_\mathbb{C} \mathcal{X} < \infty$ by hypothesis. Choose a basis $X_1, \ldots X_k$ of $\mathcal{X}$. Then

$$\mathcal{R} = \cap_{X \in \mathcal{X}} (\ker \varepsilon \cap \ker X) = \cap_{j=1}^{k} (\ker \varepsilon \cap \ker X_j).$$

Now the codimension of $\ker \varepsilon \cap \ker X_j$ in $\ker \varepsilon$ is either 0 or 1 since X_j is a functional. [Actually, it cannot be 0 since by condition (a) we would then have $X_j \equiv 0$, contradicting the hypothesis that it is a basis element. Hence, the codimension of $\ker \varepsilon \cap \ker X_j$ in $\ker \varepsilon$ is 1.] In any event, the codimension of $\mathcal{R}$ in $\ker \varepsilon$ is then $\leq k$. But this codimension is also the dimension of

$$\frac{\ker \varepsilon}{\mathcal{R}} \cong {}_{\text{inv}}\Gamma \cong \mathcal{X}_\Gamma. \tag{6.44}$$

Therefore, $\dim_\mathbb{C} \mathcal{X}_\Gamma \leq k = \dim_\mathbb{C} \mathcal{X} < \infty$. But we have already shown that $\mathcal{X} \subset \mathcal{X}_\Gamma$. Since these are finite-dimensional spaces, this forces $\mathcal{X} = \mathcal{X}_\Gamma$.

The uniqueness assertion is a consequence of $\mathcal{X} = \mathcal{X}_\Gamma$ and (6.44). ∎

6.6 Examples

Recall from Section 3.5 the Hopf algebras $\tilde{U}_q(sl_2)$ and $SL_q(2)$. Also recall that there we defined a duality pairing of Hopf algebras, denoted by $\langle \cdot, \cdot \rangle$. This allows us to consider any element $X \in \tilde{U}_q(sl_2)$ as a linear functional $SL_q(2) \to \mathbb{C}$, namely, $\alpha \mapsto \langle X, \alpha \rangle$ for all $\alpha \in SL_q(2)$. We slightly abuse notation by writing this functional as

$$X(\alpha) := \langle X, \alpha \rangle.$$

A curious example is obtained by taking $X = 1 \in \tilde{U}_q(sl_2)$. Then we have

$$1(\alpha) = \langle 1, \alpha \rangle = \varepsilon(\alpha)$$

for all $\alpha \in SL_q(2)$, where $\varepsilon : SL_q(2) \to \mathbb{C}$ is the co-unit. So we conclude that $1 = \varepsilon$ as functionals on $SL_q(2)$.

Recall that $\tilde{U}_q(sl_2)$ is generated as an algebra by the elements E, F, K, and K^{-1}. We use these together with ε to define the following linear functionals on $SL_q(2)$:

$$X_0 := q^{-1/2} F K,$$

$$X_1 := (1 - q^{-2})^{-1}(\varepsilon - K^4),$$

$$X_2 := q^{1/2} E K.$$

We then can compute

$$\tilde{\phi}(X_0) = q^{-1/2}\tilde{\phi}(F)\,\tilde{\phi}(K)$$

$$= q^{-1/2}(F \otimes K + K^{-1} \otimes F)(K \otimes K)$$

$$= q^{-1/2}(FK \otimes K^2 + 1 \otimes FK)$$

$$= (q^{-1/2}FK) \otimes K^2 + 1 \otimes q^{-1/2}FK$$

$$= X_0 \otimes K^2 + 1 \otimes X_0$$

$$= X_0 \otimes K^2 + \varepsilon \otimes X_0.$$

We used $\varepsilon = 1$ as functionals in the last equality. Similarly, we have

$$\tilde{\phi}(X_2) = q^{1/2}\tilde{\phi}(E)\,\tilde{\phi}(K)$$

$$= q^{1/2}(E \otimes K + K^{-1} \otimes E)(K \otimes K)$$

$$= q^{1/2}(EK \otimes K^2 + 1 \otimes EK)$$

$$= X_2 \otimes K^2 + 1 \otimes X_2$$

$$= X_2 \otimes K^2 + \varepsilon \otimes X_2.$$

Finally, we compute the co-multiplication acting on X_1:

$$\tilde{\phi}(X_1) = (1 - q^{-2})^{-1}\tilde{\phi}(\varepsilon - K^4)$$

$$= (1 - q^{-2})^{-1}(\tilde{\phi}(1) - \tilde{\phi}(K)^4)$$

$$= (1 - q^{-2})^{-1}(1 \otimes 1 - (K \otimes K)^4)$$

$$= (1 - q^{-2})^{-1}(1 \otimes 1 - K^4 \otimes K^4)$$

$$= (1 - q^{-2})^{-1}(1 \otimes 1 - 1 \otimes K^4 + 1 \otimes K^4 - K^4 \otimes K^4)$$

$$= (1 - q^{-2})^{-1}(1 \otimes (1 - K^4) + (1 - K^4) \otimes K^4)$$

$$= 1 \otimes X_1 + X_1 \otimes K^4$$

$$= \varepsilon \otimes X_1 + X_1 \otimes K^4.$$

We define $\mathcal{X} := \text{span}_{\mathbb{C}}\{X_0, X_1, X_2\} \subset \tilde{U}_q(sl_2)$. Then the previous three calculations show for $k = 0, 1, 2$ that

$$\tilde{\phi}(X_k) - \varepsilon \otimes X_k \in \mathcal{X} \otimes \mathcal{A}^o,$$

where $\mathcal{A} = SL_q(2)$.

We also note that

$$X_0(1) = \langle q^{-1/2} F K, 1 \rangle$$
$$= \tilde{\varepsilon}(q^{-1/2} F K)$$
$$= q^{-1/2}\tilde{\varepsilon}(F)\tilde{\varepsilon}(K)$$
$$= 0$$

since $\tilde{\varepsilon}(F) = 0$. Here 1 denotes the unit element in $SL_q(2)$.

Exercise 6.18. *Show by similar calculations that* $X_1(1) = X_2(1) = 0$ *as well.*

From all these, we can conclude according to Theorem 6.14 that $\mathcal{X}$ is the space of quantum LIVFs of a left covariant FODC Γ over the Hopf algebra $\mathcal{A} = SL_q(2)$. Moreover, this FODC is unique up to isomorphism. This FODC is called the $3D$ calculus on $SL_q(2)$. While this FODC is left covariant by construction, it is notable that it is *not* bicovariant. The reader might like to figure out why this is so.

The $3D$ calculus was a seminal discovery since it was an early example of a nontrivial FODC over a nontrivial Hopf algebra. The name is due to the fact that the dimension of this FODC is 3. This is so since the dimension of Γ is by definition equal to

$$\dim_{\mathbb{C}\,\text{inv}} \Gamma = \dim_{\mathbb{C}} \mathcal{X} = 3.$$

The first equality holds because of the nondegenerate pairing $(\cdot, \cdot)$ of the two spaces. The highly astute reader may have already noted that we merely have from the definition of $\mathcal{X}$ that $\dim_{\mathbb{C}} \mathcal{X} \leq 3$.

Exercise 6.19. *Prove that* $\{X_0, X_1, X_2\}$ *is a basis of* $\mathcal{X}$ *and consequently that* $\dim_{\mathbb{C}} \mathcal{X} = 3$.

The previous exercise is critical, since now we let $\{\omega_0, \omega_1, \omega_2\}$ be the basis of *three* elements of $_{\text{inv}}\Gamma$ dual to the basis $\{X_0, X_1, X_2\}$ of $\mathcal{X}$ with respect to the pairing $(\cdot, \cdot)$ of Theorem 6.13. So we have $(X_j, \omega_k) = \delta_{jk}$, the Kronecker delta.

Now we can use formula (6.40) to write the differential [now denoted by δ to distinguish it from the element $d \in SL_q(2)$] of the left covariant FODC (Γ, δ). For each $r \in SL_q(2)$, this yields

$$\delta r = \sum_{j=0}^{2}(X_j * r)\omega_j.$$

It suffices to evaluate this formula for the four generators a, b, c, d of $SL_q(2)$. The proofs of the next two propositions are left to the reader. Or see [48].

Proposition 6.11. *We have the following identities:*

$$\delta(a) = a\omega_1 + b\omega_2,$$
$$\delta(b) = a\omega_0 - q^2 b\omega_1,$$
$$\delta(c) = c\omega_1 + d\omega_2,$$
$$\delta(d) = c\omega_0 - q^2 d\omega_1.$$

We also let $\mathcal{R}$ be the right ideal in $SL_q(2)$ generated by these six elements:

$$b^2, \quad c^2, \quad bc, \quad (a-1)b, \quad (a-1)c, \quad a + q^{-2}d + (1 + q^{-2})1.$$

Proposition 6.12. *We have $\mathcal{R} \subset \ker\varepsilon$ and $\mathcal{R}$ is the right ideal associated with the FODC Γ.*

6.7 Postscript

The reader may well be puzzled that both left covariant and right covariant FODCs arise from *right* ideals of $\mathcal{A}$ contained in $\ker\varepsilon$. What about the left ideals of $\mathcal{A}$ contained in $\ker\varepsilon$? Why don't they play a role in this theory? The answer has to do with the definition of s, which comes from [86]. That map is not really dual to r. Another map instead of s is given by

$$s'(a \otimes b) := \phi(a)(1 \otimes b) = a^{(1)} \otimes a^{(2)} b.$$

Here and in the following formulas, $a, b \in \mathcal{A}$. An equivalent expression is $s' = (id \otimes m_\mathcal{A})(\phi \otimes 1)$. Note that the maps s' and r appear in [80] (p. 167), where they are denoted by T_1 and T_2, respectively. But $s'(r^{-1}(1 \otimes a))$ does not equal $\text{ad}(a)$, as one can check by looking, for example, at the case of finite groups. There is a fourth map lurking behind the scene. We denote it by r' to emphasize the symmetries among these four maps. It is defined by $r'(a \otimes b) := \phi(a)(b \otimes 1) = a^{(1)}b \otimes a^{(2)}$.

The pair of maps r' and s', together with all the requisite structures on their co-domain spaces, give two more models for the universal FODC. And in this formalism the structure theorems of both left and right covariant FODCs will turn out to involve *left* ideals of $\mathcal{A}$ in $\ker\varepsilon$. We leave the details to the interested reader. However, here is one interesting part of that theory:

Exercise 6.20 (Optional). *Find and then prove the formula for* ad *in terms of the maps r' and s'.*

6.8 Notes

This chapter brings the study of covariant bimodules to life, as it were, by presenting the case where the bimodule is also an FODC. In some sense, this goes right back to the 19th-century greats, though in the commutative case. So this was quite a natural development in the noncommutative setting. This chapter is based on the work presented in various papers. The first three sections contain material that appears in Woronowicz's paper [86] although I have reformulated some of the results in Section 6.2 in order to clarify that there are three (isomorphic!) universal models naturally available. I also tried to emphasize more clearly what the maps r and s are doing for us. However, all this does really appear already in [86] though that might entail a bit of digging to ferret that out.

Section 6.4 is based mostly on Ðurđevich's paper [22] though some of it already appeared in his earlier paper [21]. However, I have reformulated some of this to fit in with the approach in Section 6.2. I have also changed some notation in the hope again of better clarity though that might depend on the eye of the beholder. The inevitable downside of different notation is that it may impede reading of papers with the older notation. However, this is a common impediment easily overcome with sheer diligence and an ounce or so of intelligence. A philosophic response to ever-changing notation could be to note that the maxim that "differential geometry is the study of properties invariant under change of notation" can now be extended to apply to the noncommutative setting as well.

Most of the material in Section 6.5 appears in Section 5 of [86] though we again warn the reader that the quantum germs map π used there is simply denoted as the composition Pd and is not given a name. Also, the quantum LIVFs are not given a name in [86] nor is the space of them named. The theory is further developed in [86], including a quantum generalization of the Lie bracket for quantum LIVFs. The development of this topic in [86] is used to present the Maurer–Cartan formula in the noncommutative setting of that paper. We will return to this formula in Section 10.3. However, our presentation here is based on [48], where the quantum germs map π is denoted as ω and where Theorem 6.14 can be found.

Section 6.6 is also based on [48]. As usual, I have tried to present the material at a more leisurely pace appropriate for a nonexpert. However, the motivation for the $3D$ calculus seems to be mainly historical; namely, this is an early example of a noncommutative differential calculus over a Hopf algebra that itself arises "naturally" from mathematical physics. How the original researchers actually found it is not clear, at least not to me. Maybe they were playing around looking for something—anything!—that would be a nontrivial example of a developing general abstract theory. References are [81] and [85]. The method of using quantum LIVFs has its charm, no doubt, but the actual choice of the three functionals seems more like the work of a magician than a mathematician. The charm has its limits if one wishes to construct bicovariant FODCs, since one only knows in general that the resultant FODC is left covariant. However, the method does have the power to construct left covariant FODCs that, as in this case, are not bicovariant.

More examples of left covariant FODCs are constructed using quantum LIVFs in [48], including the $4D_\pm$-calculi on $SL_q(2)$ and some nonclassical FODCs on classical groups. The $4D_\pm$-calculi are given in [86] by specifying their corresponding right ideals $\mathcal{R}_\pm$, which are defined explicitly in terms of generators. These are bicovariant FODCs, a property that must be proved after the construction.

Structure theorems not only provide us with a panoramic view of all the isomorphism classes of the objects under study. They also give us a tool for constructing specific, new examples of these objects as well as allow us to recognize known objects as being examples of something more general. Thus, in Section 13.4, we construct certain FODCs for finite groups using the structure theory in this chapter plus that which appears in Section 11.2. So we refer the reader to Section 13.4 for an ample supply of examples of bicovariant FODCs.

Chapter 7
The Braid Groups

7.1 Motivation

The braid group enters into the construction of a higher-order differential calculus (HODC) that is known as the braided exterior calculus. This higher-order differential calculus (which is not unique as an HODC, but is functorial) is, strictly speaking, not necessary for the theory of quantum principal bundles (QPBs) as given in Chapter 12. One can do everything just using the universal differential calculus. However, we wish to present this alternative calculus for at least two reasons. First, it gives a nontrivial and quite practical example of an HODC. Second, it uses the braid group, which has played an important role in the development of noncommutative geometry. Simply put, if you want to play this game, then the braid group is a must to learn!

To construct the braided exterior calculus with its differentiation, one first constructs the braided exterior algebra, which is everything except for the differentiation. To do this, we will follow the classical case (to some extent, but not totally) by first constructing the tensor algebra and then passing to the appropriate quotient. That quotient will be constructed using representations of the braid groups. This will not be a short story. Even so, we will only see the most basic aspects of braid groups.

One major difference from the classical case is that we start with a first-order differential calculus (FODC) (Γ, d) over $\mathcal{A}$. In other words, we already have the space (namely, Γ), which corresponds to the classical 1-forms. As we have emphasized before, the FODC is not uniquely determined by $\mathcal{A}$. However, given an FODC, the tensor algebra and the braided exterior algebra will be functorial constructions of it.

© Springer International Publishing Switzerland 2015   107
S.B. Sontz, *Principal Bundles*, Universitext, DOI 10.1007/978-3-319-15829-7_7

Actually, for the moment we do not need an FODC. We simply start with an $\mathcal{A}$-bimodule Γ and then construct the tensor algebra based on it in the standard way. So we recall the usual notation:

$$\otimes^0 \Gamma = \mathcal{A},$$

$$\otimes^1 \Gamma = \Gamma,$$

$$\otimes^k \Gamma = \Gamma \otimes_{\mathcal{A}} \cdots \otimes_{\mathcal{A}} \Gamma \qquad k \text{ times, } k \geq 2,$$

$$\otimes^* \Gamma = \otimes \Gamma = \oplus_{k=0}^{\infty} \left(\otimes^k \Gamma \right).$$

Sometimes we will use the alternative notations $\Gamma^{\otimes k} = \otimes^k \Gamma$ and $\Gamma^{\otimes} = \otimes \Gamma$. Notice that the tensor products are over $\mathcal{A}$, and so here we are using the hypothesis that Γ is an $\mathcal{A}$-bimodule. The *tensor algebra* of Γ is $\otimes^* \Gamma$. It is indeed a graded algebra generated by the elements in degree 1, $\otimes^1 \Gamma = \Gamma$; this is well known. Moreover, in this context $\otimes^k \Gamma$ is an $\mathcal{A}$-bimodule for *all* $k \geq 0$. The case $k = 0$ is left to the reader to ponder, while for $k \geq 1$, the left action of $\mathcal{A}$ is defined by acting on the first factor of Γ in $\otimes^k \Gamma$ from the left; that is,

$$a \cdot (\omega_1 \otimes_{\mathcal{A}} \omega_2 \otimes_{\mathcal{A}} \cdots \otimes_{\mathcal{A}} \omega_k) := (a\omega_1) \otimes_{\mathcal{A}} \omega_2 \otimes_{\mathcal{A}} \cdots \otimes_{\mathcal{A}} \omega_k,$$

where $a \in \mathcal{A}$ and $\omega_1, \ldots, \omega_k \in \Gamma$. Similarly, the right action of $\mathcal{A}$ is defined by acting on the last factor of Γ in $\otimes^k \Gamma$ from the right. (The formula is left to the reader.) It follows immediately that $\otimes^* \Gamma$ is also an $\mathcal{A}$-bimodule.

In the case of classical differential geometry, Γ is $\Gamma(T^*M)$, the smooth sections of the cotangent bundle; these are also called the 1-forms on M. And $\Gamma(T^*M)$ is an $\mathcal{A}$-bimodule, where $\mathcal{A} = C^{\infty}(M)$, the commutative algebra of C^{∞}-functions $M \to \mathbb{R}$. But $\otimes^k \Gamma(T^*M) \cong \Gamma(\otimes^k T^*M)$ for $k \geq 2$ does *not* give us the k-forms on M. Rather, the space of k-forms is $\wedge^k \Gamma(T^*M) \cong \Gamma(\wedge^k T^*M)$ for all $k \geq 0$. As is usual in these sorts of constructions on vector bundles in classical differential geometry, we do the construction fiberwise; that is, we deal with operations (functors, to be precise) of vector spaces. For the example at hand, the question is how does one go from the vector space $\otimes^k V$ to the vector space $\wedge^k V$ for any (say, finite-dimensional) vector space V over $\mathbb{R}$? And the answer, a bit too long and technical for some tastes, begins by defining a left action of the symmetric group on k "letters," denoted S_k, on $\otimes^k V$. This is defined for $\tau \in S_k$ and $v_1 \otimes \cdots \otimes v_j \otimes \cdots \otimes v_k \in \otimes^k V$ by

$$\tau(v_1 \otimes \cdots \otimes v_j \otimes \cdots \otimes v_k) := v_{\tau(1)} \otimes \cdots \otimes v_{\tau(j)} \otimes \cdots \otimes v_{\tau(k)}.$$

(Here $\otimes$ means $\otimes_{\mathbb{R}}$.) Next, one defines the *antisymmetrization operator* $\alpha_k : \otimes^k V \to \otimes^k V$ by

$$\alpha_k = \frac{1}{k!} \sum_{\tau \in S_k} \text{sign}(\tau) \, \tau, \tag{7.1}$$

where sign(τ), the sign or parity of τ, is either $+1$ or -1. (See any text on elementary group theory for the definition and basic properties of the sign of a permutation.) This same construction works for any field of characteristic zero.

Exercise 7.1. *Prove that α_k is a projection for all $k \geq 0$. Remember that a part of the definition of a projection is that it is linear. (Be very careful with the cases $k = 0$ and $k = 1$. For example, what is the symmetric group on zero letters?)*

If we eliminate the factor $1/k!$ from the definition of α_k, show that the resulting map is linear, but not a projection.

Now we have several equivalent ways to define $\wedge^k V$. One is $\wedge^k V := \operatorname{Ran} \alpha_k$, the range (or image) of α_k. By a standard isomorphism, this is the "same" as (i.e., isomorphic to) $\otimes^k V / \ker \alpha_k$. Finally, since α_k is a projection, the nonzero elements in its range are exactly the eigenvectors associated to the eigenvalue 1 of α_k. So we have $\wedge^k V = \{w \in \otimes^k V \mid \alpha_k(w) = w\} = \ker(id - \alpha_k)$. We would like simply to repeat this procedure in the noncommutative case.

But we run into a serious snag! This will not work out for even the first nontrivial case, namely, $k = 2$. Recall that S_2 has two elements, the identity and the interchange τ of the two letters. Say $\omega_1, \omega_2 \in \Gamma$. Then

$$\omega_1 \otimes_A \omega_2 \in \Gamma \otimes_A \Gamma = \otimes^2 \Gamma,$$

and we would like to define

$$\tau(\omega_1 \otimes_A \omega_2) := \omega_2 \otimes_A \omega_1. \tag{7.2}$$

Since we are dealing with tensor products of bimodules over A, we have the identity

$$(\omega_1 a) \otimes_A \omega_2 = \omega_1 \otimes_A (a\omega_2)$$

for all $a \in A$ and all $\omega_1, \omega_2 \in \Gamma$. We next apply τ as defined in (7.2) to both sides of the previous equation. Now we have from the left side

$$\tau((\omega_1 a) \otimes_A \omega_2) = \omega_2 \otimes_A (\omega_1 a),$$

while from the right side we get

$$\tau(\omega_1 \otimes_A (a\omega_2)) = (a\omega_2) \otimes_A \omega_1.$$

We conclude from these simple considerations that

$$\omega_2 \otimes_A (\omega_1 a) = (a\omega_2) \otimes_A \omega_1 \tag{7.3}$$

for all $a \in \mathcal{A}$ and all $\omega_1, \omega_2 \in \Gamma$. However, notice that this last equation is not an identity in this theory *even if* $\mathcal{A}$ is commutative! (The example Γ_p with $p \neq x$ from Section 2.2 will show this.) This shows that the noncommutative theory brings in new structures and results even for the case when the algebra is commutative. Neat, huh?

So this attempt to define an action of S_2 on $\otimes^2 \Gamma$ using (7.2) leads to a contradiction for many bimodules Γ, though not for all of them. But another way to look at this situation is to ask: For a given bimodule Γ, what are the elements $\omega_1, \omega_2 \in \Gamma$ for which (7.3) does hold? For a trivial example, let $\omega_1 = 0$ and let ω_2 be arbitrary.

Now an idea here is to use the interchange on *some* elements of the form $\omega_1 \otimes \omega_2$ and then try to extend this definition to all of $\otimes^2 \Gamma$. With just our current hypotheses, one could focus on those elements ω_1 satisfying $\omega_1 = \omega_1 a$ for all $a \in \mathcal{A}$ and on those elements ω_2 satisfying $\omega_2 = a\omega_2$ for all $a \in \mathcal{A}$. For such elements ω_1, ω_2, we see that (7.3) does hold, but the vector subspace of $\otimes^2 \Gamma$ spanned by the corresponding elements $\omega_1 \otimes_{\mathcal{A}} \omega_2$ could be very small. Even the subbimodule of $\otimes^2 \Gamma$ spanned by such elements could be different than $\otimes^2 \Gamma$. The moral is that this idea does not work in the present context. However, we are interested in FODCs that have more structure with which to play.

7.2   Definitions

So using the previous section as a bit of motivation, we now return to the particular case when Γ is the $\mathcal{A}$-bimodule of a *bicovariant* FODC (Γ, d). Then Γ has left and right co-actions defined in terms of the co-multiplication ϕ of $\mathcal{A}$ and the differential d of the FODC. Using these co-actions, we recall that the definition of the vector space of left invariant elements in Γ is given by

$$_{\mathrm{inv}}\Gamma = \{\omega \in \Gamma \mid \Phi_\Gamma(\omega) = 1 \otimes \omega\}.$$

Similarly, we recall that the vector space of right invariant elements in Γ is

$$\Gamma_{\mathrm{inv}} = \{\eta \in \Gamma \mid {}_\Gamma\Phi(\eta) = \eta \otimes 1\}.$$

Now, the insight given by Woronowicz (see [86]) is to define a linear map

$$\sigma : \Gamma \otimes_{\mathcal{A}} \Gamma \to \Gamma \otimes_{\mathcal{A}} \Gamma$$

by first defining it as the interchange on elements of the form $\omega \otimes_{\mathcal{A}} \eta$, where $\omega \in {}_{\mathrm{inv}}\Gamma$ and $\eta \in \Gamma_{\mathrm{inv}}$ and then extending to all of $\Gamma \otimes_{\mathcal{A}} \Gamma$ by using $\mathcal{A}$-linearity (with respect to the left action of $\mathcal{A}$ on $\Gamma \otimes_{\mathcal{A}} \Gamma$, say). And an important part of this insight is that we forget all about (7.3).

Recall from Theorem 5.3 that we can find vector space bases $\{\omega_i \mid i \in I\}$ of $_{\text{inv}}\Gamma$ and $\{\eta_i \mid i \in I\}$ of Γ_{inv} with various nice properties. (And, yes, it is the same index set for the two bases! This is one of those nice properties.) The next little step forward is

Lemma 7.1. *Let $\gamma \in \Gamma^{\otimes 2} = \Gamma \otimes_{\mathcal{A}} \Gamma$ be arbitrary. Then there are unique elements $a_{ij}, b_{ij} \in \mathcal{A}$ for $i, j \in I$ such that only finitely many of then are nonzero and*

$$\gamma = \sum_{ij \in I} a_{ij} \omega_i \otimes_{\mathcal{A}} \eta_j, \tag{7.4}$$

$$\gamma = \sum_{ij \in I} b_{ij} \eta_i \otimes_{\mathcal{A}} \omega_j. \tag{7.5}$$

Remark. This result allows us to define σ on *all* of $\Gamma^{\otimes 2}$, as we will do after presenting the proof.

Proof. For the existence of these two expansions, it suffices to consider the case when $\gamma = \alpha \otimes_{\mathcal{A}} \beta$ for $\alpha, \beta \in \Gamma$. But we can write α as a finite linear combination of the ω_i's with coefficients in $\mathcal{A}$ on the left. And we can write β as a finite linear combination of the η_j's with coefficients in $\mathcal{A}$ on the left. The latter coefficients can be passed to the left of the tensor product $\otimes_{\mathcal{A}}$ and so act on the ω_i's from the right. Now these terms of the form $\omega_i a$ can in turn be written as a linear combination of the ω_i's with coefficients in $\mathcal{A}$ on the *left*; there is even an explicit formula for this. This is all by Theorems 5.1 and 5.2. Putting all this together gives the first part of the lemma. We interrupt this proof for a brief aside.

Exercise 7.2. *If this last paragraph is too terse for you to fully grasp, then your mission is to expand it with the requisite formulas.*

Since Γ is a free left $\mathcal{A}$-module with basis $\{\eta_j \mid j \in I\}$, we have a canonical isomorphism

$$\Gamma \cong \oplus_{j \in I} \mathcal{A} \eta_j$$

as left $\mathcal{A}$-modules. Let $c_j : \Gamma \to \mathcal{A}$ be the map that maps $\omega \in \Gamma$ to its coefficient in front of η_j. Clearly, c_j is a left $\mathcal{A}$-module morphism. Also, $c_j(\eta_k) = \delta_{jk}$, the Kronecker delta. Simply put, c_j is nothing but the dual basis to the basis η_j. Then

$$id \otimes_{\mathcal{A}} c_j : \Gamma \otimes_{\mathcal{A}} \Gamma \to \Gamma \otimes_{\mathcal{A}} \mathcal{A} \cong \Gamma.$$

Next, we apply $id \otimes_{\mathcal{A}} c_k$ to (7.4) and get

$$
(id \otimes c_k)\gamma = (id \otimes c_k)(\sum_{ij \in I} a_{ij}\omega_i \otimes_{\mathcal{A}} \eta_j)
$$

$$
= \sum_{ij \in I} a_{ij}\omega_i \otimes_{\mathcal{A}} c_k(\eta_j)
$$

$$
= \sum_{ij \in I} a_{ij}\omega_i \delta_{kj}
$$

$$
= \sum_{i \in I} a_{ik}\omega_i .
$$

Now, for each $k \in I$, this is an element in the left $\mathcal{A}$-module Γ and so has a unique expansion in its basis ω_i. This means that for each k, the coefficients a_{ik} are unique. And this shows that the coefficients in (7.4) are unique.

Our established tradition dictates that the quite similar proof for (7.5) is left to the reader. ∎

We deliver the promised definition.

Definition 7.1. *Let* $\gamma \in \Gamma^{\otimes 2} = \Gamma \otimes_{\mathcal{A}} \Gamma$ *be arbitrary. Let vector space bases* $\{\omega_i \mid i \in I\}$ *of* $_{inv}\Gamma$ *and* $\{\eta_i \mid i \in I\}$ *of* Γ_{inv} *be given. Define*

$$
\sigma(\gamma) := \sum_{ij \in I} a_{ij}\eta_j \otimes_{\mathcal{A}} \omega_i ,
$$

where we have used the previous lemma to write $\gamma = \sum_{ij \in I} a_{ij}\omega_i \otimes_{\mathcal{A}} \eta_j$ *for unique elements* $a_{ij} \in \mathcal{A}$. *So* $\sigma : \Gamma^{\otimes 2} \to \Gamma^{\otimes 2}$ *is a linear map.*

Since we want σ to be a left $\mathcal{A}$-module morphism and since, following Woronowicz's idea, we also want

$$
\sigma(\omega_i \otimes_{\mathcal{A}} \eta_j) = \eta_j \otimes_{\mathcal{A}} \omega_i ,
$$

then this definition of σ is forced on us. But the trick is to see what good it does for us! Well, as it turns out, a lot of good. But the story unfolds slowly. First, we make some (more!) definitions.

Definition 7.2. *Let* (Γ, d) *be a bicovariant FODC over* $\mathcal{A}$ *with left (resp., right) co-action* Φ_Γ *(resp.,* $_\Gamma\Phi$*). For each integer* $n \geq 2$, *we define a linear map* $\Phi_{\Gamma^{\otimes n}} : \Gamma^{\otimes n} \to \mathcal{A} \otimes \Gamma^{\otimes n}$ *by*

$$
\Phi_{\Gamma^{\otimes n}}(\theta_1 \otimes_{\mathcal{A}} \cdots \otimes_{\mathcal{A}} \theta_n) := (a_1^{(0)} \cdots a_n^{(0)}) \otimes (\theta_1^{(1)} \otimes_{\mathcal{A}} \cdots \otimes_{\mathcal{A}} \theta_n^{(1)}), \qquad (7.6)
$$

where $\theta_1, \ldots, \theta_n \in \Gamma$ *and* $\Phi_\Gamma(\theta_j) = a_j^{(0)} \otimes \theta_j^{(1)} \in \mathcal{A} \otimes \Gamma$ *for each* $j = 1, \ldots, n$ *is Sweedler's notation for the left co-action.*

Warning: *It is our convention that Sweedler's notation suppresses the summations of arbitrary tensors as sums of decomposable tensors. Rather, the decomposable tensor that appears in this notation represents the generic term in the suppressed summation. Moreover, we are not using the Einstein summation convention, so that there is no summation over the repeated index j in our notation for $\Phi_\Gamma(\theta_j)$. Also, recall that $\otimes$ in (7.6) means $\otimes_{\mathbb{C}}$.*

Similarly, for each integer $n \geq 2$, we define $_{\Gamma^{\otimes n}}\Phi : \Gamma^{\otimes n} \to \Gamma^{\otimes n} \otimes \mathcal{A}$ by

$$_{\Gamma^{\otimes n}}\Phi(\theta_1 \otimes_{\mathcal{A}} \cdots \otimes_{\mathcal{A}} \theta_n) := (\theta_1^{(0)} \otimes_{\mathcal{A}} \cdots \otimes_{\mathcal{A}} \theta_n^{(0)}) \otimes (b_1^{(1)} \cdots b_n^{(1)}), \qquad (7.7)$$

where $\theta_1, \ldots, \theta_n \in \Gamma$ and $_\Gamma\Phi(\theta_j) = \theta_j^{(0)} \otimes b_j^{(1)} \in \Gamma \otimes \mathcal{A}$ for each $j = 1, \ldots, n$ is Sweedler's notation for the right co-action, which has nothing to do with Sweedler's notation as used in the first part in Eq. (7.6).

For $n = 0$, we define

$$\Phi_{\Gamma^{\otimes 0}} := \phi \quad and \quad _{\Gamma^{\otimes 0}}\Phi := \phi.$$

For $n = 1$, we define

$$\Phi_{\Gamma^{\otimes 1}} := \Phi_\Gamma \quad and \quad _{\Gamma^{\otimes 1}}\Phi := _\Gamma\Phi.$$

The following result justifies using the notation of co-actions for these maps.

Proposition 7.1. *The linear map $\Phi_{\Gamma^{\otimes n}} : \Gamma^{\otimes n} \to \mathcal{A} \otimes \Gamma^{\otimes n}$ is a left co-action of $\mathcal{A}$ on $\Gamma^{\otimes n}$ for all $n \geq 0$. Similarly, $_{\Gamma^{\otimes n}}\Phi : \Gamma^{\otimes n} \to \Gamma^{\otimes n} \otimes \mathcal{A}$ is a right co-action of $\mathcal{A}$ on $\Gamma^{\otimes n}$ for all $n \geq 0$.*

Proof. We prove the first statement and leave the similar proof of the second statement to the reader (see Exercise 7.3). First, we note that $\Phi_{\Gamma^{\otimes 0}} = \phi$ and $\Phi_{\Gamma^{\otimes 1}} = \Phi_\Gamma$ are left co-actions on degree-0 and degree-1 elements, respectively. So we now consider the remaining case when $n \geq 2$. We have to check that $\Phi_{\Gamma^{\otimes n}}$ satisfies the two defining properties of a left co-action. One of these properties is $(\varepsilon \otimes id)\, \Phi_{\Gamma^{\otimes n}} = id$. To see this, we use the notation from (7.6) to get

$$(\varepsilon \otimes id)\, \Phi_{\Gamma^{\otimes n}}(\theta_1 \otimes_{\mathcal{A}} \cdots \otimes_{\mathcal{A}} \theta_n)$$

$$= (\varepsilon \otimes id)\left((a_1^{(0)} \cdots a_n^{(0)}) \otimes (\theta_1^{(1)} \otimes_{\mathcal{A}} \cdots \otimes_{\mathcal{A}} \theta_n^{(1)})\right)$$

$$= \varepsilon\left((a_1^{(0)} \cdots a_n^{(0)})\right) \otimes (\theta_1^{(1)} \otimes_{\mathcal{A}} \cdots \otimes_{\mathcal{A}} \theta_n^{(1)})$$

$$= \varepsilon(a_1^{(0)}) \cdots \varepsilon(a_n^{(0)}) \otimes (\theta_1^{(1)} \otimes_{\mathcal{A}} \cdots \otimes_{\mathcal{A}} \theta_n^{(1)})$$

$$= 1 \otimes \left(\varepsilon(a_1^{(0)})\theta_1^{(1)} \otimes_{\mathcal{A}} \cdots \otimes_{\mathcal{A}} \varepsilon(a_n^{(0)})\theta_n^{(1)}\right)$$

$$= 1 \otimes (\theta_1 \otimes_{\mathcal{A}} \cdots \otimes_{\mathcal{A}} \theta_n)$$

$$\cong \theta_1 \otimes_{\mathcal{A}} \cdots \otimes_{\mathcal{A}} \theta_n.$$

The other property of a left co-action is left to the reader to prove. See Exercise 7.3. ∎

Exercise 7.3. *Prove that* $(\Gamma^{\otimes n}, \Phi_{\Gamma^{\otimes n}}, {}_{\Gamma^{\otimes n}}\Phi)$ *is a bicovariant bimodule for all integers* $n \geq 0$. *The things still to be verified are that* $\Phi_{\Gamma^{\otimes n}}$ *satisfies the second property of a left co-action,* ${}_{\Gamma^{\otimes n}}\Phi$ *is a right co-action (though this should be obvious at this stage), and these two co-actions are compatible.*

Exercise 7.4. *The notation* ${}_{inv}\Gamma^{\otimes n}$ *means what it has to mean, namely, the space of all left invariant elements of* $\Gamma^{\otimes n}$ *with respect to its left co-action* $\Phi_{\Gamma^{\otimes n}}$. *Prove* ${}_{inv}\Gamma^{\otimes n} = {}_{inv}\Gamma \otimes \cdots \otimes {}_{inv}\Gamma$. *Translate this result into a statement about a relation between the bases of* ${}_{inv}\Gamma$ *and* ${}_{inv}\Gamma^{\otimes n}$.

Make yourself aware of the corresponding statements for right invariant elements. Optionally (if you think this will help you), write everything out in detail and provide complete proofs although it is always best to find your own personal intermediate level of required detail and do that.

We will resist the temptation to form the direct sum over n of these bicovariant bimodules for the time being. We will do this later, of course. Instead, we get down to the business of establishing the properties of σ. The result of Exercise 7.3 for $n = 2$ will be used in proving part 3 of Theorem 7.1.

7.3   Basic Theorem

The braid group begins to come into evidence in the next, basic theorem.

Theorem 7.1. *Consider the map* $\sigma : \Gamma^{\otimes 2} \to \Gamma^{\otimes 2}$ *from Definition 7.1. Then* σ *has the following properties:*

1. σ *is an* $\mathcal{A}$-*bimodule isomorphism.*
2. $\sigma(\omega \otimes_A \eta) = \eta \otimes_A \omega$ *for all* $\omega \in {}_{inv}\Gamma$ *and all* $\eta \in \Gamma_{inv}$.
3. σ *is a left and a right* $\mathcal{A}$-*co-module morphism, which means that these diagrams commute:*

$$
\begin{array}{ccc}
\Gamma^{\otimes 2} & \xrightarrow{\sigma} & \Gamma^{\otimes 2} \\
\Phi_{\Gamma^{\otimes 2}} \downarrow & & \downarrow \Phi_{\Gamma^{\otimes 2}} \\
\mathcal{A} \otimes \Gamma^{\otimes 2} & \xrightarrow{id \otimes \sigma} & \mathcal{A} \otimes \Gamma^{\otimes 2},
\end{array}
\tag{7.8}
$$

$$
\begin{array}{ccc}
\Gamma^{\otimes 2} & \xrightarrow{\sigma} & \Gamma^{\otimes 2} \\
{}_{\Gamma^{\otimes 2}}\Phi \downarrow & & \downarrow {}_{\Gamma^{\otimes 2}}\Phi \\
\Gamma^{\otimes 2} \otimes \mathcal{A} & \xrightarrow{\sigma \otimes id} & \Gamma^{\otimes 2} \otimes \mathcal{A}.
\end{array}
\tag{7.9}
$$

4. *Define*

$$\sigma_{12} := \sigma \otimes_A id_\Gamma : \Gamma^{\otimes 3} \to \Gamma^{\otimes 3}$$

$$\text{and} \quad \sigma_{23} := id_\Gamma \otimes_A \sigma : \Gamma^{\otimes 3} \to \Gamma^{\otimes 3}.$$

Then these maps satisfy the celebrated braid equation (also known as the Yang–Baxter equation), which says

$$\sigma_{23}\sigma_{12}\sigma_{23} = \sigma_{12}\sigma_{23}\sigma_{12}. \tag{7.10}$$

Remark. In the symmetric groups S_k for $k \geq 3$, one has the identity

$$(23)(12)(23) = (12)(23)(12),$$

where (ab) denotes the 2-cycle that interchanges a and b for $a, b \in \{1, \ldots, k\}$ with $a \neq b$. And this is, of course, the same braid equation (7.10). However, $(ab)^2 = e$ for every 2-cycle (ab), where $e \in S_k$ is the identity element. But $\sigma \in \text{End}(\Gamma^{\otimes 2})$ does not necessarily satisfy $\sigma^2 = id_{\Gamma^{\otimes 2}}$.

Proof. Lemma 7.1 shows that σ is injective and surjective. Also, it is immediate that σ is linear. To complete the proof of part 1, we must show that it is an A-bimodule morphism. Let $\alpha \in \Gamma^{\otimes 2}$ and write $\alpha = \sum_{ij \in I} a_{ij}\omega_i \otimes_A \eta_j$ as in Definition 7.1. Then for any $c \in A$, we have

$$c\alpha = \sum_{ij \in I} ca_{ij}\omega_i \otimes_A \eta_j,$$

which implies that

$$\sigma(c\alpha) = \sum_{ij \in I} ca_{ij}\eta_j \otimes_A \omega_i = c \sum_{ij \in I} a_{ij}\eta_j \otimes_A \omega_i = c\sigma(\alpha).$$

This proves that σ is a left A-module morphism.

Before proving that σ also is a right A-module morphism, we recall these identities for bicovariant bimodules from Theorems 5.1, 5.2, and 5.3:

$$\eta_j c = \sum_{k \in I} (c * f_{jk})\eta_k,$$

$$\omega_i c = \sum_{l \in I} (f_{il} * c)\omega_l.$$

Here $c \in \mathcal{A}$, $f_{jk} \in \mathcal{A}'$ and the convolutions products (which give elements in $\mathcal{A}$) have been defined in Theorems 5.1 and 5.2. We will also be using the associativity identity

$$(f * c) * g = f * (c * g)$$

for all $f, g \in \mathcal{A}'$ and $c \in \mathcal{A}$, which is the identity (B.18).

Now we start the proof that σ also is a right $\mathcal{A}$-module morphism. First, we note that

$$\sigma(\alpha c) = \sigma\Big(\sum_{ij \in I} a_{ij}\omega_i \otimes_{\mathcal{A}} \eta_j c\Big)$$

$$= \sum_{ij \in I} \sigma\big(a_{ij}\omega_i \otimes_{\mathcal{A}} \eta_j c\big)$$

$$= \sum_{ij \in I} a_{ij}\sigma\big(\omega_i \otimes_{\mathcal{A}} \eta_j c\big). \qquad (7.11)$$

So we continue with part of the expression on the right side, getting

$$\sigma\big(\omega_i \otimes_{\mathcal{A}} \eta_j c\big) = \sum_{k \in I} \sigma\big(\omega_i \otimes_{\mathcal{A}} (c * f_{jk})\eta_k\big)$$

$$= \sum_{k \in I} \sigma\big(\omega_i (c * f_{jk}) \otimes_{\mathcal{A}} \eta_k\big)$$

$$= \sum_{k \in I} \sum_{l \in I} \sigma\big((f_{il} * (c * f_{jk}))\omega_l \otimes_{\mathcal{A}} \eta_k\big)$$

$$= \sum_{l \in I} \sum_{k \in I} \sigma\big(((f_{il} * c) * f_{jk})\omega_l \otimes_{\mathcal{A}} \eta_k\big)$$

$$= \sum_{l \in I} \sum_{k \in I} ((f_{il} * c) * f_{jk})\eta_k \otimes_{\mathcal{A}} \omega_l$$

$$= \sum_{l \in I} \Big(\sum_{k \in I} ((f_{il} * c) * f_{jk})\eta_k\Big) \otimes_{\mathcal{A}} \omega_l$$

$$= \sum_{l \in I} \eta_j (f_{il} * c) \otimes_{\mathcal{A}} \omega_l$$

$$= \sum_{l \in I} \eta_j \otimes_{\mathcal{A}} (f_{il} * c)\omega_l$$

$$= \eta_j \otimes_{\mathcal{A}} \sum_{l \in I} (f_{il} * c)\omega_l$$

$$= \eta_j \otimes_A (\omega_i c)$$

$$= (\eta_j \otimes_A \omega_i)c.$$

Putting this back into (7.11), we find that

$$\sigma(\alpha c) = \sum_{ij \in I} a_{ij} \sigma(\omega_i \otimes_A \eta_j c)$$

$$= \sum_{ij \in I} a_{ij}((\eta_j \otimes_A \omega_i)c)$$

$$= \sum_{ij \in I} (a_{ij}(\eta_j \otimes_A \omega_i))c$$

$$= \left(\sum_{ij \in I} a_{ij} \eta_j \otimes_A \omega_i \right)c$$

$$= \sigma(\alpha)c.$$

And this ends the proof that σ is a right A-module morphism.

For part 2, we write $\omega = \sum_i \lambda_i \omega_i$ and $\eta = \sum_j \mu_j \eta_j$, where $\lambda_i, \mu_j \in \mathbb{C}$. Then we simply calculate

$$\sigma(\omega \otimes_A \eta) = \sigma\left(\sum_i \lambda_i \omega_i \otimes_A \sum_j \mu_j \eta_j \right)$$

$$= \sum_{ij} \lambda_i \mu_j \sigma(\omega_i \otimes_A \eta_j)$$

$$= \sum_{ij} \lambda_i \mu_j (\eta_j \otimes_A \omega_i)$$

$$= \sum_j \mu_j \eta_j \otimes_A \sum_i \lambda_i \omega_i$$

$$= \eta \otimes_A \omega.$$

Notice that elements in $\mathbb{C}$ float back and forth over the elements in Γ as well as over the tensor product $\otimes_A$. Expanding ω and η as above, except with coefficients in A, we will not get a similar result.

To show that the diagram (7.8) commutes, we take an element $\alpha \in \Gamma^{\otimes 2}$ and chase it through the diagram. So we first chase $\alpha = \sum a_{ij} \omega_i \otimes_A \eta_j$ horizontally with σ to get

$$\sigma\left(\sum a_{ij} \omega_i \otimes_A \eta_j \right) = \sum a_{ij} \eta_j \otimes_A \omega_i.$$

Then we chase this element down to

$$(\Phi_{\Gamma\otimes2} \circ \sigma)(\alpha) = \Phi_{\Gamma\otimes2}\Big(\sum_{ij} a_{ij}\eta_j \otimes_{\mathcal{A}} \omega_i\Big)$$

$$= \sum_{ij} \Phi_{\Gamma\otimes2}(a_{ij}\eta_j \otimes_{\mathcal{A}} \omega_i)$$

$$= \sum_{ij} \phi(a_{ij})\, \Phi_{\Gamma\otimes2}(\eta_j \otimes_{\mathcal{A}} \omega_i).$$

To use the definition of $\Phi_{\Gamma\otimes2}$, we have to calculate $\Phi_\Gamma(\eta_j)$ and $\Phi_\Gamma(\omega_i)$. First, since ω_i is left invariant, we have $\Phi_\Gamma(\omega_i) = 1\otimes\omega_i$ for all $i \in I$. Second, for the co-action of η_j, we will use the identity

$$\Phi_\Gamma(\eta_j) = \sum_{k\in I} \kappa(R_{kj}) \otimes \eta_k \in \mathcal{A} \otimes \Gamma$$

for all $j \in I$, which was established in Theorem 5.3. Here $R_{kj} \in \mathcal{A}$ for all $k, j \in I$ are the elements introduced in Theorem 5.3. Using these calculations and the definition of $\Phi_{\Gamma\otimes2}$, we obtain

$$\Phi_{\Gamma\otimes2}(\eta_j \otimes_{\mathcal{A}} \omega_i) = \sum_{k} \kappa(R_{kj}) \otimes (\eta_k \otimes_{\mathcal{A}} \omega_i).$$

Putting this all together, we see that the result of chasing across and then down is

$$(\Phi_{\Gamma\otimes2} \circ \sigma)(\alpha) = \sum_{ijk} \phi(a_{ij})\big(\kappa(R_{kj}) \otimes (\eta_k \otimes_{\mathcal{A}} \omega_i)\big).$$

Now we have the necessary information to chase α down the left vertical arrow. This is

$$\Phi_{\Gamma\otimes2}(\alpha) = \Phi_{\Gamma\otimes2}\Big(\sum_{ij} a_{ij}\omega_i \otimes_{\mathcal{A}} \eta_j\Big)$$

$$= \sum_{ij} \phi(a_{ij})\Phi_\Gamma^{\otimes2}(\omega_i \otimes_{\mathcal{A}} \eta_j)$$

$$= \sum_{ijk} \phi(a_{ij})\big(\kappa(R_{kj}) \otimes (\omega_i \otimes_{\mathcal{A}} \eta_k)\big).$$

Finally, chasing this across, we get

$$(id \otimes \sigma)\Phi_{\Gamma\otimes2}(\alpha) = \sum_{ijk} \phi(a_{ij})\big(\kappa(R_{kj}) \otimes (\eta_k \otimes_{\mathcal{A}} \omega_i)\big),$$

which is equal to the across–down chase. So the diagram (7.8) commutes.

It will come as no surprise that the commutativity of the diagram (7.9) is left to the reader. So we now move on to the proof of part 4.

We first remark that we have a result whose statement and proof follow the pattern set in Lemma 7.1. The statement is that every element in $\Gamma^{\otimes 3}$ can be written as a linear combination with coefficients in $\mathcal{A}$ acting from the left of elements of the form $\omega_1 \otimes \omega_2 \otimes \eta$, where $\omega_1, \omega_2 \in \ _{\text{inv}}\Gamma$, that is, they are left invariant, and $\eta \in \Gamma_{\text{inv}}$, that is, it is right invariant.

Exercise 7.5. *Prove this statement in the manner indicated.*

Suppose that $\omega_1, \omega_2 \in \ _{\text{inv}}\Gamma$. This simply means that $\Phi_\Gamma(\omega_1) = 1 \otimes \omega_1$ and $\Phi_\Gamma(\omega_2) = 1 \otimes \omega_2$. Then by the definition of $\Phi_{\Gamma^{\otimes 2}}$, we immediately see that

$$\Phi_{\Gamma^{\otimes 2}}(\omega_1 \otimes_{\mathcal{A}} \omega_2) = 1 \otimes (\omega_1 \otimes_{\mathcal{A}} \omega_2).$$

So $\omega_1 \otimes_{\mathcal{A}} \omega_2 \in \Gamma^{\otimes 2}$ is left invariant with respect to the co-action $\Phi_{\Gamma^{\otimes 2}}$. Since σ is a left $\mathcal{A}$-co-module morphism, it follows that $\sigma(\omega_1 \otimes_{\mathcal{A}} \omega_2)$ in $\Gamma^{\otimes 2}$ is also left invariant with respect to the co-action $\Phi_{\Gamma^{\otimes 2}}$. Let $\{\omega_i \mid i \in I\}$ be a vector space basis of $_{\text{inv}}\Gamma$. Then the set $\{\omega_j \otimes_{\mathcal{A}} \omega_k \mid j, k \in I\}$ is a vector space basis of $_{\text{inv}}(\Gamma^{\otimes 2})$. (See the next exercise.) This implies that we have an expansion with $c_{jk} \in \mathbb{C}$ as follows:

$$\sigma(\omega_1 \otimes_{\mathcal{A}} \omega_2) = \sum_{jk \in I} c_{jk}\, \omega_j \otimes_{\mathcal{A}} \omega_k. \qquad (7.12)$$

A comment on notation may be in order. The index set I is some elegant abstract set and as such does not contain lowly beasts such as the numbers 1 and 2. So the elements ω_1 and ω_2 are not to be confused with the ω_i's.

To prove the braid equation (7.10), we first evaluate $\sigma_{23}\sigma_{12}\sigma_{23}$ acting on $\omega_1 \otimes_{\mathcal{A}} \omega_2 \otimes_{\mathcal{A}} \eta$, where $\omega_1, \omega_2 \in \ _{\text{inv}}\Gamma$, and $\eta \in \Gamma_{\text{inv}}$, and then $\sigma_{12}\sigma_{23}\sigma_{12}$ acting on the same element. If these turn out to be equal, then by our previous comments, this suffices to prove the braid equation (7.10). So we evaluate

$$\sigma_{23}\sigma_{12}\sigma_{23}(\omega_1 \otimes_{\mathcal{A}} \omega_2 \otimes_{\mathcal{A}} \eta) = \sigma_{23}\sigma_{12}(\omega_1 \otimes_{\mathcal{A}} \eta \otimes_{\mathcal{A}} \omega_2)$$

$$= \sigma_{23}(\eta \otimes_{\mathcal{A}} \omega_1 \otimes_{\mathcal{A}} \omega_2)$$

$$= \eta \otimes_{\mathcal{A}} \left(\sum c_{jk}\, \omega_j \otimes_{\mathcal{A}} \omega_k \right)$$

$$= \sum c_{jk}\, (\eta \otimes_{\mathcal{A}} \omega_j \otimes_{\mathcal{A}} \omega_k).$$

In the last equality, we used the fact that $c_{jk} \in \mathbb{C}$. Now, on the other hand,

$$\sigma_{12}\sigma_{23}\sigma_{12}(\omega_1 \otimes_{\mathcal{A}} \omega_2 \otimes_{\mathcal{A}} \eta) = \sigma_{12}\sigma_{23}\left(\sum c_{jk}\, \omega_j \otimes_{\mathcal{A}} \omega_k \otimes_{\mathcal{A}} \eta \right)$$

$$= \sigma_{12}\left(\sum c_{jk}\, \omega_j \otimes_{\mathcal{A}} \eta \otimes_{\mathcal{A}} \omega_k \right)$$

$$= \sum c_{jk}\, (\eta \otimes_{\mathcal{A}} \omega_j \otimes_{\mathcal{A}} \omega_k).$$

This completes the proof of the braid equation (7.10). And that was the last remaining thing to be proved. ∎

Exercise 7.6. *Let $\{\omega_i \mid i \in I\}$ be a vector space basis of $_{\text{inv}}\Gamma$. Then the set*

$$\{\omega_j \otimes_A \omega_k \mid j, k \in I\}$$

is a vector space basis of $_{\text{inv}}(\Gamma^{\otimes 2})$. (**Hint:** *You should have already done this for another exercise.*)

7.4 Basic Representation

So the moral of this story so far is that the cumbersome symmetric group is not powerful enough to get a working noncommutative theory of the exterior algebra. We now want to get into what we will use instead of the symmetric group. This will turn out to be another, larger group. First, we discuss the symmetric group S_k for $k \geq 2$, but in no great detail. We will not prove all of our assertions; basic texts such as [41] can be consulted for proofs and more theory.

For every integer j satisfying $1 \leq j < k$, we define $\tau_j \in S_k$ to be the 2-cycle that interchanges j and $j+1$, namely, $\tau_j := (j, j+1)$. For example, in S_2 we have only one such element: $\tau_1 = (1, 2)$, while in S_3, we have two such elements: $\tau_1 = (1, 2)$ and $\tau_2 = (2, 3)$. (At this point the reader should realize that the notation τ_j is somewhat ambiguous.) Now these 2-cycles in S_k satisfy the following equations for $1 \leq i < k$ and $1 \leq j < k$:

$$\tau_i \tau_j = \tau_j \tau_i \quad \text{for all } |i - j| > 2,$$
$$\tau_i \tau_j \tau_i = \tau_j \tau_i \tau_j \quad \text{for all } |i - j| = 1, \qquad (7.13)$$
$$\tau_i^2 = e.$$

Now an important result here, due to Moore in [58], is that S_k is isomorphic to the abstract group generated by $k-1$ symbols $\tau_1, \ldots, \tau_{k-1}$ satisfying the above relations (7.13). (For a somewhat more recent treatment, see Theorem 4.1 in [46].) It is also known that S_k is a finite group with exactly $k!$ elements. Moreover, S_k is nonabelian for $k \geq 3$, while S_2 is the cyclic group with two elements, often also denoted by $\mathbb{Z}_2$.

Now the *braid group* B_k for $k \geq 2$ is defined as the group generated by $k - 1$ symbols $\sigma_1, \ldots, \sigma_{k-1}$ satisfying these relations, which are known as the *braid relations*:

$$\sigma_i \sigma_j = \sigma_j \sigma_i \quad \text{for all } |i - j| > 2,$$
$$\sigma_i \sigma_j \sigma_i = \sigma_j \sigma_i \sigma_j \quad \text{for all } |i - j| = 1.$$

It turns out that $B_2 \cong \mathbb{Z}$ is infinite and abelian, while for all $k \geq 3$, we have that B_k is infinite and nonabelian. By general principles about groups defined by generators and relations, we have that there exists a unique group homomorphism

$$p_k : B_k \to S_k$$

for all $k \geq 2$ that satisfies $p_k(\sigma_i) = \tau_i$ for all $i = 1, \ldots, k-1$. Clearly, this homomorphism is surjective. The braid group has scads of applications, many more than I could possibly hope to mention. Braid groups are originally due to E. Artin in a paper [2] published in 1925. Moreover, the braid group actually has something to do with physical braids in three-dimensional space, which then motivated a topic in topology. A nice place to read about that application is [45].

Exercise 7.7 (Optional). *We have given presentations of S_k and B_k, each of which has $k-1$ generators. Prove that each of these groups is generated by two elements.*

Part 4 of the previous theorem can now be interpreted as saying that the assignments $\sigma_1 \mapsto \sigma_{12}$ and $\sigma_2 \mapsto \sigma_{23}$ induce a homomorphism of groups $\phi_3 : B_3 \to GL(\Gamma^{\otimes 3})$, the general linear group of $\Gamma^{\otimes 3}$. This is also known as a *representation* of the group B_3 on the vector space $\Gamma^{\otimes 3}$. The representation ϕ_3 factors as $\phi_3 = \psi_3 \circ p_3$ to give us a representation ψ_3 of S_3 on $\otimes^3 \Gamma$ if and only if $\ker p_3 \subset \ker \phi_3$. This follows from the theory of groups.

A rather more trivial statement is that the assignment $\sigma_1 \mapsto \sigma$ induces a homomorphism of groups $\phi_2 : B_2 \cong \mathbb{Z} \to GL(\Gamma^{\otimes 2})$. And ϕ_2 factors through p_2 if and only if $\ker p_2 \subset \ker \phi_2$. Actually, the range $\mathrm{Ran}\, \phi_2$ is restricted to lie in the subgroup of $GL(\Gamma^{\otimes 2})$ of $\mathcal{A}$-bimodule morphisms that are also left and right $\mathcal{A}$-co-module morphisms. In short, the range of ϕ_2, and hence its kernel as well, depends a lot on all of the structures on $\Gamma^{\otimes 2}$, not just the vector space structure. And these extra structures come from the fact that Γ is a bicovariant $\mathcal{A}$-bimodule and not merely a vector space.

Cutting to the well-anticipated chase, we want to define representations $\phi_k : B_k \to GL(\Gamma^{\otimes k})$ for *all* $k \geq 2$. Once that is done, we will have that ϕ_k factors through p_k if and only if $\ker p_k \subset \ker \phi_k$. So sometimes the quantum situation reduces to the classical situation, that is, a representation of the symmetric group. But, in general, the quantum situation will involve the representation ϕ_k of the much larger group B_k. Now let's see how this works. Eventually, this helps us arrive at a reasonable definition of the braided exterior algebra.

We will use σ acting on $\Gamma^{\otimes 2}$ to construct a representation of the braid group B_k acting on $\Gamma^{\otimes k}$ for all $k \geq 2$. Of course, we already have noted that we have a representation of $B_2 \cong \mathbb{Z}$ acting on $\Gamma^{\otimes 2}$; we just mapped the abstract generator $\sigma_1 \in B_2$ to σ. Now, for $k \geq 3$, we map the generators σ_i for $i = 1, \ldots, k-1$ of B_k to the linear maps $l_i : \Gamma^{\otimes k} \to \Gamma^{\otimes k}$ defined by

$$l_i := id \otimes_{\mathcal{A}} \cdots \otimes_{\mathcal{A}} id \otimes_{\mathcal{A}} \sigma \otimes_{\mathcal{A}} id \otimes_{\mathcal{A}} \cdots \otimes_{\mathcal{A}} id,$$

where there are at first $i - 1$ factors of the identity map of $\mathcal{A}$ followed by one occurrence σ (which acts on two factors of Γ) and finally $k - i - 1$ occurrences of the identity map of $\mathcal{A}$.

Exercise 7.8. *Show the following:*

- *Show that the linear map l_i is invertible by finding its inverse.*
- *For each integer $k \geq 3$, the linear maps $l_i \in \mathrm{GL}(\Gamma^{\otimes k})$ for $i = 1, \ldots, k - 1$ satisfy the braid relations for B_k.* (**Hint:** *Do the case $k = 3$ first.*)
- *Consequently, for all $k \geq 3$, there is a unique group homomorphism $\phi_k : B_k \to \mathrm{GL}(\Gamma^{\otimes k})$ such that $\phi_k(\sigma_i) = l_i$ for $1 = 1, \ldots, k - 1$.*

Of course, the subgroup $\mathrm{Ran}\,\phi_k$ of $\mathrm{GL}(\Gamma^{\otimes k})$ must be isomorphic to some quotient of the braid group B_k, the exact quotient depending on the structure of the bicovariant bimodule Γ. So it is not necessarily true that $\mathrm{Ran}\,\phi_k$ is isomorphic to B_k. Nonetheless, we will write σ_i instead of l_i from now on. This is not really an abuse of notation. We are merely saying that the action of the group generators σ_i of B_k is given by the linear maps l_i.

This rather long chapter has been a rather brief introduction to braid groups, including just those topics needed to construct the braided exterior algebra of an FODC. But yet another, albeit short, chapter of material must be covered.

7.5   Notes

This chapter is about how braid groups enter into this one particular nook of noncommutative geometry. We present just enough of the general theory of braid groups to make this application understandable. In contrast with our general philosophy, we sometimes only stated a result without giving a proof. Many fine references for braid groups exist to help you fill in those gaps. More than any other I have relied on [46]. Braid groups have been used in numerous applications in many areas of mathematics, far too many to cite here. As far as noncommutative geometry per se is concerned, there are many, quite fine introductions to that theory which never mention braid groups or their associated braidings. For example, see [13, 47, 54], and [82]. On the other hand, there are many books where braid groups are used, such as [11, 45, 55], and [48]. Of course, all of these are texts on quantum groups and so have a different point of view. In this book we are looking at a theory that relates both to noncommutative geometry and to quantum groups. A rather interesting application of the braid group to the theory of quantum principal bundles is given in [30], where it is shown that every QPB has a natural braiding.

The particular application that interests us in this chapter is presented in Woronowicz's paper [86]. This topic will be continued in Chapter 9, on braided exterior algebras. But first we will take a short interlude.

Chapter 8
An Interlude: Some Abstract Nonsense

8.1 Bicovariant Algebras

At this point we take a brief interlude to indulge ourselves in some more abstract nonsense. This will give us language to describe the relation between a bicovariant first-order differential calculus (FODC) and its associated braided exterior algebra. The stuff that will be served up next looks an awful like what we have already seen. And that is true! However, it is not exactly the same. So heads up!

The idea behind the next definition applies directly to our current concern where we start with a bicovariant FODC (Γ, d) over $\mathcal{A}$. We want $d : \mathcal{A} \to \Gamma$ to become the beginning of a graded complex that generalizes the exterior algebra over a differential manifold together with its de Rham differential. So we are looking for a complex

$$\mathcal{A} \xrightarrow{d_0} \Gamma \xrightarrow{d_1} \Gamma^2 \xrightarrow{d_2} \cdots \xrightarrow{d_{k-1}} \Gamma^k \xrightarrow{d_k} \cdots$$

so that $\oplus_{j=0}^{\infty} \Gamma^j$ is a graded algebra. Of course, we put $\Gamma^0 = \mathcal{A}$ and $\Gamma^1 = \Gamma$. Also, $d_0 = d$, the differential of the FODC. Since both $\mathcal{A}$ and Γ are bicovariant bimodules, it is natural also to require this structure on every Γ^k.

Definition 8.1. *A bicovariant algebra over $\mathcal{A}$, a Hopf algebra, is a triple $(G, \Phi_G, {}_G\Phi)$ satisfying the following four conditions:*

1. *G is an algebra with identity 1.*
2. *$\Phi_G : G \to \mathcal{A} \otimes G$ is a left co-action of $\mathcal{A}$ on G and an algebra morphism.*
3. *${}_G\Phi : G \to G \otimes \mathcal{A}$ is a right co-action of $\mathcal{A}$ on G and an algebra morphism.*
4. *The left and right co-actions commute. This means that the following diagram commutes:*

© Springer International Publishing Switzerland 2015   123
S.B. Sontz, *Principal Bundles*, Universitext, DOI 10.1007/978-3-319-15829-7_8

$$\begin{array}{ccc}
G & \xrightarrow{\Phi_G} & \mathcal{A} \otimes G \\
{\scriptstyle _G\Phi} \downarrow & & \downarrow {\scriptstyle id \otimes {}_G\Phi} \\
G \otimes \mathcal{A} & \xrightarrow{\Phi_G \otimes id} & \mathcal{A} \otimes G \otimes \mathcal{A}.
\end{array} \tag{8.1}$$

Compare this with the diagram (5.1).

Moreover, if $G = \oplus_{k=0}^{\infty} G^k$ realizes G as a graded algebra *(meaning that $1 \in G^0$ and $G^j G^k \subset G^{j+k}$ for all $j, k \geq 0$), then we say that it is a* graded bicovariant algebra *over $\mathcal{A}$ if, furthermore, the co-actions preserve degree for every $k \geq 0$; that is,*

$$\Phi_G(G^k) \subset \mathcal{A} \otimes G^k \quad \text{and} \quad {}_G\Phi(G^k) \subset G^k \otimes \mathcal{A}.$$

Intuitively, this can be interpreted as saying that all the elements in $\mathcal{A}$ have degree 0. In this case, we denote the restriction of Φ_G (resp., ${}_G\Phi$) to G^k by Φ_{G^k} (resp., ${}_{G^k}\Phi$).

Remark. The notation G for an algebra may seem unusual. Well, so it is! The problem is that the notation $\mathcal{A}$ already is being used for something else, while A is too similar. Think algebra = al+gebra (which is etymologically justifiable) to understand from where the G comes. As it turns out, G is also used by some authors to denote a group or even a quantum group. For the time being, we will not follow that convention.

In our application of these ideas, we will have $G^0 = \mathcal{A}$ and $\Phi_{G^0} = {}_{G^0}\Phi = \phi$, the co-multiplication of $\mathcal{A}$. But this is not always so. This special case deserves its own name.

Definition 8.2. *A graded bicovariant algebra over $\mathcal{A}$, say $G = \oplus_{k=0}^{\infty} G^k$, is said to be* connected *if $G^0 = \mathcal{A}$ and $\Phi_{G^0} = {}_{G^0}\Phi = \phi$.*

Exercise 8.1. *Let $G = \oplus_{k=0}^{\infty} G^k$ be a connected graded bicovariant algebra over $\mathcal{A}$. Show for every $k \geq 0$ that $(G^k, \Phi_{G^k}, {}_{G^k}\Phi)$ is a bicovariant bimodule over $\mathcal{A}$.*

Later we will discuss the full problem of constructing a complex starting from an FODC. But for right now, in order to construct a graded algebra (without a differential), we only use these ingredients: a bicovariant bimodule $(\Gamma, \Phi_\Gamma, {}_\Gamma\Phi)$ over $\mathcal{A}$. So here is the definition:

Definition 8.3. *Let $(\Gamma, \Phi_\Gamma, {}_\Gamma\Phi)$ be a bicovariant bimodule over $\mathcal{A}$. Then we say that a graded bicovariant algebra $(G, \Phi_G, {}_G\Phi)$ with $G = \oplus_{k=0}^{\infty} G^k$ is* based on *$(\Gamma, \Phi_\Gamma, {}_\Gamma\Phi)$ if*

- *$(G, \Phi_G, {}_G\Phi)$ is connected.*
- *$(G^1, \Phi_{G^1}, {}_{G_1}\Phi) = (\Gamma, \Phi_\Gamma, {}_\Gamma\Phi)$ as bicovariant bimodules.*
- *$\text{span}_{\mathbb{C}}(G_1)^k = G^k$, where we use the usual notation*

$$(G^1)^k := \{g_1 g_2 \cdots g_k \mid g_i \in G^1 \text{ for all } i = 1, \dots, k\}.$$

Remark. In [86] Woronowicz uses the expression "built over" where I have used "based on" instead. There is really nothing wrong with his usage. For reasons I am not able to identify, I prefer my usage. This sort of phenomenon reliably dates back to Pygmalion.

We have already seen one example of this definition as detailed now:

Exercise 8.2. *We now request the reader to succumb to the temptation we described just after Exercise 7.3. Referring back to the tensor algebra* $\Gamma^{\otimes}$ *introduced earlier, we define* $\Phi_{\Gamma\otimes} := \oplus_k \Phi_{\Gamma\otimes k}$ *and* $_{\Gamma\otimes}\Phi := \oplus_k {_{\Gamma\otimes k}}\Phi$. *Show that* $(\Gamma^{\otimes}, \Phi_{\Gamma\otimes}, {_{\Gamma\otimes}}\Phi)$ *is a graded bicovariant algebra based on* $(\Gamma, \Phi_{\Gamma}, {_{\Gamma}}\Phi)$.

We will soon construct the braided exterior algebra associated to an FODC (Γ, d) over $\mathcal{A}$. This will also be graded bicovariant bimodule based on $(\Gamma, \Phi_{\Gamma}, {_{\Gamma}}\Phi)$. But before doing that, we yield to the categorical imperative. The next two exercises are for readers familiar with the language of category theory.

Exercise 8.3 (Optional). *We have defined three mathematical structures in this section. For each of them, define what a morphism is and then show that we get a category for each of the three structures.*

Exercise 8.4 (Optional). *Show that the assignment*

$$(\Gamma, \Phi_{\Gamma}, {_{\Gamma}}\Phi) \mapsto (\Gamma^{\otimes}, \Phi_{\Gamma\otimes}, {_{\Gamma\otimes}}\Phi)$$

extends to a functor. The point is that we have only defined an assignment for the objects. So the assignment for the morphisms must first be defined, and then it must be proved that the properties of a functor are satisfied.

8.2 Notes

This chapter has turned out to be a brief discussion of some definitions, all of which can be found in Woronowicz's paper [86]. Their importance will become clearer as this story unfolds.

The earliest usage of the expression "abstract nonsense" seems to predate category theory (early 1940s), but nowadays it usually concerns a straightforward translation of some elementary considerations into the terminology of category theory. While it is not meant as a derogatory term, I think that it is deliberately self-mocking. In any event, it has become part of the colloquial vocabulary of mathematicians.

Chapter 9
The Braided Exterior Algebra

9.1 Antisymmetrization

Using the representation ϕ_k of the braid group B_k, we next define (in some sort of analogy with the case of the symmetric group) an *antisymmetrization operator*

$$A_k : \Gamma^{\otimes k} \to \Gamma^{\otimes k}$$

for all $k \geq 2$. (When you come to think of it, it would really be better called an *antibraidization operator*.) Then the elements of degree $k \geq 2$ of the braided exterior algebra associated to Γ will be defined by

$$\wedge^k \Gamma := \Gamma^{\otimes k} / \ker A_k \cong \operatorname{Ran} A_k. \tag{9.1}$$

Recall that we have already decided on the definitions $\wedge^0 \Gamma := \mathcal{A}$ as well as $\wedge^1 \Gamma := \Gamma$. These two definitions are equivalent to defining A_0 and A_1 to be the appropriate identity maps. So it comes down to defining A_k—and understanding it—for $k \geq 2$. Unless mentioned otherwise, we take $k \geq 2$ throughout this section. Actually, some of the statements we will make are false for $k = 1$.

Recall that the symmetric group S_k of the set $\{1, 2, \ldots, k\}$ is generated by the 2-cycles $\tau_s = (s, s+1)$ for $s = 1, \ldots, k-1$. Also recall that these generators satisfy the braid equations as well as the relations $\tau_s^2 = e$, the identity element of S_k. Denote the set of these generators by

$$T := \{\tau_1, \ldots, \tau_{k-1}\}.$$

For every $\tau \in S_k$, we define $I(\tau)$ to be the number of inversions of pairs induced by τ. What we mean explicitly by this is

$$I(\tau) := \operatorname{card}\{ (i, j) \mid 1 \leq i < j \leq k \text{ such that } \tau(i) > \tau(j) \}. \tag{9.2}$$

For example, $I(\tau) = 0$ if and only if $\tau = e$.

© Springer International Publishing Switzerland 2015
S.B. Sontz, *Principal Bundles*, Universitext, DOI 10.1007/978-3-319-15829-7_9

Proposition 9.1. *Let* $\tau \in S_k$. *Put* $m := I(\tau)$. *Then* τ *can be written as a product of* m *factors, each of which is an element in* T. *In other words, there exist indices* $s_1, \ldots, s_m$, *not necessarily distinct, in* $\{1, 2, \ldots, k-1\}$ *such that*

$$\tau = \tau_{s_1} \tau_{s_2} \cdots \tau_{s_m}. \tag{9.3}$$

This representation is not unique for $k \geq 3$.

Remarks. For example, taking $\tau = e$ gives $m = 0$. Then the set of indices s_l is empty so that the product on the right side of (9.3) has zero factors. Such a product is defined actually and by definition is e. Also, $m = 0$ implies that $\tau = e$. So this proposition is rather trivially true for the case $m = 0$.

Since T is a set of generators of S_k, there is always some integer m such that (9.3) holds. The point here is that we can take m to be the particular value $I(\tau)$.

The fact that the representation (9.3) is not unique for $k \geq 3$ is a trivial consequence of the braid equations.

The material needed to prove Proposition 9.1 runs on for some dozen pages or so, depending on how much one can assume as known. So we will not follow our usual policy of explaining everything in great detail. Instead, we refer to Lemma 4.7(a) in [46], where not only is this result proved, but it is also shown that $m = I(\tau)$ is the smallest integer for which a factorization of the form (9.3) holds.

Next, we define a set function $\pi_k : S_k \to B_k$ by

$$\pi_k(\tau) := \sigma_{s_1} \sigma_{s_2} \cdots \sigma_{s_m}, \tag{9.4}$$

where τ is factorized as in (9.3). We warn the reader that π_k is not a homomorphism of groups. For example, $I(\tau_s) = 1$, and the factorization in (9.3) is unique in this case and is given by $\tau_s = \tau_s$. (This equation is not quite as tautological as one might think at first sight. It says that a factorization of the left-hand side is given by the product of terms, one term actually, on the right side.) So, by definition, $\pi_k(\tau_s) = \sigma_s$. Also, by the remark given above, we see that $\pi_k(e) = e$, where now e denotes the identity element of any and every group. So we have

$$\pi_k(\tau_s^2) = \pi_k(e) = e,$$
$$\pi_k(\tau_s)^2 = \sigma_s^2 \neq e.$$

A basic fact about the braid group is that all of its generators (as well as all other elements not equal to e) have infinite order. Please consult a text such as [46] for a proof of this fact. This is how we know that $\sigma_s^2 \neq e$. Consequently, the result that $\pi_k(\tau_s^2) \neq \pi_k(\tau_s)^2$ for all $s = 1, \ldots, k-1$ shows that π_k is not a homomorphism of groups.

Exercise 9.1. *What happens to this argument for the case* $k = 1$? *What is the correct result in that case?*

But we have not yet completed the definition of π_k since the representation (9.3) is not unique for $k \geq 3$. [By the way, the representation (9.3) is unique if $k = 2$, as the reader should verify.] Here is an example of what is happening for any $k \geq 3$. Let $\tau = (13)$ be the 2-cycle that interchanges 1 and 3 and leaves fixed all the other points in $\{1, 2, \ldots, k\}$. Then one can calculate that $I(\tau) = 3$. Now there are two representations of (13) in the form of (9.3). They are

$$(13) = (23)(12)(23) = (12)(23)(12) \qquad (9.5)$$

or, equivalently,

$$(13) = \tau_2 \tau_1 \tau_2 = \tau_1 \tau_2 \tau_1. \qquad (9.6)$$

Of course, we are seeing a braid relation here. Notice that the 2-cycle (13) is not an element of the generating set T. In any event, we readily calculate that

$$\tau(13) = \sigma_2 \sigma_1 \sigma_2 = \sigma_1 \sigma_2 \sigma_1, \qquad (9.7)$$

where now we are using the braid relation for three of the generators of the braid group B_k. So this shows that the definition of $\tau(13)$ does not depend on the particular choice of factorization in (9.5).

This discussion motivates the next result.

Proposition 9.2. *The definition of $\pi_k(\tau)$ does not depend on the particular choice of the factorization given in Eq. (9.3).*

Remark. Again, a proof of this would lead us very far afield. We simply note that this result is an immediate consequence of Lemma 4.11 in [46].

And while π_k is not a homomorphism of groups, we do have the next curious result.

Proposition 9.3. *The set function $\pi_k : S_k \to B_k$ satisfies*

* $\pi_k(e) = e$.
* *If $\alpha, \beta \in S_k$ satisfy $I(\alpha\beta) = I(\alpha) + I(\beta)$, then $\pi_k(\alpha\beta) = \pi_k(\alpha)\pi_k(\beta)$.*

Remark. Notice that the hypothesis in the second part fails to hold if we take $\alpha = \beta = \tau_s$. This is consistent with our earlier calculation showing that π_k is not a homomorphism.

Proof. We have already proved the first part. For the second part, we put $m = I(\alpha)$ and $n = I(\beta)$. Then we can write α and β as

$$\alpha = \tau_{t_1} \cdots \tau_{t_m},$$
$$\beta = \tau_{s_1} \cdots \tau_{s_n}$$

for appropriate subindices. This immediately implies

$$\alpha\beta = \tau_{t_1}\cdots\tau_{t_m}\tau_{s_1}\cdots\tau_{s_n}, \tag{9.8}$$

which is a factorization of $\alpha\beta$ as a product of $m + n = I(\alpha) + I(\beta) = I(\alpha\beta)$ factors in T. (We just used the hypothesis.) So (9.8) is in the form (9.3). Therefore, by using the definition of π_k three times, we obtain

$$\pi_k(\alpha\beta) = \sigma_{t_1}\cdots\sigma_{t_m}\sigma_{s_1}\cdots\sigma_{s_m}$$
$$= \pi_k(\alpha)\pi_k(\beta). \quad\blacksquare$$

Another result is quite straightforward but merits enough attention to be stated in its own proposition. Its proof follows immediately from definitions.

Proposition 9.4. *The set function $\pi_k : S_k \to B_k$ is a section of the canonical projection (which is a group homomorphism) $p_k : B_k \to S_k$, that is, $p_k \circ \pi_k = id_{S_k}$. In particular, π_k is a one-to-one function.*

Remark. The function $\pi_k : S_k \to B_k$ is not a homomorphism. Actually, the short exact sequence

$$\{e\} \to \ker p_k \to B_k \xrightarrow{p_k} S_k \to \{e\}$$

in the category of groups does not split since if it did split, we would have a subgroup of B_k isomorphic to S_k and thus B_k would have nontrivial elements of finite order. But it is known that all nontrivial elements of B_k have infinite order.

The group S_k is finite while B_k is infinite. So $\ker p_k$ is an infinite group, and accordingly the fiber $p_k^{-1}(\tau)$ above each element $\tau \in S_k$ is a coset of $\ker p_k$, which therefore also has infinitely many elements. The set function $\pi_k : S_k \to B_k$ picks out one very special element in each of these fibers.

We now have enough theory in place in order to be able to define the antisymmetrization operator.

Definition 9.1. *Let $k \geq 2$ be an integer. We define the antisymmetrization operator $A_k : \Gamma^{\otimes k} \to \Gamma^{\otimes k}$ by*

$$A_k := \sum_{\alpha \in S_k} \text{sign}(\alpha)\, \pi_k(\alpha), \tag{9.9}$$

where we are considering that the element $\pi_k(\alpha) \in B_k$ acts on the vector space $\otimes^k \Gamma$ via the representation ϕ_k. A more precise notation for this definition is, of course, given by

$$A_k := \sum_{\alpha \in S_k} \text{sign}(\alpha)\, \phi_k(\pi_k(\alpha)). \tag{9.10}$$

For completeness of notation, we also let $A_0 = id_A$ and $A_1 = id_\Gamma$, the identity operators on the appropriate spaces.

We introduced the sign of a permutation earlier. We note here in passing that $\text{sign}(\alpha) = (-1)^{I(\alpha)}$. Note that we are summing over the *finite* set S_k, not over B_k, which is infinite. However, if we prefer, we can think of this definition as a sum over the image of π_k in B_k, which is in bijection with S_k. We then have that

$$A_k = \sum_{\sigma \in \pi_k(S_k)} \text{sgn}(\sigma)\,\phi_k(\sigma), \tag{9.11}$$

where sgn $: B_k \to \{-1, +1\}$ is defined to be the composition

$$B_k \xrightarrow{p_k} S_k \xrightarrow{\text{sign}} \{-1, +1\}.$$

Using the formula (9.1) given earlier, we have the definition of $\wedge^k \Gamma$ as a vector space for all $k \geq 2$. But we also want this to be a graded algebra. So we need to know that the kernel we are dividing out is a two-sided ideal. This is our next task.

9.2 Shuffles

Anyone who has played any of the many games of cards is familiar with a special class of permutations of 52 objects. These are the shuffles. These permutations are defined as certain elements of the symmetric group S_k for any $k \geq 2$. We are thinking in the following definition of dividing the "deck of k cards" into two "halves," the first with h elements and the second with the remaining $k - h$ elements. The "cards" in each "half" maintain the order relation among themselves during the shuffle. Of course, the "cards" that we are permuting are simply the integers $1, 2, \ldots, k$. This should be enough chit-chat to motivate this mathematical definition:

Definition 9.2. *Let h be an integer $0 \leq h \leq k$. We say that $\tau \in S_k$ is an h-shuffle if $\tau(i) < \tau(j)$ whenever either $1 \leq i < j \leq h$ or $h < i < j \leq k$. We denote the set of all h-shuffles in S_k as*

$$SH_{k,h} := \{\tau \in S_k \mid \tau \text{ is an } h-\text{shuffle}\}.$$

We say that $\tau \in S_k$ is a shuffle if it is an h-shuffle for some integer h with $0 \leq h \leq k$.

Exercise 9.2. *If $\tau \in S_k$ is a shuffle, then it is an h-shuffle for some h. But is h unique? (**Hint:** Think carefully about all possible values of h.)*

Exercise 9.3. *A magician fans out a standard deck of 52 playing cards, faces down of course, and says to you, "Pick a card, any card." You look at the face of the card* *you picked and then (following the magician's request) you put the card back into the deck of the remaining 51 cards wherever you like. This operation clearly is a permutation in S_{52}. Is it a shuffle?*

Exercise 9.4. *Identify $SH_{k,0}$ and $SH_{k,k}$.*

Exercise 9.5 (Optional). *Find a formula for the cardinality of $SH_{k,h}$.*

Exercise 9.6. *Is $SH_{k,h}$ a subgroup of S_k?*

Proposition 9.5. *Let $\tau \in S_k$. Then for any $0 \leq h \leq k$, there exist unique elements $\alpha, \beta \in S_k$ and a unique $\upsilon \in SH_{k,h}$ such that*

- *$\tau = \upsilon \alpha \beta$.*
- *α leaves $\{h+1, \ldots, k\}$ pointwise fixed. So $\alpha \in S_h \times \{e\} \cong S_h$.*
- *β leaves $\{1, \ldots, h\}$ pointwise fixed. So $\beta \in \{e\} \times S_{k-h} \cong S_{k-h}$.*

Proof. At first glance, this may seem to be asking for the moon in a teacup. But the idea is actually simple. The point is that τ is going to rearrange the elements in the first half into a new order, similarly with the second half. Thus, α and β (which commute because each is the identity on the half where the other is not necessarily the identity) just produce these rearrangements, but each in its own half respectively. Next, we shuffle those two halves, putting the elements where τ wants them to go; this shuffle is denoted by υ. Then the composition of these three permutations gives us τ. This shows existence. Uniqueness follows by thinking just a wee bit more about this construction of α, β, and υ. ∎

Exercise 9.7 (Only for the more formally inclined). *Okay, I understand. That was not a real proof, and you want a real proof. So write out a detailed real proof with formulas, now that you have the idea behind it.*

Corollary 9.1. *Using the notation in Proposition 9.5, we have*

$$I(\tau) = I(\upsilon) + I(\alpha) + I(\beta),$$
$$\pi_k(\tau) = \pi_k(\upsilon) \, \pi_k(\alpha) \, \pi_k(\beta).$$

Exercise 9.8. *Prove Corollary 9.1 in the manner to which you have grown accustomed.*

The definition of A_k in (9.9) can be modified by taking the sum over any subset of S_k instead of all of S_k. We do exactly that now.

Definition 9.3. *The h-shuffle antisymmetrization operator is defined as*

$$A_{k,h} := \sum_{\alpha \in SH_{k,h}} \text{sign}(\alpha) \, \pi_k(\alpha). \tag{9.12}$$

So $A_{k,h} : \Gamma^{\otimes k} \to \Gamma^{\otimes k}$ is a linear map.

Exercise 9.9. *Identify $A_{k,0}$ and $A_{k,k}$.*

Proposition 9.6. *For any $0 \le h \le k$, we have the following identity:*

$$A_k = A_{k,h}(A_h \otimes_{\mathcal{A}} A_{k-h}). \tag{9.13}$$

Proof. Note that A_l maps (linearly, of course) $\Gamma^{\otimes l}$ to itself. So $A_h \otimes_{\mathcal{A}} A_{k-h}$ maps $\Gamma^{\otimes h} \otimes_{\mathcal{A}} \Gamma^{\otimes (k-h)} \cong \Gamma^{\otimes k}$ to itself. Hence, both sides of (9.13) are linear maps of $\Gamma^{\otimes k}$ to itself. First, we evaluate $A_h \otimes_{\mathcal{A}} A_{k-h}$. In the following, we let sign_l denote the sign function for the symmetric group S_l. Also, e denotes the identity element of the appropriate group.

$$
\begin{aligned}
A_h \otimes_{\mathcal{A}} A_{k-h} &= \sum_{\alpha \in S_h} \text{sign}_h(\alpha)\, \pi_h(\alpha) \otimes_{\mathcal{A}} \sum_{\beta \in S_{k-h}} \text{sign}_{h-k}(\beta)\, \pi_{h-k}(\beta) \\
&= \sum_{\alpha \in S_h} \sum_{\beta \in S_{k-h}} \text{sign}_h(\alpha)\text{sign}_{h-k}(\beta)(\pi_h(\alpha) \otimes_{\mathcal{A}} id)(id \otimes_{\mathcal{A}} \pi_{h-k}(\beta)) \\
&= \sum_{\alpha \in S_h} \sum_{\beta \in S_{k-h}} \text{sign}_k(\alpha \times e)\text{sign}_k(e \times \beta)\pi_k(\alpha \times e)\pi_k(e \times \beta) \\
&= \sum_{\alpha \in S_h \times e} \sum_{\beta \in e \times S_{k-h}} \text{sign}_k(\alpha)\text{sign}_k(\beta)\, \pi_k(\alpha)\pi_k(\beta) \\
&= \sum_{\alpha \in S_h \times e} \sum_{\beta \in e \times S_{k-h}} \text{sign}_k(\alpha\beta)\, \pi_k(\alpha\beta).
\end{aligned}
$$

So we continue using Proposition 9.5 and its Corollary 9.1:

$$
\begin{aligned}
A_{k,h}(A_h \otimes_{\mathcal{A}} A_{k-h}) &= \sum_{v \in SH_{k,h}} \text{sign}_k(v)\, \pi_k(v) \sum_{\alpha \in S_h \times e} \sum_{\beta \in e \times S_{k-h}} \text{sign}_k(\alpha\beta)\, \pi_k(\alpha\beta) \\
&= \sum_{v \in SH_{k,h}} \sum_{\alpha \in S_h \times e} \sum_{\beta \in e \times S_{k-h}} \text{sign}_k(v\alpha\beta)\, \pi_k(v\alpha\beta) \\
&= \sum_{\tau \in S_k} \text{sign}_k(\tau)\, \pi_k(\tau) \\
&= A_k.
\end{aligned}
$$

Notice that this argument is valid for any h satisfying $0 \le h \le k$. ∎

Another way of writing this result is

$$A_{n+m} = A_{n+m,n}(A_n \otimes_{\mathcal{A}} A_m) = A_{n+m,m}(A_m \otimes_{\mathcal{A}} A_n). \tag{9.14}$$

Theorem 9.1. *Define* $K_n := \ker A_n \subset \Gamma^{\otimes n}$. *Then* $\oplus_n K_n$ *is a graded two-sided ideal in* $\Gamma^\otimes$.

Moreover, $K_0 = 0$ *and* $K_1 = 0$.

Proof. By (9.14), we have that $\omega \in K_n \subset \Gamma^{\otimes n}$ and $\eta \in \Gamma^{\otimes m}$ imply that $\omega \otimes_A \eta \in K_{n+m}$ as well as $\eta \otimes_A \omega \in K_{n+m}$.

Since $A_0 = id_A$ and $A_1 = id_\Gamma$ by definition, we immediately get $K_0 = 0$ and $K_1 = 0$. ∎

9.3   Wedge Product

The previous section gives us the next, most important result.

Theorem 9.2. *In the above notation,* $K := \oplus_n K_n$ *is a two-sided ideal in* $\Gamma^\otimes := \oplus_n \Gamma^{\otimes n}$. *Consequently, the quotient*

$$\frac{\Gamma^\otimes}{K} \cong \oplus_n \left(\frac{\Gamma^{\otimes n}}{K_n} \right) = \oplus_n (\wedge^n \Gamma)$$

is a graded algebra with the quotient product.

Moreover, $\wedge^0 \Gamma = A$ *and* $\wedge^1 \Gamma = \Gamma$.

Proof. We are simply encoding our already-established results using the definition $\wedge^n \Gamma := \Gamma^{\otimes n} / \ker A_n = \Gamma^{\otimes n} / K_n$.

The last assertion follows from $K_0 = 0$ and $K_1 = 0$. ∎

Definition 9.4. *If* $\omega_1 \in \wedge^k \Gamma$ *and* $\omega_2 \in \wedge^l \Gamma$, *then we denote their product as given in Theorem 9.2 by*

$$\omega_1 \wedge \omega_2 \in \wedge^{k+l} \Gamma.$$

This is called the (braided) wedge product.

We also call $\wedge \Gamma \equiv \wedge^* \Gamma := \oplus_n (\wedge^n \Gamma)$ *the braided exterior algebra associated with* Γ.

Corollary 9.2. *As an algebra,* $\wedge \Gamma$ *is generated by* $\wedge^1 \Gamma = \Gamma$. *In particular, any* $\omega \in \wedge^k \Gamma$ *can be written as a linear combination of elements of the form* $\theta_1 \wedge \cdots \wedge \theta_k$ *with each* $\theta_i \in \Gamma$.

Proof. This follows from the fact that $\Gamma^\otimes$ is generated by $\Gamma^{\otimes 1} = \Gamma$. ∎

Remarks 9.1. *If we write* $\omega_1 = \eta_1 + \ker A_k$ *and* $\omega_2 = \eta_2 + \ker A_l$, *where* $\eta_1 \in \otimes^k \Gamma$ *and* $\eta_2 \in \otimes^l \Gamma$, *then we have by this definition that*

$$\omega_1 \wedge \omega_2 = \eta_1 \otimes_A \eta_2 + \ker A_{k+l}.$$

Notice that even though the wedge product $\wedge$ comes from the tensor product $\otimes_{\mathcal{A}}$, the notation for the wedge product does not have any subscript. Also, this is not the wedge product of the de Rham theory of a manifold.

As noted earlier, $\wedge^n \Gamma = \Gamma^{\otimes n} / \ker A_n \cong \operatorname{Ran} A_n$. Using this identification, we obtain the next result, which is a handy formula for doing calculations.

Corollary 9.3. *Let $\omega_1 \in \wedge^k \Gamma$ and $\omega_2 \in \wedge^l \Gamma$. Then*

$$\omega_1 \wedge \omega_2 = A_{k+l,k}(\omega_1 \otimes_{\mathcal{A}} \omega_2).$$

Proof. Using the above identification, we write $\omega_1 = A_k \eta_1$ and $\omega_2 = A_l \eta_2$, where $\eta_1 \in \Gamma^{\otimes k}$ and $\eta_2 \in \Gamma^{\otimes l}$. Then using the definition of the wedge product, we obtain

$$\omega_1 \wedge \omega_2 = \eta_1 \otimes_{\mathcal{A}} \eta_2 + \ker A_{k+l} \in \Gamma^{\otimes(k+l)} / \ker A_{k+l},$$

a coset in the quotient space. But this then gets identified with its image under A_{k+l}, namely, with $A_{k+l}(\eta_1 \otimes_{\mathcal{A}} \eta_2)$. Next, we calculate

$$
\begin{aligned}
\omega_1 \wedge \omega_2 &= A_{k+l}(\eta_1 \otimes_{\mathcal{A}} \eta_2) \\
&= A_{k+l,k}(A_k \otimes_{\mathcal{A}} A_l)(\eta_1 \otimes_{\mathcal{A}} \eta_2) \\
&= A_{k+l,k}(A_k \eta_1 \otimes_{\mathcal{A}} A_l \eta_2) \\
&= A_{k+l,k}(\omega_1 \otimes_{\mathcal{A}} \omega_2),
\end{aligned}
$$

which finishes the proof. ∎

9.4 Bicovariance

We now examine the relation between the antisymmetrization operator A_n and the left $\mathcal{A}$-co-action $\Phi_{\Gamma^{\otimes n}}$ on $\Gamma^{\otimes n}$. We recall that the diagram (7.8) says that σ is a left $\mathcal{A}$-co-module morphism on $\Gamma^{\otimes 2}$. We claim that this together with the definitions of A_n and $\Phi_{\Gamma^{\otimes n}}$ is enough to prove that A_n is a left $\mathcal{A}$-co-module morphism on $\otimes^n \Gamma$; that is, we have the following commutative diagram:

$$
\begin{array}{ccc}
\Gamma^{\otimes n} & \xrightarrow{\ A_n\ } & \Gamma^{\otimes n} \\
{\scriptstyle \Phi_{\Gamma^{\otimes n}}} \downarrow & & \downarrow {\scriptstyle \Phi_{\Gamma^{\otimes n}}} \\
\mathcal{A} \otimes \Gamma^{\otimes n} & \xrightarrow{\ id \otimes A_n\ } & \mathcal{A} \otimes \Gamma^{\otimes n}.
\end{array}
\qquad (9.15)
$$

Exercise 9.10. *Prove that this diagram commutes.*
Hint: *Prove first that the diagram*

$$\begin{array}{ccc}
\Gamma^{\otimes n} & \xrightarrow{\phi_n(\sigma_j)} & \Gamma^{\otimes n} \\
\Phi_{\Gamma^{\otimes n}} \downarrow & & \downarrow \Phi_{\Gamma^{\otimes n}} \\
\mathcal{A} \otimes \Gamma^{\otimes n} & \xrightarrow{id \otimes \phi_n(\sigma_j)} & \mathcal{A} \otimes \Gamma^{\otimes n}
\end{array}$$

commutes for every $j = 1, \ldots n - 1$, *where* ϕ_n *is the representation of the braid group* B_n *on* $\Gamma^{\otimes n}$.

Now we pick some element $\alpha \in \ker A_n$ and chase it in the diagram (9.15) from the upper left corner to the lower right corner. First, going across, we get zero, and so chasing across–down gives zero. Next, chasing this element down, we get $\Phi_{\Gamma^{\otimes n}}(\alpha) \in \ker(id \otimes A_n) = \mathcal{A} \otimes \ker A_n$. In short,

$$\Phi_{\Gamma^{\otimes n}}(\ker A_n) \subset \mathcal{A} \otimes \ker A_n.$$

Thus, $K_n = \ker A_n$ is left invariant under the co-action $\Phi_{\Gamma^{\otimes n}}$, which implies that $\Phi_{\Gamma^{\otimes n}}$ passes to the quotient $\wedge^n \Gamma$, where it is denoted by $\Phi_{\wedge^n \Gamma}$. It then follows that $\Phi_{\wedge^n \Gamma}$ is a left co-action of $\mathcal{A}$ on $\wedge^n \Gamma$.

Exercise 9.11. *Prove the last assertion using basic ring theory.*

Similarly, using the diagram (7.9), which says that σ is a right $\mathcal{A}$-co-module morphism on $\Gamma^{\otimes 2}$, we deduce that K_n is a right invariant subspace under the co-action $_{\Gamma^{\otimes n}}\Phi$. And this implies that $_{\Gamma^{\otimes n}}\Phi$ passes to the quotient $\wedge^n \Gamma$, where it is denoted by $_{\wedge^n \Gamma}\Phi$. And similarly, it follows that $_{\wedge^n \Gamma}\Phi$ is a right co-action of $\mathcal{A}$ on $\wedge^n \Gamma$.

Exercise 9.12. *Pause as long as needed in order to ponder until you are sure you understand the last paragraph thoroughly.*

We write $\Phi_{\wedge \Gamma} := \oplus_n \Phi_{\wedge^n \Gamma}$ (resp., $_{\wedge \Gamma}\Phi := \oplus_n {}_{\wedge^n \Gamma}\Phi$), which is then a left (resp., right) co-action of $\mathcal{A}$ on $\wedge \Gamma := \oplus_n \wedge^n \Gamma$.

The formulas for these co-actions evaluated on decomposable elements of degree n are obtained by taking the formulas for $\Phi_{\Gamma^{\otimes n}}$ and $_{\Gamma^{\otimes n}}\Phi$ and projecting them to the quotient spaces. First, for the left co-action $\Phi_{\wedge^n \Gamma}$ on $\wedge^n \Gamma$, we have

$$\Phi_{\wedge^n \Gamma}(\theta_1 \wedge \cdots \wedge \theta_n) = (a_1^{(0)} \cdots a_n^{(0)}) \otimes (\theta_1^{(1)} \wedge \cdots \wedge \theta_n^{(1)}),$$

where $\theta_1, \ldots, \theta_n \in \Gamma$, and $\Phi_\Gamma(\theta_j) = a_j^{(0)} \otimes \theta_j^{(1)} \in \mathcal{A} \otimes \Gamma$ for each $j = 1, \ldots, n$ is Sweedler's notation for the left co-action on Γ. (**Warning**: We do not use Einstein's summation notation. So there is no sum on j here.) Compare this with Eq. (7.6). Notice that $\otimes$ continues to mean what it has meant all along: the tensor product over $\mathbb{C}$ of complex vector spaces.

Second, for the right co-action $_{\wedge^n \Gamma}\Phi$ on $\wedge^n \Gamma$, we have

$$_{\wedge^n \Gamma}\Phi(\theta_1 \wedge \cdots \wedge \theta_n) = (\theta_1^{(0)} \wedge \cdots \wedge \theta_n^{(0)}) \otimes (b_1^{(1)} \cdots b_n^{(1)}),$$

where $\theta_1, \ldots, \theta_n \in \Gamma$, and ${}_\Gamma\Phi(\theta_j) = \theta_j^{(0)} \otimes b_j^{(1)} \in \Gamma \otimes \mathcal{A}$ for each $j = 1, \ldots, n$ is Sweedler's notation for the right co-action on Γ. And compare this with Eq. (7.7).

We now leave to the reader the proof of this important result:

Theorem 9.3 $(\wedge\Gamma, \Phi_{\wedge\Gamma}, {}_{\wedge\Gamma}\Phi)$. *is a bicovariant graded algebra based on the bicovariant bimodule* $(\Gamma, \Phi_\Gamma, {}_\Gamma\Phi)$.

Remark. When we refer to $\wedge\Gamma$ as the braided exterior algebra based on Γ, we really mean the bicovariant graded algebra $(\wedge\Gamma, \Phi_{\wedge\Gamma}, {}_{\wedge\Gamma}\Phi)$ based on the bicovariant bimodule $(\Gamma, \Phi_\Gamma, {}_\Gamma\Phi)$.

Exercise 9.13 (Optional). *The assignment*

$$(\Gamma, \Phi_\Gamma, {}_\Gamma\Phi) \mapsto (\wedge\Gamma, \Phi_{\wedge\Gamma}, {}_{\wedge\Gamma}\Phi)$$

extends to a functor from the category of bicovariant bimodules over $\mathcal{A}$ to the category of bicovariant graded algebras over $\mathcal{A}$.

9.5 Notes

The material from this chapter on the braided exterior algebra comes from Woronowicz's paper [86]. However, we have slightly changed terminology, preferring to say "exterior" instead of "external." This chapter is a sequel to the chapter on the braid groups as well as being a prelude to the first section of the next chapter on the braided exterior calculus, which will be about the braided exterior algebra together with a canonical differential.

I am not aware of who was the first researcher to introduce the braid group into noncommutative geometry. However, the paper [86] by Woronowicz must be one of the earliest, if not the first. That paper also highlights the importance of the braid group in the study of noncommutative spaces.

Chapter 10
Higher-Order Differential Calculi

10.1 Braided Exterior Calculus

In this section, we will construct a higher-order differential calculus (HODC) provided that we are given a first-order differential calculus (FODC). This construction will be a functor. What we will construct are the differentials in the braided exterior algebra that are analogous to the de Rham differentials in the classical case.

Here is the theorem we have all been waiting for ever since the definition in Chapter 2 of an FODC.

Theorem 10.1. *Let (Γ, d) be a bicovariant FODC over $\mathcal{A}$. Let $(\wedge\Gamma, \Phi_{\wedge\Gamma}, {}_{\wedge\Gamma}\Phi)$ denote the braided exterior algebra with its canonical co-actions based on $(\Gamma, \Phi_\Gamma, {}_\Gamma\Phi)$, where Φ_Γ (resp., ${}_\Gamma\Phi$) is the canonical left (resp., right) co-action of $\mathcal{A}$ on Γ.*

Then there exists a unique linear map $d : \wedge\Gamma \to \wedge\Gamma$, the differential of the braided exterior algebra with these properties:

1. *(Degree-1 differential) $d : \wedge^k\Gamma \to \wedge^{k+1}\Gamma$ for all $k \geq 0$.*
2. *(Extension) $d : \wedge^0\Gamma \to \wedge^1\Gamma$ coincides with the differential d in the FODC.*
3. *(Graded Leibniz rule) $d(\omega_1 \wedge \omega_2) = d\omega_1 \wedge \omega_2 + (-1)^{|\omega_1|}\omega_1 \wedge d\omega_2$ for homogeneous elements ω_1.*
4. *(Chain complex) $d^2 = 0$.*
5. *(Bicovariant differential)*

$$\Phi_{\wedge\Gamma} d = (id \otimes d)\, \Phi_{\wedge\Gamma}$$

$$_{\wedge\Gamma}\Phi d = (d \otimes id)\, _{\wedge\Gamma}\Phi.$$

Remark. This proof will involve something the reader might think is a very clever trick. But it is a method. The (nonmathematical) definition is that a method is a trick used at least three times. This seems to me to be a reasonable definition, but people of good faith do differ on these sorts of matters. Anyway, if the reader is

© Springer International Publishing Switzerland 2015
S.B. Sontz, *Principal Bundles*, Universitext, DOI 10.1007/978-3-319-15829-7_10

seeing this method for the first (or even second) time, it is natural enough to think of it as a trick. Be patient! In time, through more experience you will realize that it is a method. A corollary of this definition of method is that all tricks are (or soon will be) methods.

Proof. We consider $\mathcal{A}$ to be a left $\mathcal{A}$-module in the usual way. Next, we form the direct sum of left $\mathcal{A}$-modules

$$\Gamma' := \mathcal{A} \oplus \Gamma, \tag{10.1}$$

which is itself a left $\mathcal{A}$-module. We introduce the notation $X := (1,0) \in \Gamma'$. So any element $\omega' \in \Gamma'$ can be written uniquely as $\omega' = cX + \omega$, where $c \in \mathcal{A}$ and $\omega \in \Gamma$. We also have the equivalent notation $\omega' = (c, \omega) \in \mathcal{A} \oplus \Gamma$. The explicit formulas for the left action of $\mathcal{A}$ on ω' in each of these notations are

$$a\,(cX + \omega) = acX + a\omega,$$

$$a\,(c, \omega) = (ac, a\omega)$$

for all $a \in \mathcal{A}$.

We use the injection $\Gamma \to \Gamma'$ given by $\omega \mapsto (0, \omega)$ to identify Γ with a subspace of Γ'. Hereafter, we shall write $\Gamma \subset \Gamma'$ to indicate this inclusion.

Now we define for any $b \in \mathcal{A}$ and any $\omega' \in \Gamma'$

$$\omega' b := cbX + c\,db + \omega b, \tag{10.2}$$

where $\omega' = cX + \omega$ as above. In the equivalent pair notation, this reads as

$$(c, \omega)\,b = (cb, c\,db + \omega b).$$

We claim that this is a right action. To show this, we first observe that

$$(c, \omega)\,1 = (c1, c\,d1 + \omega 1) = (c, \omega)$$

since $d1 = 0$. Next, we check that

$$
\begin{aligned}
((c, \omega)\,b_1)\,b_2 &= (cb_1, c\,db_1 + \omega b_1)\,b_2 \\
&= ((cb_1)b_2, (cb_1)db_2 + (c\,db_1 + \omega b_1)\,b_2) \\
&= (c(b_1 b_2), c(b_1 db_2) + (c\,db_1)\,b_2 + (\omega b_1)\,b_2) \\
&= (c(b_1 b_2), c(b_1 db_2) + c\,(db_1\,b_2) + \omega(b_1\,b_2)) \\
&= (c(b_1 b_2), cd(b_1 b_2) + \omega(b_1\,b_2)) \\
&= (c, \omega)(b_1 b_2).
\end{aligned}
$$

Note that we have been rather careful (some might say too careful) to show all the associativity equalities using the sufficient quantity of parentheses. Anyway, the upshot is that (10.2) does define a right action.

Now that we have both a left and a right action of $\mathcal{A}$ on Γ', the natural question is whether this gives us an $\mathcal{A}$-bimodule structure on Γ'. So we take $a, b \in \mathcal{A}$ and $\omega' = (c, \omega) \in \Gamma'$, where $c \in \mathcal{A}$ and $\omega \in \Gamma$.

We then compute

$$a\,(\omega'\,b) = a(cb, c\,db + \omega b)$$
$$= (a(cb), a(c\,db) + a(\omega b))$$
$$= ((ac)b, (ac)db + (a\omega)b)$$
$$= (ac, a\omega)\,b$$
$$= (a\,(c, \omega))\,b$$
$$= (a\,\omega')\,b.$$

This shows, again with an excessive amount of parentheses, that the left and right actions of $\mathcal{A}$ on Γ' commute. So Γ' is an $\mathcal{A}$-bimodule.

Now what good does all this do for us? Well, the commutator is an essential ingredient in the theory of noncommutative structures. We consider any $a \in \mathcal{A}$ and compute its commutator with the element $(1, 0) \in \Gamma'$ using the above definitions of the left and right actions. This gives

$$(1, 0)\,a - a\,(1, 0) = (a, 1\,da + 0\,a) - (a, 0) = (0, da).$$

Writing this identity in the other notation for elements in Γ' yields

$$da = Xa - aX,$$

since $X = (1, 0)$. The "trick" is that we have represented the differential d in any arbitrary FODC as a commutator. Pretty, huh? This is known as the *extended bimodule method* in [85]. We also call Γ' with this $\mathcal{A}$-bimodule structure the *extended bimodule of* Γ.

Now we have a commutator representation for the differential $d : \mathcal{A} \to \Gamma$. But we can think of this as $d : \wedge^0 \Gamma \to \wedge^1 \Gamma$. So the idea now is to *define* $d : \wedge^k \Gamma \to \wedge^{k+1} \Gamma$ for $k \geq 1$ as a commutator with X (but using at first the slightly larger space Γ') and then show that everything works out in the subspace Γ. So now that we have the "trick" well in hand, we just have to chug through the details.

And there are details! We start by putting left and right co-actions of $\mathcal{A}$ on the extended bimodule Γ'. (This might be called an instance of the *bicovariance imperative*, as we shall see.) For $\omega' = (c, \omega)$ as above, we define

$$\Phi_{\Gamma'}(\omega') = \Phi_{\Gamma'}(c, \omega) := \phi(c)(1 \otimes X) + \Phi_{\Gamma}(\omega) \in \mathcal{A} \otimes \Gamma'.$$

In the expression $\phi(c)(1 \otimes X)$, we are using the left action of $\mathcal{A} \otimes \mathcal{A}$ on $\mathcal{A} \otimes \Gamma'$. Similarly, we define

$$_{\Gamma'}\Phi(\omega') = {}_{\Gamma'}\Phi(c, \omega) := \phi(c)(X \otimes 1) + {}_{\Gamma}\Phi(\omega) \in \Gamma' \otimes \mathcal{A}.$$

We now leave the next part of this proof as an exercise for the reader.

Exercise 10.1. *Show that $\Phi_{\Gamma'}$ (resp., ${}_{\Gamma'}\Phi$) is a left (resp., right) co-action of $\mathcal{A}$ on Γ'. Then show that $(\Gamma', \Phi_{\Gamma'}, {}_{\Gamma'}\Phi)$ is a bicovariant bimodule over $\mathcal{A}$.*

Because of this exercise, we automatically get the braided exterior algebra $(\wedge\Gamma', \Phi_{\wedge\Gamma'}, {}_{\wedge\Gamma'}\Phi)$ based on $(\Gamma', \Phi_{\Gamma'}, {}_{\Gamma'}\Phi)$. The inclusion $\Gamma \subset \Gamma'$ induces an inclusion of braided exterior algebras $\wedge\Gamma \subset \wedge\Gamma'$. Now we have enough to be able to define d, but for the moment in the braided exterior algebra $\wedge\Gamma'$. We take $\theta \in \wedge^k \Gamma'$ and define

$$d\theta := X \wedge \theta - (-1)^k \theta \wedge X. \tag{10.3}$$

We note that in $\wedge\Gamma'$ we can define a *graded commutator* for any homogeneous elements ρ, θ of degree j, k, respectively, by

$$[\rho, \theta] := \rho \wedge \theta - (-1)^{jk} \theta \wedge \rho.$$

Notice that $X \in \Gamma' = \wedge^1 \Gamma'$, so that the degree of X is 1, which we write as $\deg X = 1$. With this definition we have an alternative formula for the differential d in the braided exterior algebra $\wedge\Gamma'$, namely,

$$d\theta = [X, \theta].$$

This fulfills the wish of defining the differential d as a commutator with X. Before checking that the five conditions listed in the theorem are satisfied, we present some auxiliary material.

Next, we take $\omega \in \Gamma \subset \Gamma'$ and calculate

$$\Phi_{\Gamma'}(\omega) = \Phi_{\Gamma'}(0, \omega) = \phi(0)(1 \otimes X) + \Phi_{\Gamma}(\omega) = \Phi_{\Gamma}(\omega) \in \mathcal{A} \otimes \Gamma.$$

This tells us that the subspace Γ is invariant under the left co-action $\Phi_{\Gamma'}$ and that the restriction of $\Phi_{\Gamma'}$ to Γ is Φ_{Γ}. A similar calculation shows that Γ is invariant under the right co-action ${}_{\Gamma'}\Phi$ and that the restriction of ${}_{\Gamma'}\Phi$ to Γ is ${}_{\Gamma}\Phi$.

Now let's consider the left co-action $\Phi_{\Gamma'}$ of the element $X = (1, 0) \in \Gamma'$. This is

$$\Phi_{\Gamma'}(X) = \Phi_{\Gamma'}(1, 0) = \phi(1)(1 \otimes X) + \Phi_{\Gamma}(0) = 1 \otimes X.$$

So X is left invariant. A similar calculation shows that X is also right invariant. Let σ' denote the braiding operation of $\Gamma' \otimes \Gamma' = \Gamma'^{\otimes 2}$. Then by the definition of σ', we see that

$$\sigma'(X \otimes_A X) = X \otimes_A X.$$

Before doing the next calculation, we recall that the symmetric group on two letters S_2 has two elements, namely, the identity e, with $\mathrm{sign}(e) = 1$, and the interchange τ, with $\mathrm{sign}(\tau) = -1$. Also, $\pi_2(e) = e$ and $\pi_2(\tau) = \sigma'$, as we have seen earlier. Recall that $\pi_2 : S_2 \to B_2$ is defined in (9.4). Applying the antisymmetrization operator A_2 to $X \otimes X$, we find that

$$A_2(X \otimes_A X) = \sum_{\alpha \in S_2} \mathrm{sign}(\alpha)\pi_2(\alpha)(X \otimes_A X) = \sum_{\alpha \in \{e,\tau\}} \mathrm{sign}(\alpha)\pi_2(\alpha)(X \otimes_A X)$$

$$= \mathrm{sign}(e)\pi_2(e)(X \otimes_A X) + \mathrm{sign}(\tau)\pi_2(\tau)(X \otimes_A X)$$

$$= (X \otimes_A X) - \sigma'(X \otimes_A X)$$

$$= 0.$$

In short, this says that $X \otimes_A X \in \ker A_2$. So it follows immediately that $X \wedge X = 0$ from the definition of $\wedge$.

We now verify the five conditions listed in the theorem. Since

$$d\theta = X \wedge \theta - (-1)^k \theta \wedge X$$

and $\deg X = 1$, we immediately have that $d\theta \in \wedge^{k+1}\Gamma'$ if $\theta \in \wedge^k \Gamma'$. This is condition 1.

Taking $\theta = a \in \mathcal{A} = \wedge^0 \Gamma$ in the definition of d gives

$$da = X \wedge a - a \wedge X,$$

since $\deg a = 0$. But the wedge products with the degree-0 element a are by construction the left or right $\mathcal{A}$-actions. So we now have

$$da = Xa - aX,$$

which we have already shown is the differential in the original FODC. So condition 2 is shown.

Condition 3 follows from this explicit calculation. We start by taking elements $\omega_1 \in \wedge^k \Gamma$ and $\omega_2 \in \wedge^l \Gamma$. Then, by the definition of d, we have that

$$d\omega_1 = X \wedge \omega_1 - (-1)^k \omega_1 \wedge X,$$

$$d\omega_2 = X \wedge \omega_2 - (-1)^l \omega_2 \wedge X,$$

$$d(\omega_1 \wedge \omega_2) = X \wedge \omega_1 \wedge \omega_2 - (-1)^{k+l} \omega_1 \wedge \omega_2 \wedge X.$$

If we use the associativity of the wedge product, it then follows that

$$d\omega_1 \wedge \omega_2 + (-1)^k \omega_1 \wedge d\omega_2$$

$$= X \wedge \omega_1 \wedge \omega_2 - (-1)^k \omega_1 \wedge X \wedge \omega_2$$

$$+ (-1)^k \omega_1 \wedge X \wedge \omega_2 - (-1)^{k+l} \omega_1 \wedge \omega_2 \wedge X$$

$$= X \wedge \omega_1 \wedge \omega_2 - (-1)^{k+l} \omega_1 \wedge \omega_2 \wedge X$$

$$= d(\omega_1 \wedge \omega_2).$$

And so condition 3 is proved.

Next, using the associativity of the wedge product, we see that

$$d(d\theta) = X \wedge (d\theta) - (-1)^{k+1}(d\theta) \wedge X$$

$$= X \wedge (X \wedge \theta - (-1)^k \theta \wedge X) - (-1)^{k+1}(X \wedge \theta - (-1)^k \theta \wedge X) \wedge X$$

$$= (X \wedge X) \wedge \theta - \theta \wedge (X \wedge X)$$

$$= 0$$

since $X \wedge X = 0$. This establishes condition 4.

We will check the first statement in condition 5 and leave the other as an exercise for the reader. So

$$\Phi_{\wedge\Gamma} d\theta = \Phi_{\wedge\Gamma}\left(X \wedge \theta - (-1)^k \theta \wedge X\right)$$

$$= \Phi_{\wedge\Gamma}(X \wedge \theta) - (-1)^k \Phi_{\wedge\Gamma}(\theta \wedge X).$$

To proceed further, we take $\theta = \theta_1 \wedge \cdots \wedge \theta_k$ with each $\theta_j \in \Gamma$. Then we write $\Phi_\Gamma(\theta_j) = a_j^{(0)} \wedge \theta_j^{(0)}$, using Sweedler's notation. Also, we already have established that X is left invariant; that is, $\Phi_\Gamma(X) = 1 \otimes X$. So we have

$$\Phi_{\wedge\Gamma}(X \wedge \theta) = \left(a_1^{(0)} \cdots a_k^{(0)}\right) \otimes \left(X \wedge \theta_1^{(0)} \wedge \cdots \wedge \theta_k^{(0)}\right),$$

$$\Phi_{\wedge\Gamma}(\theta \wedge X) = \left(a_1^{(0)} \cdots a_k^{(0)}\right) \otimes \left(\theta_1^{(0)} \wedge \cdots \wedge \theta_k^{(0)} \wedge X\right),$$

$$\Phi_{\wedge\Gamma}(\theta) = \left(a_1^{(0)} \cdots a_k^{(0)}\right) \otimes \left(\theta_1^{(0)} \wedge \cdots \wedge \theta_k^{(0)}\right).$$

Continuing the above calculation, we find that

$$\Phi_{\wedge\Gamma} d\theta = \Phi_{\wedge\Gamma}(X \wedge \theta) - (-1)^k \Phi_{\wedge\Gamma}(\theta \wedge X)$$

$$= \left(a_1^{(0)} \cdots a_k^{(0)}\right) \otimes \left(X \wedge \theta_1^{(0)} \wedge \cdots \wedge \theta_k^{(0)} - (-1)^k \theta_1^{(0)} \wedge \cdots \wedge \theta_k^{(0)} \wedge X\right)$$

$$= \left(a_1^{(0)} \cdots a_k^{(0)}\right) \otimes [X, \theta_1^{(0)} \wedge \cdots \wedge \theta_k^{(0)}].$$

But, on the other hand,

$$
\begin{aligned}
(id \otimes d)\Phi_{\wedge\Gamma}(\theta) &= (id \otimes d)\big((a_1^{(0)} \cdots a_k^{(0)}) \otimes (\theta_1^{(0)} \wedge \cdots \wedge \theta_k^{(0)})\big) \\
&= (a_1^{(0)} \cdots a_k^{(0)}) \otimes d\,(\theta_1^{(0)} \wedge \cdots \wedge \theta_k^{(0)}) \\
&= (a_1^{(0)} \cdots a_k^{(0)}) \otimes [X, \theta_1^{(0)} \wedge \cdots \wedge \theta_k^{(0)}].
\end{aligned}
$$

And that proves the first identity in condition 5.

But we still have not finished the proof since we have to show that d leaves $\wedge\Gamma$ invariant and that it is the unique linear map satisfying the five conditions. So suppose $\theta \in \wedge\Gamma$. But the standard form for elements in Γ implies that we may assume with no loss of generality that

$$
\theta = a_0\, da_1 \wedge \cdots \wedge da_k,
$$

where $a_j \in \mathcal{A}$ for $j = 0, \ldots, k$. Then using the properties already shown for d, we calculate that

$$
d\theta = da_0 \wedge da_1 \wedge \cdots \wedge da_k.
$$

This formula shows that the linear map d is uniquely determined by these properties. Also, the expression on the right side clearly lies in $\wedge\Gamma$, thereby showing the invariance of $\wedge\Gamma$ under d. And now we indeed have finished the proof. ∎

Exercise 10.2 (Optional). *Show that this assignment of an HODC to any FODC is a functor.*

10.2   Universal Differential Calculus

The material in this section is taken from [21], Appendix B, where the terminology "universal differential envelope" is used. This turns out to be another functor of an FODC that produces a differential calculus. It is not the same functor as presented in the previous section though there is a relation with it, as we will see. Many of the results of [86] will be used in this section.

The point here is that an FODC does not uniquely determine an enveloping differential calculus, which is defined axiomatically, not functorially. (Does that sound familiar?) So what we have seen in the previous section is simply one possible functor for going from an FODC to a graded, differential algebra that contains (or envelops) the given FODC. In this section, we will see yet another way to do this. The latter way is a functor with a universal property. Both functors have their interest. And there are concepts in common involved in the construction of these two functors.

To start this section, we assume that $\mathcal{A}$ is an associative algebra with 1.

Definition 10.1. *Let* (Γ, d) *be an FODC over* $\mathcal{A}$. *Define the* universal differential calculus $\Gamma^\wedge$ *of the FODC to be the quotient algebra*

$$\Gamma^\wedge := \otimes^* \Gamma / \mathcal{Q}, \tag{10.4}$$

where $\otimes^* \Gamma$ *is the tensor algebra of* Γ,

$$\mathcal{Q} := \left\langle \left\{ Q = \sum_i da_i \otimes_\mathcal{A} db_i \,\middle|\, a_i, b_i \in \mathcal{A}, \ \sum_i a_i \, db_i = 0 \right\} \right\rangle, \tag{10.5}$$

and $\langle S \rangle$ *denotes the two-sided ideal generated by the set* S.

Since $\otimes^* \Gamma$ is a graded algebra and $\mathcal{Q}$ is a graded two-sided ideal, the quotient $\Gamma^\wedge$ is a graded algebra, which we write as $\Gamma^\wedge = \oplus_n \Gamma^{\wedge n}$, where $\Gamma^{\wedge n}$ is the subspace of elements of degree $n \geq 0$. We write $\deg \rho = n$ if $\rho \in \Gamma^{\wedge n}$.

Similarly, we write $\mathcal{Q} = \oplus_n \mathcal{Q}^n$. Notice that $\mathcal{Q}^0 = \mathcal{Q}^1 = 0$, so that $\Gamma^{\wedge 0} = \mathcal{A}$ and $\Gamma^{\wedge 1} = \Gamma$. In this notation, we have $d : \Gamma^{\wedge 0} \to \Gamma^{\wedge 1}$.

The multiplication of the elements $\alpha, \beta \in \Gamma^\wedge$ is written as juxtaposition: $\alpha\beta$. We certainly do not want to write $\wedge$ for this product since $\Gamma^\wedge$ is not the braided exterior algebra $\wedge\Gamma$.

Exercise 10.3. *This definition has an apparent asymmetry in the condition* $\sum a_i \, db_i = 0$ *on* $a_i, b_i \in \mathcal{A}$. *Why don't we impose the condition* $\sum da_i \, b_i = 0$ *instead? Or* $\sum db_i \, a_i = 0$? *Or* $\sum b_i \, da_i = 0$?

Theorem 10.2. *Let* (Γ, d) *be an FODC over* $\mathcal{A}$. *There exists a unique complex of linear maps*

$$\Gamma^{\wedge 0} \xrightarrow{d_0} \Gamma^{\wedge 1} \xrightarrow{d_1} \Gamma^{\wedge 2} \xrightarrow{d_2} \cdots \xrightarrow{d_{n-1}} \Gamma^{\wedge n} \xrightarrow{d_n} \cdots$$

such that the following hold:

1. *Enveloping property:* $d_0 = d : \mathcal{A} \to \Gamma$.
2. *Chain complex:* $d_n d_{n-1} = 0$ *for all* $n \geq 1$.
3. *Graded Leibniz rule:* $d(\rho\sigma) = d(\rho)\,\sigma + (-1)^k \rho \, d\sigma$ *for* $\rho \in \Gamma^{\wedge k}$, *which we write as* $k = \deg \rho$.

Remark. One says that the complex envelops the FODC (Γ, d). In the third property, we are already using the standard convention of dropping the subscript from the notation d_n. So the second property is usually written as $d^2 = 0$. However, we will still use the notation d_0 once in the following proof.

Proof. First, we note that we simply define $d_0 := d$ in order to have the enveloping property.

Take any $\theta \in \Gamma = \Gamma^{\wedge 1}$ and write it in standard form $\theta = \sum_i a_i \, db_i$ with $a_i, b_i \in \mathcal{A}$. Then define

$$d\theta := \sum_i da_i \, db_i \in \Gamma^{\wedge 2}.$$

We must show that this definition does not depend on the choice of standard form of θ. Thus, suppose that $\sum_i a_i db_i = 0$. It suffices to show $\sum_i da_i db_i = 0$. But by the definition of $\mathcal{Q}$, we have $\sum_i da_i \otimes_{\mathcal{A}} db_i \in \mathcal{Q}$. Passing to the quotient yields $\sum_i da_i db_i = 0$ in $\Gamma^{\wedge 2}$.

Let $a \in \mathcal{A}$. Then $d(da) = d(1da) = dl \, da = 0$, since $d1 = 0$. So this verifies $d^2 = 0$ for the only case where it has been defined so far.

Also, let $\theta = \sum_i a_i db_i \in \Gamma$. Then we calculate

$$d(a \, \theta) = d\left(a \sum_i a_i \, db_i\right)$$

$$= d\left(\sum_i a \, a_i \, db_i\right)$$

$$= \sum_i d(a \, a_i) \, db_i$$

$$= \sum_i \left((da) \, a_i \, db_i + a \, (da_i) \, db_i\right)$$

$$= (da) \, \theta + a \, d\theta. \tag{10.6}$$

This is a special case of the graded Leibniz rule.

Exercise 10.4. *Verify this other special case of the graded Leibniz rule:*

$$d(\theta \, a) = (d\theta) \, a - \theta \, da$$

for $u \in \mathcal{A}$ and $\theta \in \Gamma$.

The graded Leibniz rule for $a, b \in \mathcal{A}$ says $d(ab) = (da) \, b + a \, db$. We already know that this is true since it is part of the definition of an FODC. At this point we have verified the graded Leibniz rule for all the cases for which we have defined d.

Let $t = \theta_1 \otimes_{\mathcal{A}} \cdots \otimes_{\mathcal{A}} \theta_j \otimes_{\mathcal{A}} \cdots \otimes_{\mathcal{A}} \theta_n \in \otimes^n \Gamma$, where $\theta_j \in \Gamma$ for $1 \leq j \leq n$ and $n \geq 1$. Then using the above definition of $d\theta_j$, we define

$$dt := \sum_{j=1}^{n} (-1)^{j-1} [\theta_1]_{\mathcal{Q}} \cdots [\theta_{j-1}]_{\mathcal{Q}} \, d\theta_j \, [\theta_{j+1}]_{\mathcal{Q}} \cdots [\theta_n]_{\mathcal{Q}} \in \Gamma^{\wedge(n+1)}, \tag{10.7}$$

where $[\cdot]_{\mathcal{Q}}$ denotes the quotient map $\otimes^* \Gamma \to \Gamma^{\wedge}$. Of course, $[\theta_j]_{\mathcal{Q}} = \theta_j$, so that the notation $[\cdot]_{\mathcal{Q}}$ is not really needed for now. Note that for $n = 1$, this definition is consistent with the previous definition. For $n = 0$, we take $d : \otimes^0 \Gamma = \mathcal{A} \to \Gamma^{\wedge 1} = \Gamma$ to be the differential of the FODC (Γ, d).

Also, take $u = \eta_1 \otimes_A \cdots \otimes_A \eta_k \otimes_A \cdots \otimes_A \eta_m \in \otimes^m \Gamma$, where $\eta_k \in \Gamma$ for $1 \le k \le m$ and $m \ge 1$. Using definitions and the fact that $[\cdot]_Q : \otimes^* \Gamma \to \Gamma^\wedge$ is an algebra morphism, we obtain

$$d(t \otimes_A u) = d(\theta_1 \otimes_A \cdots \otimes_A \theta_n \otimes_A \eta_1 \otimes_A \cdots \otimes_A \eta_m)$$

$$= \sum_{j=1}^{n} (-1)^{j-1} [\theta_1]_Q \cdots d\theta_j \cdots [\theta_n]_Q [\eta_1]_Q \cdots [\eta_m]_Q$$

$$+ (-1)^n [\theta_1]_Q \cdots [\theta_n]_Q \sum_{k=1}^{m} (-1)^{k-1} [\eta_1]_Q \cdots d\eta_k \cdots [\eta_m]_Q$$

$$= dt \, [\eta_1]_Q \cdots [\eta_m]_Q + (-1)^n [\theta_1]_Q \cdots [\theta_n]_Q \, du$$

$$= dt \, [\eta_1 \otimes_A \cdots \otimes_A \eta_m]_Q + (-1)^n [\theta_1 \otimes_A \cdots \otimes_A \theta_n]_Q \, du$$

$$= dt \, [u] + (-1)^n [t] \, du.$$

This is itself the graded Leibniz rule. But what about the three cases that this calculation does not cover? Those cases are (i) $n = m = 0$, (ii) $n = 0$ and $m \ge 1$, and (iii) $n \ge 1$ and $m = 0$. But these are *related* to the three special cases for which we have already shown the graded Leibniz rule. Case (i) was considered above, and the graded Leibniz rule holds. So let's consider case (ii) here. We will use the graded Leibniz rule established above for the case $n = 0$ and $m = 1$ given in (10.6).

So we have that

$$d(au) = d(a\eta_1 \otimes_A \cdots \otimes_A \eta_m)$$

$$= d(a\eta_1)\eta_2 \cdots \eta_m + \sum_{k=2}^{m} (-1)^{k-1} a\eta_1 \cdots d\eta_k \cdots \eta_m$$

$$= (da \, \eta_1 + a \, d\eta_1)\eta_2 \cdots \eta_m + \sum_{k=2}^{m} (-1)^{k-1} a\eta_1 \cdots d\eta_k \cdots \eta_m$$

$$= (da)(\eta_1\eta_2 \cdots \eta_m) + a \sum_{k=1}^{m} (-1)^{k-1} \eta_1 \cdots d\eta_k \cdots \eta_m$$

$$= (da) [u]_Q + [a]_Q (du).$$

Exercise 10.5. *Show that the graded Leibniz rule holds in case (iii).*

So we have shown that the maps $d : \otimes^n \Gamma \to \Gamma^{\wedge(n+1)}$ for $n \ge 0$ satisfy the graded Leibniz rule.

Now we want to factor this through the quotient map $\otimes^n \Gamma \to \Gamma^{\wedge n}$. This can be done if and only if $Q \subset \ker d$. So take a generator $Q = \sum_i da_i \otimes_A db_i$ of Q and calculate using the graded Leibniz rule:

$$d(Q \otimes_A \rho) = d\Big(\sum_i da_i \otimes_A db_i \otimes_A \rho\Big)$$

$$= \sum_i d^2 a_i \, db_i \, [\rho] - \sum_i da_i \, d^2 b_i \, [\rho] + \sum_i da_i \, db_i \, d\rho$$

$$= \Big(\sum_i da_i \, db_i\Big) d\rho$$

since $d^2 a = 0$ for $a \in A$. But since Q is a generator of Q, we have that $\sum_i a_i db_i = 0$, which in turn implies by applying d that $\sum_i da_i \, db_i = 0$. So $Q \otimes_A \rho \in \ker d$. Similarly, the reader can show that $\rho \otimes_A Q \in \ker d$. Therefore, the two-sided ideal generated by the Qs is contained in $\ker d$, that is, $Q \subset \ker d$.

We now factor the maps $d : \otimes^n \Gamma \to \Gamma^{\wedge(n+1)}$ through the quotient to get maps, again denoted by d,

$$d : \Gamma^{\wedge n} \to \Gamma^{\wedge(n+1)}$$

for all $n \geq 0$. Properties 1 and 3 are immediate consequences of what we have already proved. Property 2 is a standard exercise.

This shows the existence part of the theorem. We stop here and let the reader continue. ∎

Exercise 10.6. *Prove the uniqueness part of this theorem by showing that the three properties of d determine the formulas for d.*

Exercise 10.7. *Do the standard exercise to prove property 2. **Hint:** Take the formula for the definition of dt given in (10.7) and then apply d to it using (what else?) Eq. (10.7) for the case of elements of degree $n + 1$.*

This exercise turns out to be a well-known result in homological algebra.

We now discuss various universal properties of $(\Gamma^{\wedge}, d)$. Here is the first.

Theorem 10.3. *Suppose (Ω, d_Ω) is a differential algebra and (Γ, d) is an FODC over A.*

Let $\varphi^0 : A = \Gamma^{\wedge 0} \to \Omega$ be an algebra morphism and $\varphi^1 : \Gamma = \Gamma^{\wedge 1} \to \Omega$ be a linear map, which are compatible in the sense that

$$\varphi^1(a \, db) = \varphi^0(a)\big(d_\Omega \varphi^0(b)\big) \tag{10.8}$$

for all $a, b \in A$.

Then there exist unique linear maps $\varphi^n : \Gamma^{\wedge n} \to \Omega$ for every integer $n \geq 2$ such that $\varphi := \oplus_{n\geq 0} \varphi^n : \oplus_{n\geq 0} \Gamma^{\wedge n} \to \Omega$ is a morphism of differential algebras.

Remark. To understand the compatibility condition (10.8) a little better, we use φ^0 to define a left A-module structure on Ω by

$$a\omega := \varphi^0(a)\,\omega,$$

where the right side is the product of two elements in Ω. Let's consider these two conditions on φ^0 and φ^1:

1. φ^1 is a left $\mathcal{A}$-module morphism; that is, $\varphi^1(a\rho) = \varphi^0(a)\varphi^1(\rho)$ for $a \in \mathcal{A}$ and $\rho \in \Gamma$.
2. The pair φ^0, φ_1 intertwines differentials; that is, $\varphi^1 d = d_\Omega \varphi^0$.

Then these conditions are sufficient for (10.8) to hold, as this computation shows:

$$\varphi^1(a\, db) = \varphi^0(a)\varphi^1(db) = \varphi^0(a)\big(d_\Omega\varphi^0(b)\big).$$

Now suppose that Ω has a unit element 1 and that $\varphi^0(1) = 1$. Then these two conditions are also necessary for (10.8) to hold. To show this, we assume (10.8) as a hypothesis and put $a = 1$ in it to get

$$\varphi^1(db) = \varphi^1(1\, db) = \varphi^0(1)\big(d_\Omega\varphi^0(b)\big) = d_\Omega\varphi^0(b)$$

for all $b \in \mathcal{A}$. This proves condition 2. Next, we take $a, b, c \in \mathcal{A}$ and put $\rho := b\, dc \in \Gamma$. Then using (10.8) twice, we calculate

$$\begin{aligned}
\varphi^1(a\rho) &= \varphi^1(ab\, dc) \\
&= \varphi^0(ab)d_\Omega\varphi^0(c) \\
&= \varphi^0(a)\varphi^0(b)d_\Omega\varphi^0(c) \\
&= \varphi^0(a)\varphi^1(b\, dc)) \\
&= \varphi^0(a)\varphi^1(\rho),
\end{aligned}$$

which suffices for showing condition 1, since any element in Γ can be written in standard form.

Proof. First, we define $\varphi^{\otimes 0}(a) := \varphi^0(a)$ for all $a \in \mathcal{A}$. Next, we define $\varphi^{\otimes 1}(\theta) := \varphi^1(\theta)$ for all $\theta \in \Gamma$. Finally, for each integer $n \geq 2$, we define $\varphi^{\otimes n} : \otimes^n \Gamma \to \Omega$ by

$$\varphi^{\otimes n}(t) := \varphi^{\otimes 1}(\theta_1) \cdots \varphi^{\otimes 1}(\theta_n),$$

where $t = \theta_1 \otimes_{\mathcal{A}} \cdots \otimes_{\mathcal{A}} \theta_n \in \otimes^n \Gamma$ with $\theta_j \in \Gamma$ for $1 \leq j \leq n$ and $n \geq 2$. Put $\varphi^\otimes = \oplus_{n \geq 0}\varphi^{\otimes n} : \oplus_{n \geq 0} \otimes^n \Gamma \to \Omega$.

Also, let $u = \eta_1 \otimes_{\mathcal{A}} \cdots \otimes_{\mathcal{A}} \eta_m \in \otimes^m \Gamma$, where $\eta_k \in \Gamma$ for $1 \leq k \leq m$ and $m \geq 1$. Then

$$\begin{aligned}
\varphi^\otimes(t)\,\varphi^\otimes(u) &= \varphi^{\otimes n}(\theta_1 \otimes_{\mathcal{A}} \cdots \otimes_{\mathcal{A}} \theta_n)\varphi^{\otimes m}(\eta_1 \otimes_{\mathcal{A}} \cdots \otimes_{\mathcal{A}} \eta_m) \\
&= \varphi^{\otimes 1}(\theta_1) \cdots \varphi^{\otimes 1}(\theta_n)\varphi^{\otimes 1}(\eta_1) \cdots \varphi^{\otimes 1}(\eta_m) \\
&= \varphi^{\otimes(n+m)}(t \otimes_{\mathcal{A}} u) \\
&= \varphi^\otimes(t \otimes_{\mathcal{A}} u),
\end{aligned}$$

which shows that $\varphi^\otimes$ is multiplicative in all but three cases, the same three cases as before! For $n = m = 0$, the multiplicativity of $\varphi^\otimes$ comes down to that of φ^0. For the case $n = 0$ and $m \geq 1$, we compute as follows for $\eta_1 = b\,dc$ and $a, b, c \in \mathcal{A}$:

$$
\begin{aligned}
\varphi^\otimes(au) &= \varphi^{\otimes m}(a\eta_1 \otimes_A \cdots \otimes_A \eta_m) \\
&= \varphi^{\otimes 1}(a\eta_1)\varphi^{\otimes 1}(\eta_2)\cdots\varphi^{\otimes 1}(\eta_m) \\
&= \varphi^{\otimes 1}(ab\,dc)\varphi^{\otimes 1}(\eta_2)\cdots\varphi^{\otimes 1}(\eta_m) \\
&= \varphi^{\otimes 0}(ab)\,d_\Omega\varphi^{\otimes 0}(c)\,\varphi^{\otimes 1}(\eta_2)\cdots\varphi^{\otimes 1}(\eta_m) \\
&= \varphi^{\otimes 0}(a)\,\varphi^{\otimes 0}(b)\,d_\Omega\varphi^{\otimes 0}(c)\,\varphi^{\otimes 1}(\eta_2)\cdots\varphi^{\otimes 1}(\eta_m) \\
&= \varphi^{\otimes 0}(a)\,\varphi^{\otimes 1}(b\,dc)\,\varphi^{\otimes 1}(\eta_2)\cdots\varphi^{\otimes 1}(\eta_m) \\
&= \varphi^{\otimes 0}(a)\,\varphi^{\otimes 1}(\eta_1)\,\varphi^{\otimes 1}(\eta_2)\cdots\varphi^{\otimes 1}(\eta_m) \\
&= \varphi^{\otimes 0}(a)\,\varphi^{\otimes m}(u) \\
&= \varphi^\otimes(a)\,\varphi^\otimes(u),
\end{aligned}
$$

where the compatibility condition between φ^0 and φ^1 was used in the fourth and sixth equalities.

Exercise 10.8. *Verify that $\varphi^\otimes(ta) = \varphi^\otimes(t)\varphi^\otimes(a)$ for $\deg t = n \geq 1$ and $a \in \mathcal{A}$.*

To show that the map $\varphi^\otimes$ passes to the quotient space $\Gamma^\wedge$, we will show that $\mathcal{Q} \subset \ker\varphi^\otimes$. So let $Q = \sum_i da_i \otimes_A db_i$ with $\sum_i a_i\,db_i = 0$ be a typical generator of $\mathcal{Q}$. Then we obtain

$$
\begin{aligned}
\varphi^\otimes(Q) &= \sum_i \varphi^{\otimes 2}\big(da_i \otimes_A db_i\big) \\
&= \sum_i \varphi^{\otimes 1}(da_i)\varphi^{\otimes 1}(db_i) \\
&= \sum_i \varphi^{\otimes 1}(1\,da_i)\varphi^{\otimes 1}(1\,db_i) \\
&= \sum_i \varphi^{\otimes 0}(1)\,d_\Omega\varphi^{\otimes 0}(a_i)\,\varphi^{\otimes 0}(1)\,d_\Omega\varphi^{\otimes 0}(b_i) \\
&= \sum_i d_\Omega\varphi^{\otimes 0}(a_i)\,d_\Omega\varphi^{\otimes 0}(b_i) \\
&= \sum_i d_\Omega\big(\varphi^{\otimes 0}(a_i)\,d_\Omega\varphi^{\otimes 0}(b_i)\big) \\
&= d_\Omega\Big(\sum_i \varphi^{\otimes 0}(a_i)\,d_\Omega\varphi^{\otimes 0}(b_i)\Big)
\end{aligned}
$$

$$= d_\Omega \left(\sum_i \varphi^{\otimes 1}(a_i \, db_i) \right)$$

$$= d_\Omega \varphi^{\otimes 1} \left(\sum_i a_i \, db_i \right)$$

$$= 0.$$

So $\mathcal{Q} \in \ker \varphi^\otimes$ indeed holds, which in turn implies $\mathcal{Q} \subset \ker \varphi^\otimes$. Thus, the map $\varphi^\otimes$ passes to the quotient space $\Gamma^\wedge$ and is denoted by φ. Since it is the quotient of an algebra morphism, it is itself an algebra morphism.

To finish this proof, one must show that $\varphi = \oplus_{n \geq 0} \varphi^n$ is a morphism of differential algebras (so it remains to show it commutes with the differentials) and that it is unique. ∎

Exercise 10.9. *Finish the proof as indicated above. That the maps φ^n are unique follows from formulas for them. These formulas (which the reader must find) are consequences of the properties of these maps.*

Corollary 10.1. *Let $\wedge \Gamma$ be the braided exterior algebra associated to the FODC (Γ, d). Then there is a unique morphism of differential algebras $\phi : \Gamma^\wedge \to \wedge \Gamma$, which is the identity in degrees 0 and 1. Moreover, ϕ is an epimorphism; that is, it is surjective.*

Remark. In general, ϕ is not an isomorphism. Also, in general, $\wedge \Gamma$ is not given as the quotient of the tensor algebra by a two-sided ideal generated by homogeneous quadratic elements. Recall that $\Gamma^\wedge$ does have this property.

Here is a second universal property of $\Gamma^\wedge$. But we will use the notation $\Gamma^\flat := \Gamma^\wedge$ for this result. Just in case, we recall that a vector space morphism is nothing other than a linear map.

Theorem 10.4. *Suppose (Ω, d_Ω) is a differential algebra and (Γ, d) is an FODC over $\mathcal{A}$.*

Let $\varphi^0 : \mathcal{A} = \Gamma^{\flat 0} \to \Omega$ be an antimultiplicative *vector space morphism and $\varphi^1 : \Gamma = \Gamma^{\flat 1} \to \Omega$ be a linear map, which are compatible in the sense that*

$$\varphi^1(a \, db) = \big(d_\Omega \varphi^0(b) \big) \varphi^0(a)$$

for all $a, b \in \mathcal{A}$.

Then there exist unique linear maps $\varphi^{\flat n} : \Gamma^{\flat n} \to \Omega$ for every integer $n \geq 2$ such that $\varphi^\flat := \oplus_{n \geq 0} \varphi^{\flat n} : \oplus_{n \geq 0} \Gamma^{\flat n} \to \Omega$ is a vector space morphism, where we put $\varphi^{\flat i} := \varphi^i$ for $i = 0, 1$, such that these properties hold:

- *Intertwining differentials: $\varphi^\flat d = d_\Omega \varphi^\flat$.*
- *Graded antimultiplicative: $\varphi^\flat(\sigma \theta) = (-1)^{jk}(\varphi^\flat(\theta)\varphi^\flat(\sigma))$ for all homogeneous elements $\sigma, \theta \in \Gamma^\flat$ with $\deg \sigma = j$ and $\deg \theta = k$.*

Proof. The proof follows an outline quite parallel to the proof of the previous theorem. And to keep notation locally under control, we will use the same notation here for some things that are actually *different*. Here is such a reuse (but not abuse!) of notation. We again let $t = \theta_1 \otimes_A \cdots \otimes_A \theta_n \in \otimes^n \Gamma$, where $\theta_j \in \Gamma$ for $1 \leq j \leq n$ and $n \geq 1$. We define

$$\varphi^{\otimes n}(t) := \text{sign}(\zeta_n)\, \varphi^1(\theta_n) \cdots \varphi^1(\theta_1),$$

where ζ_n is the permutation of the set $\{1, 2, \ldots, n\}$ that reverses order (that is, $1 \mapsto n, 2 \mapsto n-1, \ldots, n-1 \mapsto 2, n \mapsto 1$) and sign is the sign of a permutation. It turns out that $\text{sign}(\zeta_n) = (-1)^{n(n-1)/2}$. We also define $\varphi^{\otimes 0}(a) = \varphi^0(a)$ for all $a \in A$.

And again we let $u = \eta_1 \otimes_A \cdots \otimes_A \eta_m \in \otimes^m \Gamma$, where $\eta_k \in \Gamma$ for $1 \leq k \leq m$ and $m \geq 1$. Now here comes the key computation that distinguishes this result from the previous one. We get

$$\varphi^{\otimes n}(t)\varphi^{\otimes m}(u) = \text{sign}(\zeta_n)\, \text{sign}(\zeta_m)\, \varphi^1(\theta_n) \cdots \varphi^1(\theta_1)\varphi^1(\eta_m) \cdots \varphi^1(\eta_1)$$

$$= \text{sign}(\zeta_n)\, \text{sign}(\zeta_m)\, \text{sign}(\zeta_{n+m}) \cdot$$

$$\varphi^{\otimes(n+m)}(\varphi^1(\eta_1) \otimes_A \cdots \otimes_A \varphi^1(\eta_m) \otimes_A \varphi^1(\theta_1) \otimes_A \cdots \otimes_A \varphi^1(\theta_n))$$

$$= \text{sign}(\zeta_n)\, \text{sign}(\zeta_m)\, \text{sign}(\zeta_{n+m})\, \varphi^{\otimes(n+m)}(u \otimes_A t).$$

So now it is a question of calculating a sign, and this comes down to finding the parity of

$$\frac{1}{2}\big(n(n-1) + m(m-1) + (n+m)(n+m-1)\big)$$

$$= n^2 - n + m^2 - m + nm$$

$$= n(n-1) + m(m-1) + nm.$$

But the factors $n(n-1)$ and $m(m-1)$ are even, and so the parity is determined by nm. In short,

$$\text{sign}(\zeta_n)\, \text{sign}(\zeta_m)\, \text{sign}(\zeta_{n+m}) = (-1)^{nm},$$

so that we obtain

$$\varphi^{\otimes(n+m)}(u \otimes_A t) = (-1)^{nm}\varphi^{\otimes n}(t)\varphi^{\otimes m}(u).$$

And this is the graded antimultiplicative property for the case $n \geq 1$ and $m \geq 1$. Three cases remain! And those cases will be shown by using the hypotheses on φ^0 and φ^1.

Exercise 10.10. *Do those three cases in detail.*

Next, a slight variant of the previous argument shows that $\varphi^{\otimes n}$ passes to the quotient space to give a map there, which we denote φ^{bn}. It clearly is graded antimultiplicative.

We still need to prove the intertwining formula for the differentials. ∎

Exercise 10.11. *Prove by a direct calculation the intertwining formula for the differentials.*

Now that we have two universal properties under the belt, a couple more seems like no big deal. The claim is that each of the previous two universal properties has a partner where the map φ^0 is antilinear. Do you see that?

Exercise 10.12. *Keep working on these two new universal properties until they become clear.*

Theorem 10.5. *Let (Γ, d) be a left covariant FODC over a Hopf algebra $\mathcal{A}$. Denote the left co-action of $\mathcal{A}$ on Γ by*

$$\Phi_\Gamma : \Gamma \to \mathcal{A} \otimes \Gamma.$$

Then there exists a unique algebra morphism and left co-action of $\mathcal{A}$ on $\Gamma^\wedge$, which we denote by

$$\Phi_{\Gamma^\wedge} : \Gamma^\wedge \to \mathcal{A} \otimes \Gamma^\wedge,$$

such that $\Phi_{\Gamma^{\wedge 0}} = \phi$ and $\Phi_{\Gamma^{\wedge 1}} = \Phi_\Gamma$, and $\Phi_{\Gamma^\wedge}$ intertwines the differential in the sense that $\Phi_{\Gamma^\wedge} d = (id \otimes d) \Phi_{\Gamma^\wedge}$.

Exercise 10.13. *Prove this theorem. In part, this is an application of one of the universal properties. Optionally, also state the corresponding theorem for right covariant FODCs.*

Exercise 10.14. *If (Γ, d) is a bicovariant FODC, then by the above results we know that $(\Gamma^\wedge, d)$ is both a left and right covariant FODC. Prove that it is also bicovariant, that is, that the left and right co-actions commute.*

Here is yet another result that will be using later on.

Theorem 10.6. *The co-multiplication map $\phi : \mathcal{A} \to \mathcal{A} \otimes \mathcal{A}$ has a unique extension*

$$\hat{\phi} : \Gamma^\wedge \to \Gamma^\wedge \otimes \Gamma^\wedge,$$

which is a morphism of graded differential algebras and a morphism of left $\mathcal{A}$-modules, where the left action of $a \in \mathcal{A}$ on $\Gamma^\wedge \otimes \Gamma^\wedge$ is defined to be the usual action on the left by $\phi(a)$.

Also, for all $\theta \in \Gamma$, we have that

$$\hat{\phi}(\theta) = \Phi_\Gamma(\theta) + {}_\Gamma\Phi(\theta). \tag{10.9}$$

(This identity implies that $\hat{\phi}$ is neither a right co-action nor a left co-action. Of course, ϕ is both a right and left co-action.) Also, ${}_{\mathrm{inv}}\Gamma^\wedge$ is invariant under $\hat{\phi}$ in the sense that

$$\hat{\phi}\big({}_{\mathrm{inv}}\Gamma^\wedge\big) \subset {}_{\mathrm{inv}}\Gamma^\wedge \otimes \Gamma^\wedge. \tag{10.10}$$

In particular, for all $\theta \in {}_{\mathrm{inv}}\Gamma$, we have

$$\hat{\phi}(\theta) = 1_{\mathcal{A}} \otimes \theta + \mathrm{ad}(\theta). \tag{10.11}$$

Here $\mathrm{ad} : {}_{\mathrm{inv}}\Gamma \to {}_{\mathrm{inv}}\Gamma \otimes \mathcal{A}$ *also denotes the restriction of* $\mathrm{ad} : \Gamma \to \Gamma \otimes \mathcal{A}$ *(the right adjoint co-action of $\mathcal{A}$ on Γ) to ${}_{\mathrm{inv}}\Gamma$.*

Remark. We will get to the proof in a moment, but first it may not be clear to the reader what the differential structure is on $\Gamma^\wedge \otimes \Gamma^\wedge$. In general, the graded tensor product $D_1 \otimes D_2$ of two graded differential algebras D_1 and D_2 with respective differentials d_1 and d_2 (each raising degree by 1) has the differential d (also raising degree by 1) defined on a decomposable element $\theta_1 \otimes \theta_2$ by

$$d(\theta_1 \otimes \theta_2) := d_1\theta_1 \otimes \theta_2 + (-1)^{|\theta_1|}\theta_1 \otimes d_2\theta_2, \tag{10.12}$$

where $\theta_1 \in D_1$ is a homogeneous element of degree $|\theta_1|$ and $\theta_2 \in D_2$.

Exercise 10.15. *Establish these elementary properties of Definition (10.12):*

1. *Suppose that both d_1 and d_2 satisfy the graded Leibniz rule. Prove that the map d satisfies the graded Leibniz rule.*
2. *Also, show that d is a differential, that is, $d^2 = 0$.*
3. *Prove that $d(d_1\eta_1 \otimes d_2\eta_2) = 0$ for all $\eta_1 \in D_1$ and $\eta_2 \in D_2$.*

Proof of Theorem 10.6: We will first assume that $\hat{\phi}$ exists and has all the desired properties. Then we will prove (10.9). So we take $\theta = a\, db$, where $a, b \in \mathcal{A}$. Since any element in Γ is a finite sum of such terms and (10.9) is linear in θ, it suffices to consider just this special case. Then we have that

$$\hat{\phi}(\theta) = \hat{\phi}(a\, db)$$
$$= \hat{\phi}(a)\,\hat{\phi}(db)$$
$$= \phi(a)\,(d \otimes id + id \otimes d)\phi(b).$$

In the third equality, we used that $\hat{\phi}$ is equal to ϕ in degree 0 and that $d \otimes id + id \otimes d$ is the differential for

$$(\Gamma^\wedge \otimes \Gamma^\wedge)^0 = \Gamma^{\wedge 0} \otimes \Gamma^{\wedge 0}$$

according to the previous remark. On the other hand, we already have a formula for the left co-action, namely,

$$\Phi_\Gamma(a\, db) = \phi(a)(id \otimes d)\phi(b),$$

while the right co-action is given similarly by

$$_\Gamma\Phi(a\, db) = \phi(a)(d \otimes id)\phi(b).$$

So these identities immediately imply that

$$\hat{\phi}(\theta) = \Phi_\Gamma(a\, db) + {}_\Gamma\Phi(a\, db) = \Phi_\Gamma(\theta) + {}_\Gamma\Phi(\theta),$$

which is Eq. (10.9), as desired.

The first calculation in the previous paragraph forces the definition of $\hat{\phi}$ if we require it to be a multiplicative, differential morphism. Namely, for $\theta = a_0\, da_1 \cdots da_n$ with $a_0, a_1, \ldots, a_n \in \mathcal{A}$, we have to define $\hat{\phi}$ by

$$\hat{\phi}(\theta) := \phi(a_0)\,(d \otimes id + id \otimes d)\phi(a_1)\cdots(d \otimes id + id \otimes d)\phi(a_n). \qquad (10.13)$$

This shows the uniqueness statement. Also, then the argument in the previous paragraph quickly converts into a rigorous argument! But we now have to prove that $\hat{\phi}$ so defined is a morphism of graded differential algebras.

Taking $\theta = a_0\, da_1 \cdots da_n$ as above, we have on the one hand that

$$\hat{\phi}(d^\wedge\theta) = \hat{\phi}\big(d^\wedge(a_0\, da_1 \cdots da_n)\big)$$

$$= \hat{\phi}(da_0\, da_1 \cdots da_n)$$

$$= ((d \otimes id + id \otimes d)\phi(a_0)\,(d \otimes id + id \otimes d)\phi(a_1)\cdots(d \otimes id + id \otimes d)\phi(a_n)).$$

On the other hand,

$$d\hat{\phi}(\theta) = d\big(\phi(a_0)\,(d \otimes id + id \otimes d)\phi(a_1)\cdots(d \otimes id + id \otimes d)\phi(a_n)\big)$$

$$= ((d \otimes id + id \otimes d)\phi(a_0)\,(d \otimes id + id \otimes d)\phi(a_1)\cdots(d \otimes id + id \otimes d)\phi(a_n)),$$

which shows that $\hat{\phi}$ is a differential morphism.

Taking $\theta = a_0\, da_1 \cdots da_n$ and $\eta = b_0\, db_1 \cdots db_m$ with the $a's$ and $b's$ in $\mathcal{A}$, we see that

$$\hat{\phi}(\theta\,\eta) = \hat{\phi}(a_0\,da_1\cdots da_n\,b_0\,db_1\cdots db_m)$$

$$= \hat{\phi}(a_0\,b_0\,da_1\cdots da_n\,db_1\cdots db_m)$$

$$= \phi(a_0\,b_0)\,(d\otimes id + id\otimes d)\phi(a_1)\cdots(d\otimes id + id\otimes d)\phi(b_m)$$

$$= \phi(a_0)\phi(b_0)\,(d\otimes id + id\otimes d)\phi(a_1)\cdots(d\otimes id + id\otimes d)\phi(b_m)$$

$$= \phi(a_0)\,(d\otimes id + id\otimes d)\phi(a_1)\cdots(d\otimes id + id\otimes d)\phi(a_n)$$

$$\cdot\,\phi(b_0)\,(d\otimes id + id\otimes d)\phi(b_1)\cdots(d\otimes id + id\otimes d)\phi(b_m)$$

$$= \hat{\phi}(\theta)\hat{\phi}(\eta).$$

And this shows that $\hat{\phi}$ is a multiplicative morphism.

To show that $\hat{\phi}$ is a left $\mathcal{A}$-morphism, take an element $c \in \mathcal{A}$ and continue with the notation $\theta = a_0\,da_1\cdots da_n$ given above. Then we note that

$$\hat{\phi}(c\theta) = \hat{\phi}(ca_0\,da_1\cdots da_n)$$

$$= \phi(ca_0)\,(d\otimes id + id\otimes d)\phi(a_1)\cdots(d\otimes id + id\otimes d)\phi(a_n)$$

$$= \phi(c)\,\phi(a_0)\,(d\otimes id + id\otimes d)\phi(a_1)\cdots(d\otimes id + id\otimes d)\phi(a_n)$$

$$= \phi(c)\hat{\phi}(\theta),$$

which proves that $\hat{\phi}$ is a left $\mathcal{A}$-morphism.

Finally, we take $\theta \in {}_{inv}\Gamma$. Then we have that

$$\Phi_\Gamma(\theta) = 1_{\mathcal{A}} \otimes \theta$$

since θ is left invariant. On the other hand, the right co-action on θ is

$$_\Gamma\Phi(\theta) = \mathrm{ad}(\theta),$$

where before we merely noted that the two sides of this equation corresponded to each other under some isomorphism. Now we use that isomorphism to identify these as the same element. Substituting these two identities into (10.9) yields

$$\hat{\phi}(\theta) = \Phi_\Gamma(\theta) + {}_\Gamma\Phi(\theta) = 1_{\mathcal{A}} \otimes \theta + \mathrm{ad}(\theta),$$

which proves (10.11). But each of the two terms on the right side of (10.11) lies in ${}_{inv}\Gamma^\wedge \otimes \Gamma^\wedge$. So this also establishes the inclusion (10.10). ∎

10.3   The Maurer–Cartan Formula

We now prove an important identity in the universal calculus of the previous section. We mentioned this identity earlier. It is called the *Maurer–Cartan formula* for this noncommutative setting. We will explain this terminology after the statement and proof of the result.

Theorem 10.7 (Maurer–Cartan formula). *For all $a \in \mathcal{A}$, we have*

$$d(\pi(a)) = -\pi(a^{(1)})\,\pi(a^{(2)}). \tag{10.14}$$

Proof. Recall that the definition of the quantum germs map $\pi : \mathcal{A} \to {}_{\text{inv}}\Gamma$ is given by $\pi(a) = \kappa(a^{(1)})d(a^{(2)})$. Then we immediately have

$$\pi(a^{(1)})\,\pi(a^{(2)}) = \kappa(a^{(11)})d(a^{(12)})\kappa(a^{(21)})d(a^{(22)}). \tag{10.15}$$

Then, using Leibniz's rule, we can write the middle two terms here as

$$d(a^{(12)})\kappa(a^{(21)}) = d\left((a^{(12)})\kappa(a^{(21)})\right) - a^{(12)}d\kappa(a^{(21)}).$$

Now it might not be obvious how to evaluate $d\left((a^{(12)})\kappa(a^{(21)})\right)$, and, frankly speaking, this is a disadvantage of the iterated Sweedler's notation being used here. However, by co-associativity, we note that

$$a^{(11)} \otimes a^{(12)} \otimes a^{(21)} \otimes a^{(22)} = a^{(1)} \otimes a^{(211)} \otimes a^{(212)} \otimes a^{(22)}.$$

So

$$d\left((a^{(12)})\kappa(a^{(21)})\right) = d\left((a^{(211)})\kappa(a^{(212)})\right) = d(\varepsilon(a^{(21)})1) = 0.$$

Putting this all back into (10.15), we obtain

$$\begin{aligned}
\pi(a^{(1)})\,\pi(a^{(2)}) &= \kappa(a^{(11)})d(a^{(12)})\kappa(a^{(21)})d(a^{(22)}) \\
&= -\kappa(a^{(11)})a^{(12)}d\kappa(a^{(21)})d(a^{(22)}) \\
&= -\varepsilon(a^{(1)})d\kappa(a^{(21)})d(a^{(22)}) \\
&= -d\kappa(\varepsilon(a^{(1)})a^{(21)})d(a^{(22)}) \\
&= -d\kappa(\varepsilon(a^{(11)})a^{(12)})d(a^{(2)}) \\
&= -d\kappa(a^{(1)})d(a^{(2)}) \\
&= -d\left(\kappa(a^{(1)})d(a^{(2)})\right) \\
&= -d(\pi(a)).
\end{aligned}$$

And this was what we wished to prove. ∎

Here is some indication of why this is called the Maurer–Cartan formula. We recall from Section 6.5 the notation X_i for the basis of the space of quantum LIVFs and ω_i for the dual basis in $_{inv}\Gamma$ with respect to the duality pairing $(\cdot, \cdot)$. From Corollary 6.5, we have for $a \in \mathcal{A}$ that

$$\pi(a) = \sum_i X_i(a)\omega_i. \qquad (10.16)$$

Since $X_i(a) \in \mathbb{C}$, this implies that

$$d\pi(a) = \sum_i X_i(a)d\omega_i,$$

which is the left side of (10.14). On the other hand, the right side of (10.14) is given by

$$-\pi(a^{(1)})\,\pi(a^{(2)}) = -\sum_{ij} X_i(a^{(1)})\,\omega_i\, X_j(a^{(2)})\,\omega_j$$

$$= -\sum_{ij} (X_i * X_j)(a)\,\omega_i\,\omega_j,$$

using (10.16) twice and the definition of the convolution product $*$. Hence, the formula (10.14) is equivalent to

$$\sum_i X_i(a)d\omega_i = -\sum_{ij} (X_i * X_j)(a)\,\omega_i\,\omega_j,$$

which is essentially Eq. (5.23) in [86], the only difference being that the calculation in [86] is done in the braided exterior calculus while here it is done in the universal differential calculus. But Eqs. (5.23) and (5.24) in [86] are equivalent, and the latter is the Maurer–Cartan formula in the noncommutative setting, as explained in further detail in [86].

10.4   Notes

Universal objects are nowadays part of what can aptly be called abstract nonsense. They are (universally!) anticipated by researchers though in some settings they do not exist.

The material in the first section on the braided exterior calculus is presented in Woronowicz's paper [86].

Our presentation in the second section follows that given in Đurđevich's paper [21], Appendix B. This material is also found in [12] by A. Connes.

These two particular HODCs are extreme opposites in a very precise sense. We consider all HODCs $\mathcal{H}$ over $\mathcal{A}$ for which there exists an extension of the co-product $\phi : \mathcal{A} \rightarrow \mathcal{A} \otimes \mathcal{A}$ to $\mathcal{H}$. As a side point, we remark that any such HODC is automatically bicovariant. Now in this class of HODCs there is a maximal one, which is the universal differential calculus. And in this class there is a minimal HODC, which is the braided exterior calculus.

The proof in the third section comes from a conversation with Micho Đurđevich. This was just one of many such helpful details for which I thank him most gratefully. However, he used the single-digit Sweedler's notation for the threefold co-product, namely, $a^{(1)} \otimes a^{(2)} \otimes a^{(3)} \otimes a^{(4)}$, while I usually prefer the iterated Sweedler's notation, which is not single-digit except for the co-product itself. So for him the proof starts with this version of (10.15):

$$\pi(a^{(1)}) \, \pi(a^{(2)}) = \kappa(a^{(1)}) d(a^{(2)}) \kappa(a^{(3)}) d(a^{(4)}).$$

Notice that Sweedler's notation on the left side here is *not* that on the right side. Frankly speaking, I find this single-digit Sweedler's notation to be clearer in this particular case, but I decided at some point not to use it very much in this book. However, note that I did use it in the definition of the adjoint co-actions of a Hopf algebra on itself.

Chapter 11
*-Structures

11.1 Generalities

We have not discussed *-structures before, because it would have made a long story even longer. Now we propose to put in the *s during a second pass through the procedure "FODC" $\to$ "HODC." This additional story will not be very short, but the upshot will be that everything works out as nicely as one could expect.

But first an unavoidable number of definitions.

Definition 11.1. *Suppose that V is a vector space over $\mathbb{C}$. We say that a map* $*$: $V \to V$, *denoted as $v \mapsto v^*$, is a *-operation (or a* conjugation*) if*

$$(v + w)^* = v^* + w^*, \quad \text{(additive)}$$

$$(\lambda v)^* = \lambda^* v^*, \quad \text{(antilinear)}$$

$$v^{**} \equiv (v^*)^* = v \quad \text{(involutive)}$$

for all $v, w \in W$ and all $\lambda \in \mathbb{C}$. Note we have used the same symbol $*$ in λ^* to denote complex conjugation.

*Let V be a vector space with *-operation. We say that $v \in V$ is a* real *element if $v^* = v$ and that it is an* imaginary *element if $v^* = -v$.*

*Let V and W be vector spaces with *-operations. Then we say that a linear map $T : V \to W$ is a *-morphism (or* hermitian*) if $(Tv)^* = T(v^*)$ for all $v \in V$.*

If $\dim_{\mathbb{C}} V > 0$, then a *-operation on V is *not* linear, but rather antilinear.

The terminology "hermitian" in this context has nothing to do with the concept of hermitian operators in Hilbert spaces.

We will always consider the complex plane $\mathbb{C}$ to be a vector space with *-operation, where its *-operation is simply the usual complex conjugation.

We leave the proof of the next result to the reader.

© Springer International Publishing Switzerland 2015

S.B. Sontz, *Principal Bundles*, Universitext, DOI 10.1007/978-3-319-15829-7_11

Proposition 11.1. *Let V and W be vector spaces with ∗-operations, each being denoted by* ∗. *We define a map* ∗ $: V \otimes W \to V \otimes W$ *by*

$$\left(\sum_k \lambda_k v_k \otimes w_k\right)^* := \sum_k \lambda_k^* v_k^* \otimes w_k^* \tag{11.1}$$

for $v_k \in V$, $w_k \in W$, and $\lambda_k \in \mathbb{C}$. Then this is a ∗-operation. In particular, $(v \otimes w)^* = v^* \otimes w^*$ for $v \in V$ and $w \in W$.

The ∗-operation defined in Proposition 11.1 will be called the *standard ∗-operation on $V \otimes W$*.

Definition 11.2. *Suppose that an algebra $\mathcal{A}$ over $\mathbb{C}$ has a ∗-operation that is an antimultiplicative map. This means $(ab)^* = b^* a^*$ for all $a, b \in \mathcal{A}$. Then we say that $\mathcal{A}$ is a ∗-algebra. Furthermore, if $\mathcal{A}$ has an identity element 1, then we also require that $1^* = 1$.*

Then $\mathbb{C}$ is a ∗-algebra, where the ∗-operation is complex conjugation. Other examples are the algebra of $n \times n$ matrices with complex entries with the ∗-operation acting on a matrix being its adjoint matrix (i.e., the transpose conjugated). More generally, the set of all bounded linear operators in any Hilbert space with the ∗-operation being the adjoint is also a ∗-algebra. An example of a commutative ∗-algebra is the vector space of all continuous functions $f : X \to \mathbb{C}$, where X is a topological space, the multiplication is defined pointwise, and the ∗-operation is defined pointwise by complex conjugation of the values of f. All of these examples are actually C^*-algebras, but we will not delve further into that rather important topic.

Here is another straightforward result whose proof is left to the reader.

Proposition 11.2. *Suppose that $\mathcal{A}$ and $\mathcal{B}$ are ∗-algebras. Then the algebra $\mathcal{A} \otimes \mathcal{B}$ is a ∗-algebra with the ∗-operation defined in Proposition 11.1.*

Definition 11.3. *Suppose that $\mathcal{A}$ is a Hopf algebra that has a ∗-operation. Suppose that it is a ∗-algebra and that the two co-algebra structure maps*

$$\phi : \mathcal{A} \to \mathcal{A} \otimes \mathcal{A}, \quad \varepsilon : \mathcal{A} \to \mathbb{C}$$

(the co-multiplication and co-unit maps, respectively) are ∗-morphisms. This means explicitly that

$$\phi(a^*) = (\phi(a))^*, \quad \varepsilon(a^*) = (\varepsilon(a))^*.$$

Then we say that $\mathcal{A}$ is a ∗-Hopf algebra. (This is also sometimes referred to as a Hopf ∗-algebra.)

Moreover, a morphism of ∗-Hopf algebras is defined, as one might expect, to be a ∗-morphism that is also a Hopf algebra morphism.

Recall that $\mathcal{A} \otimes \mathcal{A}$ has a $*$-operation by Proposition 11.1. We are using this $*$-operation in the expression $(\phi(a))^*$ above.

Note that we do *not* require that the antipode κ also be a $*$-morphism. This goes against what one might expect to be the "natural" definition. As we have noted on other matters, there are no "wrong" definitions, just as there are no "right" definitions. At least, that is a common philosophical principle, which we also adapt in our mathematical studies. While this definition may appear counterintuitive to some of us, the motivation behind it obviously comes from interesting examples, many of which appear in the works of Faddeev and the Leningrad school.

Nonetheless, the co-inverse of a $*$-Hopf algebra does have a quite specific relation with the $*$-operation. Here are two basic results about that.

Proposition 11.3. *The co-inverse κ of a $*$-Hopf algebra $\mathcal{A}$ satisfies*

$$\kappa(\kappa(a^*)^*) = a \qquad (11.2)$$

for all $a \in \mathcal{A}$. In particular, $\kappa : \mathcal{A} \to \mathcal{A}$ is a bijection with $\kappa^{-1}(a) = \kappa(a^*)^*$.

Remark. Since the co-inverse κ in a compact matrix compact group (see [84]) is defined on a $*$-Hopf algebra, this result implies that such a co-inverse κ is a bijection. For the proof of Proposition 11.3, see [80].

The next result gives a necessary and sufficient condition for the co-inverse of a $*$-Hopf algebra to be a $*$-morphism.

Corollary 11.1. *Let $\mathcal{A}$ be a $*$-Hopf algebra. Then κ is a $*$-morphism if and only if $\kappa^{-1} = \kappa$.*

Proof. First, suppose that κ is a $*$-morphism. Then $\kappa^{-1}(a) = \kappa(a^*)^* = \kappa(a)$. Conversely, suppose that $\kappa^{-1}(a) = \kappa(a)$; that is, $\kappa(a) = \kappa(a^*)^*$. Putting $b = a^*$ then gives $\kappa(b^*) = \kappa(b)^*$. $\blacksquare$

The $*$-Hopf algebra that arises from a finite group (see Chapter 13) has a co-inverse that satisfies $\kappa^{-1} = \kappa$. It seems that an important step in the development of the theory of Hopf algebras (and quantum groups) was the construction of examples where the co-inverse does not satisfy $\kappa^{-1} = \kappa$. Consequently, when a $*$-operation is introduced on such a Hopf algebra, the co-inverse (which is uniquely determined by the bi-algebra structure) cannot be a $*$-morphism.

We conclude this section with a more "natural" definition.

Definition 11.4. *Suppose that Γ is a bimodule over a $*$-algebra $\mathcal{A}$ and that Γ has a $*$-operation. Then we say that Γ is a $*$-bimodule over $\mathcal{A}$ if we have these two properties for all $a \in \mathcal{A}$ and all $\theta \in \Gamma$:*

$$(a\theta)^* = \theta^* a^*, \qquad (\theta a)^* = a^* \theta^*. \qquad (11.3)$$

Remark. We imposed two properties in this definition, out of a false sense of symmetry. But this is redundant. Using the substitutions $a \to a^*$ and $\theta \to \theta^*$ changes each identity in (11.3) into the other.

11.2   ∗-FODCs

We are now ready to discuss first-order differential calculi (FODCs).

Definition 11.5. *We say that a first-order differential calculus* (Γ, d) *over a* ∗-*algebra* $\mathcal{A}$ *is a* ∗-*FODC (or a* ∗-*calculus) if for all* $a_k, b_k \in \mathcal{A}$, *we have that*

$$\sum_k a_k d b_k = 0 \implies \sum_k d(b_k^*) a_k^* = 0. \tag{11.4}$$

Theorem 11.1. *Suppose that* (Γ, d) *is a* ∗-*FODC over the* ∗-*algebra* $\mathcal{A}$. *Then there exists a unique antilinear involution* $\Gamma \to \Gamma$, *denoted by* $\omega \mapsto \omega^*$ *for* $\omega \in \Gamma$, *which satisfies*

$$(a\omega)^* = \omega^* a^*, \tag{11.5}$$

$$(\omega a)^* = a^* \omega^*, \tag{11.6}$$

$$(da)^* = d(a^*) \tag{11.7}$$

for all $\omega \in \Gamma$ *and all* $a \in \mathcal{A}$.

Remark. Either of the redundant identities (11.5) and (11.6) says that Γ with this ∗-operation is a ∗-bimodule over the ∗-algebra $\mathcal{A}$. The identity (11.7) says that $d : \mathcal{A} \to \Gamma$ is a ∗-morphism.

Proof. If (11.5) and (11.7) hold, then for any $a_k, b_k \in \mathcal{A}$, we compute

$$\Big(\sum_k a_k d b_k\Big)^* = \sum_k (a_k d b_k)^* = \sum_k (d b_k)^* a_k^* = \sum_k d(b_k^*) a_k^*.$$

This establishes the uniqueness of the ∗-operation in Γ, given the ∗-operation in $\mathcal{A}$ and the identities (11.5) and (11.7). So, given any $\omega \in \Gamma$, we write $\omega = \sum a_k d b_k$ for some $a_k, b_k \in \mathcal{A}$ (standard form) and *define*

$$\omega^* := \sum_k d(b_k^*) a_k^* \in \Gamma.$$

Of course, there is no unique standard form, but the condition that (Γ, d) is a ∗-FODC assures that the right side of this definition is independent of the choice of standard form.

The mapping $^* : \Gamma \to \Gamma$ given by $\omega \mapsto \omega^*$ is clearly additive. Moreover (continuing with the notation above and taking $\lambda \in \mathbb{C}$), we have that

$$(\lambda \omega)^* = \Big(\lambda \sum_k a_k d b_k\Big)^* = \Big(\sum_k (\lambda a_k) d b_k\Big)^*$$

$$= \sum_k d(b_k^*)(\lambda a_k)^* = \lambda^* \sum_k d(b_k^*) a_k^*$$

$$= \lambda^* \omega^*.$$

In short, $^* : \Gamma \to \Gamma$ is antilinear. Next, to show that this is an involution, we calculate

$$(a\, db)^{**} = \left(d(b^*)a^*\right)^*$$

$$= \left(1 d(b^* a^*) - b^* d(a^*)\right)^*$$

$$= d((b^* a^*)^*)1^* - d(a^{**})b^{**}$$

$$= d(ab) - d(a)b$$

$$= a\, db,$$

where we have used Leibniz's rule in the second and last equalities. This immediately implies that $\omega^{**} = \omega$ for all $\omega \in \Gamma$; that is, * is an involution.

Next, we verify (11.5):

$$(a\omega)^* = \left(a \sum_k a_k db_k\right)^*$$

$$= \left(\sum_k a a_k db_k\right)^*$$

$$= \sum_k d(b_k^*)(a a_k)^*$$

$$= \sum_k d(b_k^*) a_k^* a^*$$

$$= \omega^* a^*.$$

As noted in the general situation, (11.6) is equivalent to (11.5).

Finally, to show (11.7), we note that the very definition of the $*$-operation in Γ gives

$$(da)^* = (1 da)^* = d(a^*)1^* = d(a^*)1 = d(a^*).$$

And that concludes the proof of the theorem. ∎

Exercise 11.1. *Suppose that $\mathcal{A}$ is a $*$-algebra and (Γ, d) is an FODC over $\mathcal{A}$ with a $*$-operation on Γ satisfying (11.5) and (11.7). Then (Γ, d) is a $*$-FODC; that is, the implication (11.4) holds.*

The moral of this exercise is that the conditions in Theorem 11.1 give an alternative way to define a $*$-FODC.

Definition 11.6. *Let* (Γ, d) *and* (Γ', d') *be* *-FODCs over a* *-algebra* $\mathcal{A}$. *Then a linear map* $\psi : \Gamma \to \Gamma'$ *is said to be a* *-FODC morphism if* ψ *is an FODC morphism and a* *-morphism.*

Respecting the categorical imperative, we say ψ *is a* *-FODC isomorphism if it has an inverse* *-FODC morphism.*

Proposition 11.4. *Let* $\mathcal{A}$ *be a* *-algebra. Suppose that* (Γ_1, d_1) *and* (Γ_2, d_2) *are* *-FODCs over* $\mathcal{A}$. *Let* $\psi : \Gamma_1 \to \Gamma_2$ *be any FODC morphism. Then* ψ *is a* *-FODC morphism. In particular, if* Γ_1 *and* Γ_2 *are isomorphic as FODCs, they are isomorphic as* *-FODCs.*

Proof. Take any $\omega \in \Gamma_1$ and write it in standard form, $\omega = \sum_k a_k \, d_1 b_k$ for some elements $a_k, b_k \in \mathcal{A}$. Now, by definition, an FODC morphism is an $\mathcal{A}$-bimodule map that also intertwines the differentials. Here the second condition means $\psi d_1 = d_2$. So we have

$$\psi(\omega) = \psi \left(\sum_k a_k \, d_1 b_k \right) = \sum_k a_k \, \psi(d_1 b_k) = \sum_k a_k \, d_2 b_k.$$

Now using the definition of the *-operations in the *-FODCs Γ_1 and Γ_2 in the second and fifth equalities, respectively, we calculate

$$\psi(\omega^*) = \psi\left(\left(\sum_k a_k \, d_1 b_k \right)^* \right)$$

$$= \psi\left(\sum_k d_1(b_k^*) \, a_k^* \right)$$

$$= \sum_k \psi(d_1(b_k^*)) \, a_k^*$$

$$= \sum_k d_2(b_k^*) \, a_k^*$$

$$= \left(\sum_k a_k \, d_2 b_k \right)^*$$

$$= (\psi(\omega))^*.$$

This calculation shows that ψ is a *-morphism. But by hypothesis, ψ is an FODC morphism. And therefore, by definition, ψ is a *-FODC morphism.

In particular, any FODC isomorphism is a *-FODC isomorphism. And this implies the last assertion in the proposition. ∎

And now, after a brief interlude proving things, even more definitions.

Definition 11.7. *Let* (Γ, Φ_Γ) *be a left covariant bimodule over* $\mathcal{A}$, *which is also a* *-bimodule. Then we say that* (Γ, Φ_Γ) *is a left covariant* *-bimodule over* $\mathcal{A}$ *provided* *that* Φ_Γ *is a* ∗-*morphism; that is,* $\Phi_\Gamma(\rho^*) = (\Phi_\Gamma(\rho))^*$ *for all* $\rho \in \Gamma$. *In this case, we say that the* ∗-*operation in* Γ *is left covariant.*

Here is the corresponding concept with "right" instead of "left."

Definition 11.8. *Let* $(\Gamma, {}_\Gamma\Phi)$ *be a right covariant bimodule over* $\mathcal{A}$, *which is also a* ∗-*bimodule. Then we say that* $(\Gamma, {}_\Gamma\Phi)$ *is a* right covariant ∗-bimodule over $\mathcal{A}$ *provided that* ${}_\Gamma\Phi$ *is a* ∗-*morphism; that is,* ${}_\Gamma\Phi(\rho^*) = ({}_\Gamma\Phi(\rho))^*$ *for all* $\rho \in \Gamma$. *In this case, we say that the* ∗-*operation in* Γ *is right covariant.*

And next we define the "bicovariant" version of this concept.

Definition 11.9. *Let* $(\Gamma, \Phi_\Gamma, {}_\Gamma\Phi)$ *be a bicovariant bimodule over* $\mathcal{A}$, *which is also a* ∗-*bimodule. Then we say that* $(\Gamma, \Phi_\Gamma, {}_\Gamma\Phi)$ *is a* bicovariant ∗-bimodule over $\mathcal{A}$ *(or more simply that it is* ∗-*covariant) provided that it is both a left and right covariant* ∗-*bimodule over* $\mathcal{A}$. *In this case, we also say that the* ∗-*operation in* Γ *is* ∗-*covariant.*

Here is a rather immediate consequence.

Proposition 11.5. *Let* (Γ, Φ_Γ) *be a left covariant* ∗-*bimodule over* $\mathcal{A}$. *As usual, denote the set of its left invariant elements by* ${}_{\text{inv}}\Gamma$. *Then* $\omega \in {}_{\text{inv}}\Gamma$ *implies that* $\omega^* \in {}_{\text{inv}}\Gamma$; *that is,* $({}_{\text{inv}}\Gamma)^* \subset {}_{\text{inv}}\Gamma$. *In other words,* ${}_{\text{inv}}\Gamma$ *is invariant under the* ∗-*operation. Moreover, using that the* ∗-*operation is involutive, we have that* $({}_{\text{inv}}\Gamma)^* = {}_{\text{inv}}\Gamma$.

Proof. Take $\omega \in {}_{\text{inv}}\Gamma$. Then

$$\Phi_\Gamma(\omega^*) = (\Phi_\Gamma(\omega))^* = (1 \otimes \omega)^* = 1^* \otimes \omega^* = 1 \otimes \omega^*,$$

which implies that $\omega^* \in {}_{\text{inv}}\Gamma$. The rest of the statements are left for the reader's consideration. ∎

Exercise 11.2. *State and prove the corresponding result for a right covariant* ∗-*bimodule over* $\mathcal{A}$.

Theorem 11.2. *Suppose that* (Γ, d) *is a* ∗-*FODC over the Hopf algebra* $\mathcal{A}$, *which is a* ∗-*algebra. Suppose that the co-multiplication map* ϕ *of* $\mathcal{A}$ *is a* ∗-*morphism. (For example, this is true if* $\mathcal{A}$ *is a* ∗-*Hopf algebra.)*

- *If* (Γ, d) *is left covariant, then* (Γ, Φ_Γ) *is a left covariant* ∗-*bimodule.*
- *If* (Γ, d) *is right covariant, then* $(\Gamma, {}_\Gamma\Phi)$ *is a right covariant* ∗-*bimodule.*
- *If* (Γ, d) *is bicovariant, then* $(\Gamma, \Phi_\Gamma, {}_\Gamma\Phi)$ *is* ∗-*covariant.*

Proof. We will use the fact that the left co-action Φ_Γ as well as the right co-action ${}_\Gamma\Phi$ are defined in terms of the maps ϕ and d and that these two maps are ∗-morphisms.

We start with the first assertion. Recall that $\Phi_\Gamma : \Gamma \to \mathcal{A} \otimes \Gamma$ is given by the formula

$$\Phi_\Gamma(\omega) = \Phi_\Gamma(a\,db) = \phi(a)(id \otimes d)\phi(b)$$

for the case when $\omega = a\,db$. (The general case follows from this simpler case by linearity.) Using $\omega^* = d(b^*)\,a^*$, we then calculate

$$
\begin{aligned}
\Phi_\Gamma(\omega^*) &= \Phi_\Gamma(d(b^*)\,a^*) \\
&= \Phi_\Gamma\Big(1d(b^*a^*) - b^*d(a^*)\Big) \\
&= \phi(1)(id \otimes d)\phi(b^*a^*) - \phi(b^*)(id \otimes d)\phi(a^*) \\
&= (1 \otimes 1)(id \otimes d)\big(\phi(b^*)\phi(a^*)\big) - \phi(b^*)(id \otimes d)\phi(a^*) \\
&= (id \otimes d)\big(\phi(b^*)\phi(a^*)\big) - \phi(b^*)(id \otimes d)\phi(a^*) \\
&= \big((id \otimes d)\phi(b^*)\big)\phi(a^*),
\end{aligned}
$$

using Leibniz's rule for $id \otimes d$. On the other hand,

$$
\begin{aligned}
\Phi_\Gamma(\omega)^* &= \big(\Phi_\Gamma(a\,db)\big)^* \\
&= \big(\phi(a)(id \otimes d)\phi(b)\big)^* \\
&= \big((id \otimes d)\phi(b)\big)^*\big(\phi(a)\big)^* \\
&= \big((id \otimes d)(\phi(b))^*\big)(\phi(a))^* \\
&= \big((id \otimes d)\phi(b^*)\big)\phi(a^*),
\end{aligned}
$$

which establishes $\Phi_\Gamma(\omega^*) = \Phi_\Gamma(\omega)^*$ in this simpler case.

A similar argument proves that $_\Gamma\Phi(\omega^*) = (_\Gamma\Phi(\omega))^*$ if (Γ, d) is a right covariant FODC. This is the second assertion. And the last assertion is just the combination of the previous two. ∎

We return to the universal FODC (U, D) but now in this new context.

Theorem 11.3. *Let $\mathcal{A}$ be a ∗-Hopf algebra. Then the universal FODC (U, D) over $\mathcal{A}$ is a ∗-FODC and $(U, \Phi_U, {}_U\Phi)$ is a ∗-covariant bimodule over $\mathcal{A}$.*

The unique ∗-operation on U as defined in Theorem 11.1 is given for $u = \sum_k a_k \otimes b_k \in U$ with $a_k, b_k \in \mathcal{A}$ by

$$
u^\dagger = \Big(\sum_k a_k \otimes b_k\Big)^\dagger = -\sum_k b_k^* \otimes a_k^*. \tag{11.8}
$$

Remark. The formula (11.8) may come as a surprise to the reader. It certainly was a surprise for me. It is not Eq. (2.64) in [86]. Also, it is not the ∗-structure that we defined in (11.1) for the tensor product $\mathcal{A} \otimes \mathcal{A}$ and then restricted to U. This is why we have used the dagger symbol † to denote it. But we have no choice in this matter since Theorem 11.1 defines the unique ∗-operation in U with the properties (11.5), (11.6), and (11.7) provided that (U, D) is a ∗-FODC over $\mathcal{A}$.

In [86], a *-operation is defined on U by Eq. (2.64); it does not satisfy (11.7). It differs from our definition by a minus sign. Consequently, some of our formulas do not agree with those in [86] though only by a sign. But let us be clear about this; the definition in [86] is not an error since the goal there was to define a *-operation satisfying only (11.5) and (11.6). If we multiply the right side of the formula (11.8) by any nonzero constant $\lambda \in \mathbb{C}$, then we get a definition of another *-operation that satisfies (11.5) and (11.6). But if $\lambda \neq 1$, then (11.7) will not be satisfied. Equation (2.64) in [86] corresponds to taking $\lambda = -1$.

However, we do consider our approach to be more transparent since we will see exactly how Eq. (11.8) arises in the following proof.

Proof. To show that (U, D) is a *-FODC, we suppose that $a_k, b_k \in \mathcal{A}$ are given such that $\sum_k a_k D b_k = 0$. Then we have to prove that $\sum_k D(b_k^*)a_k^* = 0$. Recall the definition $Db = 1 \otimes b - b \otimes 1$ for $b \in \mathcal{A}$. So we start from

$$0 = \sum_k a_k D b_k$$
$$= \sum_k a_k(1 \otimes b_k - b_k \otimes 1)$$
$$= \sum_k (a_k \otimes b_k - a_k b_k \otimes 1). \tag{11.9}$$

Then we next calculate

$$\sum_k D(b_k^*)a_k^* = \sum_k (1 \otimes b_k^* - b_k^* \otimes 1)a_k^*$$
$$= \sum_k (1 \otimes b_k^* a_k^* - b_k^* \otimes a_k^*)$$
$$= \sum_k (1 \otimes (a_k b_k)^* - b_k^* \otimes a_k^*). \tag{11.10}$$

At this point, we argue very carefully. While we will eventually put a new *-operation on U, it is nonetheless true that we also have the *-operation on $\mathcal{A} \otimes \mathcal{A}$ defined by (11.1) and denoted by *. While U is not necessarily invariant under this *-operation, that does not matter here. We now use this *-operation to continue the analysis started in (11.10). We also use the interchange map $\sigma_\mathcal{A} : \mathcal{A} \otimes \mathcal{A} \to \mathcal{A} \otimes \mathcal{A}$ given by $a \otimes b \mapsto b \otimes a$. So we obtain

$$\sum_k D(b_k^*)a_k^* = \sum_k (1 \otimes (a_k b_k)^* - b_k^* \otimes a_k^*)$$
$$= \sum_k (1 \otimes a_k b_k - b_k \otimes a_k)^*$$

$$= \left(\sum_k \left(1 \otimes a_k b_k - b_k \otimes a_k \right) \right)^*$$

$$= \left(\sum_k \sigma_A \left(a_k b_k \otimes 1 - a_k \otimes b_k \right) \right)^*$$

$$= \left(\sigma_A \left(\sum_k \left(a_k b_k \otimes 1 - a_k \otimes b_k \right) \right) \right)^*$$

$$= 0,$$

where the last equality is just an application of Eq. (11.9). We were definitely not allowed to use the relation $\sum_k a_k b_k = 0$ in this paragraph. And we did not.

Since (U, D) is a $*$-FODC over $\mathcal{A}$, we can now calculate the $*$-operation in U, which is defined in Theorem 11.1. So take any $u = \sum_k a_k \otimes b_k \in U$, that is, $\sum_k a_k b_k = 0$. Then $u = \sum_k a_k D b_k$ (as we have seen many times), which gives u in standard form. To distinguish clearly that this is a new conjugation (or $*$-operation), we will denote its action on u by $u^\dagger$. In the following calculation, the first equality is the definition of $u^\dagger$ as given in Theorem 11.1, the second equality is a repetition of (11.10) (which is valid here), and the rest is simple algebra. So we have

$$u^\dagger = \sum_k D(b_k^*) a_k^* = \sum_k \left(1 \otimes (a_k b_k)^* - b_k^* \otimes a_k^* \right)$$

$$= 1 \otimes \left(\sum_k a_k b_k \right)^* - \sum_k b_k^* \otimes a_k^* = - \sum_k b_k^* \otimes a_k^*.$$

This not only is exactly the formula (11.8), but also shows from where that formula comes.

The reader might wish to check directly that $D(a^*) = (Da)^\dagger$ for $a \in \mathcal{A}$; that is, Eq. (11.7) holds, as it must. If there were no minus sign in (11.8), then this relation would not hold.

Finally, Theorem 11.2 implies that (U, Φ_U) is a left covariant $*$-bimodule over $\mathcal{A}$ and that $(U, {}_U\Phi)$ is a right covariant $*$-bimodule over $\mathcal{A}$. But we already know that these co-actions commute. So $(U, \Phi_U, {}_U\Phi)$ is a $*$-covariant bimodule over $\mathcal{A}$. ∎

The moral of this tale is that the universal FODC (U, D) has all the $*$-structure we could possibly want!

Theorem 11.4. *Let $\mathcal{N}$ be an $\mathcal{A}$-subbimodule of U. Put $\Gamma_{\mathcal{N}} = U/\mathcal{N}$ and let $\pi_{\mathcal{N}}$: $U \to \Gamma_{\mathcal{N}}$ be the canonical quotient map. Set $d_{\mathcal{N}} = \pi_{\mathcal{N}} \circ D$. Then by Theorem 2.1, $(\Gamma_{\mathcal{N}}, d_{\mathcal{N}})$ is an FODC over $\mathcal{A}$, and, moreover, any FODC over $\mathcal{A}$ is isomorphic as an FODC to $(\Gamma_{\mathcal{N}}, d_{\mathcal{N}})$ for some $\mathcal{N}$ that is an $\mathcal{A}$-subbimodule of U. Then $(\Gamma_{\mathcal{N}}, d_{\mathcal{N}})$ is a $*$-FODC over $\mathcal{A}$ if and only if $\mathcal{N}$ is invariant under its $*$-operation $\dagger$ (i.e., $\mathcal{N}^\dagger \subset \mathcal{N}$).*

Remark. By a standard argument (take conjugates and use the involution property), the condition $\mathcal{N}^\dagger \subset \mathcal{N}$ is equivalent to $\mathcal{N}^\dagger = \mathcal{N}$.

Proof. Suppose that $(\Gamma_\mathcal{N}, d_\mathcal{N})$ is a ∗-FODC. Then we have shown that there is a ∗-operation on $\Gamma_\mathcal{N}$. Moreover, this satisfies $\pi_\mathcal{N}(u^\dagger) = (\pi_\mathcal{N}(u))^*$ for all $u \in U$. (Exercise!) If we take $u \in \mathcal{N}$, this implies $\pi_\mathcal{N}(u^\dagger) = (\pi_\mathcal{N}(u))^* = 0$, which in turn means that $u^\dagger \in \mathcal{N}$. And this shows that $\mathcal{N}$ is invariant under the ∗-operation †.

Conversely, suppose that $\mathcal{N}$ is invariant under †. To show that $(\Gamma_\mathcal{N}, d_\mathcal{N})$ is a ∗-FODC, we consider elements $a_k, b_k \in \mathcal{A}$ that satisfy $\sum_k a_k d_\mathcal{N} b_k = 0$. We must show that $\sum_k (d_\mathcal{N} b_k^*) a_k^* = 0$. First, we note

$$0 = \sum_k a_k d_\mathcal{N} b_k = \sum_k a_k \pi_\mathcal{N} D b_k = \pi_\mathcal{N} \left(\sum_k a_k D b_k \right),$$

which tells us that $\sum_k a_k D b_k \in \mathcal{N}$. Since $\mathcal{N}$ is invariant under †, we then get $\left(\sum_k a_k D b_k \right)^\dagger \in \mathcal{N}$. So $\pi_\mathcal{N} \left(\left(\sum_k a_k D b_k \right)^\dagger \right) = 0$. But then we have

$$\sum_k (d_\mathcal{N} b_k^*) a_k^* = \sum_k (\pi_\mathcal{N} D b_k^*) a_k^* = \pi_\mathcal{N} \left(\sum_k (D b_k^*) a_k^* \right) = \pi_\mathcal{N} \left(\sum_k a_k D b_k \right)^\dagger = 0,$$

which concludes the proof. ∎

Now we transport the ∗-operation on U to U_L. To do this, we use the map $r : U \to U_L$.

Definition 11.10. *Define a map* $^* : U_L \to U_L$ *by*

$$\omega^* := r((r^{-1}\omega)^\dagger) \tag{11.11}$$

for all $\omega \in U_L$.

We are reverting to the tradition of denoting all conjugations by *. The only exception to this was our use of the dagger † for the conjugation in U.

Exercise 11.3. *Show that Eq. (11.11) defines a ∗-operation on U_L and that $r : U \to U_L$ is a ∗-morphism.*

Theorem 11.5. *Let $\mathcal{R} \subset \ker \varepsilon$ be a right ideal in a ∗-Hopf algebra $\mathcal{A}$. Put $\mathcal{N} = \mathcal{A} \otimes \mathcal{R} \subset \mathcal{A} \otimes \ker \varepsilon = U_L$. Let $(\Gamma, d) = (\Gamma_{L\mathcal{N}}, d_{L\mathcal{N}})$ be the left covariant FODC associated to $\mathcal{R}$, where $\Gamma_{L\mathcal{N}} = U_L / \mathcal{N}$ (see Theorem 6.8). Then (Γ, d) is a ∗-FODC over $\mathcal{A}$ if and only if $(\kappa(\mathcal{R}))^* \subset \mathcal{R}$, where κ is the co-inverse in $\mathcal{A}$.*

Proof. By the previous theorem, (Γ, d) is a ∗-FODC if and only if $\mathcal{N}^\dagger \subset \mathcal{N}$, which in turn holds if and only if $\mathcal{N}^* \subset \mathcal{N}$.

First, we assume that $\mathcal{N}^* \subset \mathcal{N}$. So (Γ, d) is a left invariant ∗-FODC. Let $a \in \mathcal{R}$ be an arbitrary element. We must show that $(\kappa(a))^* \equiv \kappa(a)^* \in \mathcal{R}$. Now $1 \otimes a \in \mathcal{N} = \mathcal{A} \otimes \mathcal{R} \subset U_L$ is left invariant in U_L. But, in general, for any left covariant

∗-FODC with left co-action $\Phi_\Gamma : \Gamma \to \mathcal{A} \otimes \Gamma$ and any left invariant $\omega \in \Gamma$ (i.e., $\Phi_\Gamma(\omega) = 1 \otimes \omega$), we have

$$\Phi_\Gamma(\omega^*) = (\Phi_\Gamma(\omega))^* = (1 \otimes \omega)^* = 1 \otimes \omega^*,$$

which says that ω^* is again left invariant. Applying this result to our specific case where $\Gamma = U_L$ and $\omega = 1 \otimes a$, we see that $(1 \otimes a)^* \in U_L$ is left invariant. And we also have $(1 \otimes a)^* \in \mathcal{N}$ by our hypothesis.

Let's take $\omega = 1 \otimes a$, where $a \in \ker \varepsilon$, and see what happens. We claim that

$$(1 \otimes a)^* = -1 \otimes \kappa(a)^* \tag{11.12}$$

holds for all $a \in \mathcal{A}$. We will prove this in a moment. But using it for now, we get $(-1) \otimes \kappa(a)^* \in \mathcal{N} = \mathcal{A} \otimes \mathcal{R}$, which implies that $\kappa(a)^* \in \mathcal{R}$. And this proves that $(\kappa(\mathcal{R}))^* \subset \mathcal{R}$.

Before proving (11.12), let's note that it disagrees with Eq. (2.65) in [86]. This is because our definition of the ∗-operation on the left side differs by a sign from the definition of the ∗-operation used in [86]. And this difference comes from our definition of the ∗-operation on U as mentioned above.

To compute the left side of (11.12), namely, $(1 \otimes a)^*$, we apply r^{-1}, then † in U, and finally r [cf. (11.11)] to it:

$$1 \otimes a \xrightarrow{r^{-1}} (\kappa \otimes id)\phi(a) = \kappa(a^{(1)}) \otimes a^{(2)}$$

$$\kappa(a^{(1)}) \otimes a^{(2)} \xrightarrow{\dagger} -a^{(2)*} \otimes \kappa(a^{(1)})^*$$

$$-a^{(2)*} \otimes \kappa(a^{(1)})^* \xrightarrow{r} -(a^{(2)*} \otimes 1)\phi(\kappa(a^{(1)})^*).$$

Here we just used the formulas for these maps. Next, we compute

$$-(a^{(2)*} \otimes 1)\phi(\kappa(a^{(1)})^*) = -(a^{(2)} \otimes 1)^*(\phi(\kappa(a^{(1)})))^*$$
$$= -\left(\phi(\kappa(a^{(1)}))(a^{(2)} \otimes 1)\right)^*$$
$$= -(1 \otimes \kappa(a))^*$$
$$= -1 \otimes \kappa(a)^*,$$

where we used Eq. (B.14) in the next-to-last equality.

Conversely, we assume that $(\kappa(\mathcal{R}))^* \subset \mathcal{R}$. The left invariant elements in $\mathcal{N} = \mathcal{A} \otimes \mathcal{R}$ are exactly those of the form $1 \otimes b$, where $b \in \mathcal{R}$. Equivalently, $_{inv}\mathcal{N} = 1 \otimes \mathcal{R}$. Take such an element $1 \otimes b \in {}_{inv}\mathcal{N}$ and apply the formula (11.12) to it to get

$$(1 \otimes b)^* = -1 \otimes \kappa(b)^* = 1 \otimes (-\kappa(b)^*) \in 1 \otimes \mathcal{R} = {}_{inv}\mathcal{N} \subset \mathcal{N}$$

by the hypothesis. So $(_{\text{inv}}\mathcal{N})^* \subset \mathcal{N}$. But any element $\theta \in \mathcal{N}$ can be written as a linear combination with coefficients in $\mathcal{A}$ of element in $_{\text{inv}}\mathcal{N}$, namely, as $\theta = \sum a_k (1 \otimes b_k)$, where $a_k \in \mathcal{A}$ and $1 \otimes b_k \in \,_{\text{inv}}\mathcal{N}$. Then

$$\theta^* = \sum_k (1 \otimes b_k)^* a_k^* \in (_{\text{inv}}\mathcal{N})^* \mathcal{A} \subset \mathcal{N}\mathcal{A} \subset \mathcal{N}.$$

And that proves that $\mathcal{N}^* \subset \mathcal{N}$. ∎

The next result determines the ∗-operation on $\operatorname{Ran} \pi = \,_{\text{inv}}\Gamma$. It can be shown that this then determines the ∗-operation on Γ.

Proposition 11.6. *Suppose that (Γ, d) is a ∗-covariant FODC. Then for all $a \in \mathcal{A}$, the quantum germs map π satisfies*

$$\pi(a)^* = -\pi(\kappa(a)^*). \tag{11.13}$$

Proof. First, by using (11.12), $\kappa(1) = 1$, and so forth, we see that

$$\begin{aligned}
\pi(a)^* &= \left(1 \otimes (a - \varepsilon(a)1)\right)^* \\
&= -1 \otimes \left(\kappa(a - \varepsilon(a)1)\right)^* \\
&= -1 \otimes \left(\kappa(a) - \kappa(\varepsilon(a)1)\right)^* \\
&= -1 \otimes \left(\kappa(a) - \varepsilon(a)\kappa(1)\right)^* \\
&= -1 \otimes \left(\kappa(a) - \varepsilon(a)1\right)^* \\
&= -1 \otimes \left(\kappa(a)^* - \varepsilon(a)^*1\right). \tag{11.14}
\end{aligned}$$

Next, we use $\varepsilon\kappa = \varepsilon$ [see (B.11)] and the fact that ε is a ∗-morphism to obtain

$$\begin{aligned}
\pi(\kappa(a)^*) &= 1 \otimes \left(\kappa(a)^* - \varepsilon(\kappa(a)^*)1\right) \\
&= 1 \otimes \left(\kappa(a)^* - \left(\varepsilon\kappa(a)\right)^*1\right) \\
&= 1 \otimes \left(\kappa(a)^* - \varepsilon(a)^*1\right) \\
&= -\pi(a)^*,
\end{aligned}$$

where the last equality is Eq. (11.14). ∎

Remark. This result shows that, in general, the quantum germs map π is not a ∗-morphism.

We continue now with the study of quantum left invariant vector fields (LIVFs), but now in the context of *-operations.

Theorem 11.6. *Let Γ be a left covariant FODC over a *-Hopf algebra $\mathcal{A}$ with $_{\mathrm{inv}}\Gamma$ finite dimensional and with $\mathcal{X}$ being its space of quantum LIVFs. (Recall that we say in this case that Γ is finite dimensional.) Then Γ is a *-FODC over $\mathcal{A}$ if and only if $\mathcal{X}$ is *-invariant in*

$$\mathcal{A}' = \{ f : \mathcal{A} \to \mathbb{C} \mid f \text{ is linear} \},$$

*where the *-operation on $\mathcal{A}'$ is defined for all $f \in \mathcal{A}'$ by*

$$f^*(a) := \big(f(\kappa(a)^*) \big)^*$$

for $a \in \mathcal{A}$.

Remark. For all $f \in \mathcal{A}'$, we have that $f^* \in \mathcal{A}'$, and this *is* a *-operation.

Proof. First, suppose that $\mathcal{X}$ is *-invariant. Take an element $a \in \mathcal{R}$, the right ideal in $\ker \varepsilon$ associated with the FODC Γ. Then for any $X \in \mathcal{X} \subset \mathcal{A}'$, we have $X^*(a) = \big(X(\kappa(a)^*) \big)^*$ by definition. By the hypothesis, we also have $X^* \in \mathcal{X}$ so that $X^*(a) = 0$ by the definition of $\mathcal{X}$. We conclude that $X(\kappa(a)^*) = 0$ for all $a \in \mathcal{R}$. So $\kappa(a)^* \in \ker X$ for all $X \in \mathcal{X}$. But this element also satisfies

$$\varepsilon(\kappa(a)^*) = \varepsilon(\kappa(a))^* = \varepsilon(a)^* = 0$$

since $\mathcal{R} \subset \ker \varepsilon$. So we obtain $\kappa(a)^* \in \ker \varepsilon$. This implies that

$$\kappa(a)^* \in \mathcal{R}' := \cap_{X \in \mathcal{X}}(\ker \varepsilon \cap \ker X) = \cap_{j=1}^k(\ker \varepsilon \cap \ker X_j),$$

where $\{X_1, \dots, X_k\}$ is a (finite) basis of the finite-dimensional space $\mathcal{X}$. We claim that $\mathcal{R}' = \mathcal{R}$, so that $\kappa(a)^* \in \mathcal{R}$ follows. In summary, we have shown that $a \in \mathcal{R}$ implies that $\kappa(a)^* \in \mathcal{R}$. But this is precisely the condition that guarantees that Γ is a *-FODC.

Conversely, suppose that Γ is a *-FODC. Take $X \in \mathcal{X}$ and $a \in \mathcal{R}$. So we have $\kappa(a)^* \in \mathcal{R}$. Then $X^*(a) = \big(X(\kappa(a)^*) \big)^* = 0$ since $X \in \mathcal{X}$. Also, $X^*(1) = \big(X(\kappa(1)^*) \big)^* = \big(X(1) \big)^* = 0$. Then by the definition of $\mathcal{X}$, we have that $X^* \in \mathcal{X}$. We have shown that $X \in \mathcal{X}$ implies that $X^* \in \mathcal{X}$. Therefore, $\mathcal{X}$ is *-invariant. $\blacksquare$

Exercise 11.4. *In the notation of this proof, verify that $\mathcal{R}' = \mathcal{R}$. (This exercise is not profound though the reader might find the proof to be tricky.)*

11.3 *-Braided Exterior Algebras

We first define the general structure.

Definition 11.11. *Let $\mathcal{A}$ be a $*$-Hopf algebra. Suppose that $(\Omega, \Phi_\Omega, {}_\Omega\Phi)$ is a graded bicovariant algebra over $\mathcal{A}$ that is also a $*$-covariant bimodule over $\mathcal{A}$ with a $*$-operation $* : \Omega \to \Omega$. Then we say that $(\Omega, \Phi_\Omega, {}_\Omega\Phi)$ is a graded $*$-covariant algebra provided that*

1. *The $*$-operation preserves degree; that is, $(\Omega^n)^* \subset \Omega^n$ for all $n \geq 0$.*
2. *The $*$-operation is graded antimultiplicative. This means that for all $\omega_1 \in \Omega^j$ and $\omega_2 \in \Omega^k$, we have*

$$(\omega_1\omega_2)^* = (-1)^{jk}\omega_2^*\omega_1^*. \tag{11.15}$$

In this section, we'll say "exterior algebra" instead of "braided exterior algebra." The importance of the previous definition resides in the next theorem.

Theorem 11.7. *Suppose that $\mathcal{A}$ is a $*$-Hopf algebra and that $(\Gamma, \Phi_\Gamma, {}_\Gamma\Phi)$ is a $*$-covariant bimodule over $\mathcal{A}$. Then the exterior algebra $(\wedge\Gamma, \Phi_{\wedge\Gamma}, {}_{\wedge\Gamma}\Phi)$ based on $(\Gamma, \Phi_\Gamma, {}_\Gamma\Phi)$ has a unique $*$-operation such that the following hold:*

1. *On $\wedge^0\Gamma = \mathcal{A}$, the $*$-operation of $\wedge^*\Gamma$ is that of the $*$-Hopf algebra $\mathcal{A}$.*
2. *On $\wedge^1\Gamma = \Gamma$, the $*$-operation of $\wedge^*\Gamma$ is that of the $*$-bimodule Γ.*
3. *$\wedge\Gamma = \oplus_n \wedge^n \Gamma$ is a graded $*$-covariant algebra.*

Proof. Since the exterior algebra is defined as a quotient of the tensor algebra $\otimes\Gamma$, we first define a $*$-structure in the tensor algebra. But degrees 0 and 1 are easy to deal with. On $\otimes^0\Gamma = \mathcal{A}$ we put the $*$-operation of the $*$-Hopf algebra $\mathcal{A}$, while on $\otimes^1\Gamma = \Gamma$ we put the $*$-operation of the $*$-bimodule Γ. For $n \geq 2$, we define the $*$-operation on $\otimes^n\Gamma$ by

$$\left(\theta_1 \otimes_{\mathcal{A}} \theta_2 \otimes_{\mathcal{A}} \cdots \otimes_{\mathcal{A}} \theta_n\right)^* := \operatorname{sign}(\zeta_n)\left(\theta_n^* \otimes_{\mathcal{A}} \cdots \otimes_{\mathcal{A}} \theta_2^* \otimes_{\mathcal{A}} \theta_1^*\right), \tag{11.16}$$

where $\theta_1, \ldots, \theta_n \in \Gamma$ and $\zeta_n \in S_n$ is the permutation of $\{1, 2, \ldots n\}$ that reverses the order, that is, $1 \mapsto n, 2 \mapsto n - 1, \ldots, n \mapsto 1$. [In general, $\zeta_n(k) = n - k + 1$.] Clearly, the right side of (11.16) lies in $\otimes^n\Gamma$, and so this map preserves degree.

Exercise 11.5. *Show (by induction, for example) $\operatorname{sign}(\zeta_n) = (-1)^{n(n-1)/2}$. Show that (11.16) defines an antilinear involution of $\otimes^n\Gamma$ that is graded antimultiplicative [cf. (11.15)] on $\otimes\Gamma$ and $*$-covariant.*

The next step in the argument is to show that this $*$-operation passes to the quotient. Recall $\pi_n : S_n \to B_n$ from Proposition 9.4. Now we claim that for all $\tau \in S_n$, we have

$$A_n(\tau^*) = \operatorname{sign}(\zeta_n)\left((\pi_n(\zeta_n))^{-1} A_n(\tau)\right)^*. \tag{11.17}$$

Given this result, we see immediately that $A_n(\tau) = 0$ implies that $A_n(\tau^*) = 0$. In other words, $(\ker A_n)^* \subset \ker A_n$ for all $n \geq 2$. So the $*$-operation in $\otimes\Gamma$ passes to the quotient $\wedge^n\Gamma = \otimes^n\Gamma/\ker A_n$. The properties of the $*$-operation also pass to the quotient. And except for a demonstration of Eq. (11.17), that finishes the proof of the existence of the $*$-operation. ∎

Exercise 11.6. *Optionally, prove (11.17). Otherwise, read and understand its proof in [86].*

Exercise 11.7. *Prove that*

$$(a\,\theta_1 \wedge \cdots \wedge \theta_n)^* = \mathrm{sign}(\zeta_n)\theta_n^* \wedge \cdots \wedge \theta_1^* a^* = (\theta_1^* \wedge \cdots \wedge \theta_n^*)a^*,$$

where $a \in \mathcal{A}$ and $\theta_1, \ldots, \theta_n \in \Gamma$.

Exercise 11.8. *Prove the uniqueness statement of the previous theorem in the usual way by using the formula for the $*$-operation that follows from its properties.*

11.4 $*$-Braided Exterior Calculus

This higher-order differential calculus (HODC) consists of nothing other than the differential d on the braided exterior algebra associated to an FODC, as in Theorem 10.1. If we suppose that we start with a $*$-FODC (Γ, d), then by the previous section, the associated braided exterior algebra has a $*$-operation. The question is how this differential d relates to the $*$-operation. The answer is the next result.

Theorem 11.8. *We assume the hypotheses of Theorem 10.1 and that (Γ, d) is a $*$-FODC. Then d is a $*$-morphism; that is, for all $\omega \in \wedge\Gamma$, we have that*

$$d(\omega^*) = (d\omega)^*. \tag{11.18}$$

Proof. We know the result is true for elements in degrees 0 and 1, that is, in $\mathcal{A}$ and Γ. So without loss of generality, $\omega = a\,db_1 \wedge \cdots \wedge db_n$ for some $n \geq 2$, where $a, b_1, \ldots, b_n \in \mathcal{A}$. Computing the right side of (11.18), we get

$$(d\omega)^* = (da \wedge db_1 \wedge \cdots \wedge db_n)^*$$
$$= \mathrm{sign}(\zeta_{n+1})\big((db_n)^* \wedge \cdots \wedge (db_1)^* \wedge (da)^*\big).$$

Now the left side of (11.18) is

$$d(\omega^*) = d\big(\mathrm{sign}(\zeta_n)\,d(b_n)^* \wedge \cdots \wedge d(b_1)^* a^*\big)$$
$$= (-1)^n\,\mathrm{sign}(\zeta_n)\,\big(d(b_n)^* \wedge \cdots \wedge d(b_1)^* \wedge d(a)^*\big)$$
$$= (-1)^n\,\mathrm{sign}(\zeta_n)\,\big((db_n)^* \wedge \cdots \wedge (db_1)^* \wedge (da)^*\big).$$

So showing that these two expressions are equal reduces to showing that $(-1)^n \operatorname{sign}(\zeta_n) = \operatorname{sign}(\zeta_{n+1})$. ∎

Exercise 11.9. *Show that* $(-1)^n \operatorname{sign}(\zeta_n) = \operatorname{sign}(\zeta_{n+1})$ *by using the formula in Exercise 11.5.*

11.5 ∗-Universal Differential Algebra

Finally, we introduce a ∗-structure into the universal differential algebra of a ∗-FODC (Γ, d) over a ∗-algebra $\mathcal{A}$, and then we examine some of its properties. So we have already constructed the universal differential algebra $(\Gamma^\wedge, d)$. Now the ∗-operation in $\mathcal{A}$ is an antilinear, antimultiplicative map $* : \mathcal{A} \to \mathcal{A}$, which furthermore satisfies $d(a^*) = (da)^*$ for $a \in \mathcal{A}$, where the ∗-operation on the right side is that of Γ. By one of the theorems (whose statement as well as proof was an exercise) concerning a universal property of $(\Gamma^\wedge, d)$, there exists a unique antilinear, graded multiplicative map of differential algebras

$$* : \Gamma^\wedge \to \Gamma^\wedge$$

(denoted again by $*$), extending the ∗-operations in degrees 0 and 1. So this gives us the sought-after ∗-structure in the universal differential algebra $(\Gamma^\wedge, d)$.

Recall that $*$ being a morphism of differential algebras implies that

$$(d\omega)^* = d(\omega^*)$$

for all $\omega \in \Gamma^\wedge$. Also, recall that $*$ being a graded multiplicative map means that for homogeneous elements $\omega_1 \in \Gamma^{\wedge j}$ and $\omega_2 \in \Gamma^{\wedge k}$, we have

$$(\omega_1 \omega_2)^* = (-1)^{jk} \omega_2^* \, \omega_1^*.$$

Exercise 11.10. *Prove that this ∗-operation is an involution, namely, that* $\omega^{**} = \omega$ *for all* $\omega \in \Gamma^\wedge$.

Furthermore, assume that the ∗-FODC (Γ, d) is left covariant over a ∗-Hopf algebra $\mathcal{A}$. Then $(\Gamma^\wedge, d)$ has a left covariant structure that extends the left covariant structures in degrees 0 and 1 as we have seen in Theorem 10.5. (Recall that in degree 0 we have $\mathcal{A}$, which is left covariant with respect to the co-action ϕ, the comultiplication of $\mathcal{A}$. In degree 1 the left covariant structure is that of Γ.) We denote the left co-action by

$$\Phi_{\Gamma^\wedge} : \Gamma^\wedge \to \mathcal{A} \otimes \Gamma^\wedge.$$

Then for all $\omega \in \Gamma^\wedge$,

$$\Phi_{\Gamma^\wedge}(\omega^*) = (\Phi_{\Gamma^\wedge}(\omega))^*. \tag{11.19}$$

The *-operation on the right side is that introduced for the tensor product of two vector spaces that have *-operations. We then say that $\Gamma^\wedge$ is *left *-covariant*.

Exercise 11.11. *Prove the identity (11.19). Formulate the hypotheses and prove the corresponding identity for right covariant *-FODCs. One then says that $\Gamma^\wedge$ is right *-covariant.*

As usual, if $\Gamma^\wedge$ is both left *-covariant and right *-covariant, then we say that it is *-covariant*. We have seen that this is the case if the *-FODC (Γ, d) is bicovariant over a *-Hopf algebra $\mathcal{A}$.

We next continue with the notation and hypotheses of Theorem 10.10.

Theorem 11.9. *Assume that Γ is *-covariant. Then $\hat{\phi}$ is a *-morphism.*

Proof. We will use that ϕ is a *-morphism in this proof. Take $\theta \in \Gamma^{\wedge n}$ of the form $\theta = a\, db_1 \cdots db_n$ for some $a, b_1, \ldots, b_n \in \mathcal{A}$. Then, using the notation $D = d \otimes id + id \otimes d$, we calculate

$$
\begin{aligned}
\hat{\phi}(\theta^*) &= \hat{\phi}\big((a\, db_1 \cdots db_n)^*\big) \\
&= \hat{\phi}\big(d(b_n)^* \cdots d(b_1)^* a^*\big) \\
&= \hat{\phi}\big(d(b_n)^*\big) \cdots \hat{\phi}\big(d(b_1)^*\big)\hat{\phi}(a^*) \\
&= D\phi(b_n^*) \cdots D\phi(b_1^*)\phi(a^*) \\
&= D(\phi(b_n)^*) \cdots D(\phi(b_1)^*)\phi(a)^* \\
&= (D\phi(b_n))^* \cdots (D\phi(b_1))^*\phi(a)^* \\
&= \big(\phi(a)D(\phi(b_1)) \cdots D(\phi(b_n))\big)^* \\
&= \big(\hat{\phi}(\theta)\big)^*,
\end{aligned}
$$

where the *-operation in each occurrence is the one appropriate for the space to which each element belongs. We also used that D is a *-morphism. ∎

11.6   Examples

It turns out that there are various nonequivalent ways to define a *-structure on $SL_q(2)$. One of these is given for $0 \neq q \in \mathbb{R}$ on the standard generators by

$$\begin{pmatrix} a & b \\ c & d \end{pmatrix}^* = \begin{pmatrix} d & -qc \\ -q^{-1}b & a \end{pmatrix}.$$

Exercise 11.12. *Show that this consistently extends to give a *-structure on $SL_q(2)$ and this makes $SL_q(2)$ (with its previously defined Hopf algebra structure) into a *-Hopf algebra.*

With this *-structure, $SL_q(2)$ is denoted as $SU_q(2)$. For $q = 1$, this is the *-algebra of complex coefficient polynomials in the entries of the matrices in $SU(2)$, the classical Lie group of unitary, 2×2 matrices with complex entries and with determinant 1. This is a standard nontrivial example of a *-Hopf algebra with interesting *-FODCs. (See [81] and [85].)

Exercise 11.13. *Prove that $SU_q(2)$ is the *-algebra generated by the elements a and c subject to the relations*

$$cc^* = c^*c,$$

$$ac = qca,$$

$$ac^* = qc^*a,$$

$$a^*a + c^*c = 1,$$

$$aa^* + q^2 c^*c = 1.$$

Exercise 11.14. *Suppose that $q \in \mathbb{R} \setminus \{0\}$. Show that the relations in the previous exercise are equivalent to requiring that this matrix is unitary:*

$$u = \begin{pmatrix} a & -qc^* \\ c & a^* \end{pmatrix}.$$

Recall that a matrix u is unitary if $u^*u = uu^* = I$, where I is the identity matrix and u^* is the complex conjugation of the transpose of u.

Now we describe the Hopf algebra operations on $SU_q(2)$. These operations are already defined since they are the operations on $SL_q(2)$. We only need to put $b = -qc^*$ and $d = a^*$ into the definitions for $SL_q(2)$ to get the corresponding equations for $SU_q(2)$ in terms of only a and c.

Since ε is a multiplicative *-morphism, this means giving two equations:

$$\varepsilon(a) = 1 \quad \text{and} \quad \varepsilon(c) = 0.$$

Since ϕ is also a multiplicative *-morphism, this again is described by just two equations:

$$\phi(a) = a \otimes a - qc^* \otimes c \quad \text{and} \quad \phi(c) = c \otimes a + a^* \otimes c.$$

Now κ is not a *-morphism in general though it is antimultiplicative. So we have to give four equations to describe it:

$$\kappa(a) = a^*, \quad \kappa(a^*) = a, \quad \kappa(c) = -qc, \quad \kappa(c^*) = -q^{-1}c^*.$$

Next, we will introduce a *-FODC on $SU_q(2)$. But we have already defined this FODC. It is the $3D$ calculus described in Section 6.6. We merely have to show that this is also a *-FODC. The simplest way to do this is to use Theorem 11.6. Then one has to show that the space $\mathcal{X}$ of LIVFs used to define this FODC is *-invariant. And that is a straightforward exercise.

A more direct approach for proving this is to show that $\kappa(r)^* \in \mathcal{R}$ for all $r \in \mathcal{R}$, the right ideal that defines the $3D$ calculus (see Theorem 11.5). Referring again back to Section 6.6, we see that this reduces to verifying $\kappa(r)^* \in \mathcal{R}$ for the six generators of $\mathcal{R}$ that are given there. And that also is a straightforward exercise.

Exercise 11.15. *Please do at least one of these straightforward exercises.*

In much the same way the $4D_\pm$-calculi on $SL_q(2)$ are *-FODCs. See [48] for more details on this and other similar examples.

11.7 Notes

The inclusion of *-operations is not a necessary feature of noncommutative geometry. Excluding it allows one to replace the field of complex numbers by an arbitrary field. But it is an extra, much as air conditioning for a car, which can be handy. And for certain environments, it is even necessary. I have included it so that the reader can see in detail how it works. It is a feature that is easily ignored if one wishes to generalize in that direction, much as one can drive a car without using its air conditioning. However, specializing to the field of complex numbers without also specializing to objects with appropriate *-operations seems to me to be a structural sort of inconsistency. There might well be something called the *-*imperative* supporting my opinion. Anyway, in physics the basic field according to the most commonly accepted viewpoint is the complex numbers. It was with this in mind that I have chosen to develop this theory using the field of complex numbers and *-operations.

The material in this chapter is based on [86], except for Section 11.1, which is as old as the hills, and Theorem 11.6 and Section 11.6, which are based on [48].

Chapter 12
Quantum Principal Bundles

The importance of the role of principal fiber bundles in classical differential geometry and physics is well established. We now consider the generalization of this structure in the context of noncommutative geometry. The material in this chapter is based on several of Micho Đurđevich's papers (see especially [22] and [26] but also [21] and [23]).

We have developed the theory of higher-order differential calculi (HODCs) first precisely because that will play a key role in the theory of quantum principal bundles.

12.1 Definition

We start off with the basic definition. The goal of this definition is to have three algebras $\mathcal{A}$, $\mathcal{B}$, and $\mathcal{V}$ corresponding respectively to the functions on the group G, on the total space E, and on the base space M of a classical principal bundle. We also want a right co-action of $\mathcal{A}$ on $\mathcal{B}$ corresponding to the pullback of the right action of G on E.

Definition 12.1. *A quantum principal bundle (often abbreviated as QPB) $P = (\mathcal{B}, \mathcal{A}, {}_{\mathcal{B}}\Phi)$ consists of the following objects:*

- *(qpb1) A $*$-algebra $\mathcal{B}$ with unit $1_{\mathcal{B}}$ ("functions" on the total space) as well as a $*$-Hopf algebra $\mathcal{A}$ with unit $1_{\mathcal{A}}$ ("functions" on the model fiber space, the structure group).*
- *(qpb2) A right co-action ${}_{\mathcal{B}}\Phi : \mathcal{B} \to \mathcal{B} \otimes \mathcal{A}$ (of the fiber group space on the total space) that is also a unital $*$-morphism of algebras.*
- *(qpb3) The map $\beta : \mathcal{B} \otimes \mathcal{B} \to \mathcal{B} \otimes \mathcal{A}$ defined for $b_1, b_2 \in \mathcal{B}$ by*

$$\beta(b_1 \otimes b_2) := (b_1 \otimes 1_{\mathcal{A}}){}_{\mathcal{B}}\Phi(b_2) \tag{12.1}$$

is surjective. (The co-action ${}_{\mathcal{B}}\Phi$ is free.)

© Springer International Publishing Switzerland 2015   181
S.B. Sontz, *Principal Bundles*, Universitext, DOI 10.1007/978-3-319-15829-7_12

In (qb1), the "total space" is represented by the $*$-algebra $\mathcal{B}$, which is intuitively a "quantum space." On the other hand, the structure group of the bundle is represented by the $*$-Hopf algebra $\mathcal{A}$ (with invertible co-inverse, as always), which is intuitively a "quantum group."

In (qpb2), unital means that

$$_{\mathcal{B}}\Phi(1_{\mathcal{B}}) = 1_{\mathcal{B}} \otimes 1_{\mathcal{A}},$$

while $*$-morphism of algebras means that

$$_{\mathcal{B}}\Phi(b_1 b_2) = {_{\mathcal{B}}\Phi(b_1)}\, {_{\mathcal{B}}\Phi(b_2)}$$

and $\qquad _{\mathcal{B}}\Phi(b^*) = {_{\mathcal{B}}\Phi(b)}^*.$

These conditions can be expressed in terms of commutative diagrams. Here we are using the standard $*$-algebra structure of $\mathcal{B} \otimes \mathcal{A}$ that arises from the $*$-algebra structures of $\mathcal{B}$ and $\mathcal{A}$, namely, $(b \otimes a)^* = b^* \otimes a^*$.

In (qpb3), the multiplication on the right side of (12.1) is the standard multiplication in the tensor product of two algebras. The right side of (12.1) can also be thought of as a left action of $\mathcal{B}$ on $\mathcal{B} \otimes \mathcal{A}$, written as $b_1 \, {_{\mathcal{B}}\Phi(b_2)}$, where in general $b(c \otimes a) = (bc) \otimes a$ for $b, c \in \mathcal{B}$ and $a \in \mathcal{A}$. Either way, we have

$$\beta = (\mu_{\mathcal{B}} \otimes id_{\mathcal{A}}) \circ (id_{\mathcal{B}} \otimes {_{\mathcal{B}}\Phi})$$

as another expression for β, where $\mu_{\mathcal{B}}$ is the multiplication in $\mathcal{B}$.

Usually, we write all unit elements simply as 1 without any subscript and let context resolve the ambiguity of the notation.

We also define

$$\mathcal{V} := \{ b \in \mathcal{B} \mid {_{\mathcal{B}}\Phi(b)} = b \otimes 1 \} = \mathcal{B}_{\mathrm{inv}} \subset \mathcal{B},$$

the right invariant "functions" on the total space, that is, the "functions" on the base space. Notice that $\mathcal{V}$ is a sub-$*$-algebra of $\mathcal{B}$ and that $1 \in \mathcal{V}$. Also notice that we have just written 1 twice without a subscript. So interpret these two usages!

Whenever we say that P is a QPB, we not only mean that this definition holds but also that the same notation $\mathcal{B}, \mathcal{A}, {_{\mathcal{B}}\Phi}$ and so forth applies.

The notation for the right co-action $_{\mathcal{B}}\Phi$ in Đurđevich's papers is F. Using his notations, the definition (12.1) for β reads as $\beta(b_1 \otimes b_2) := b_1 F(b_2)$. The general notational convention we use is that a right co-action of the (ungraded) Hopf algebra $\mathcal{A}$ on a vector space V should be denoted $_V\Phi$. Since we never consider more than one right co-action on a given V, this notation is adequate and unambiguous. Similarly, a left co-action of $\mathcal{A}$ on V should be denoted Φ_V.

Another condition that is used in this context is

- (qpb4) *For each $a \in \mathcal{A}$, there exist finitely many elements $b_j, c_j \in \mathcal{B}$ such that*

$$1_{\mathcal{B}} \otimes a = \sum_j b_j \, _{\mathcal{B}}\Phi(c_j) = \beta\Big(\sum_j b_j \otimes c_j\Big).$$

This condition is equivalent to saying that the image of the map β defined in condition (qpb3) contains $1_{\mathcal{B}} \otimes \mathcal{A}$. But β is a morphism of left $\mathcal{B}$-modules, where the left action on each space is the multiplication of $\mathcal{B}$ on the first factor. So it follows that conditions (qpb3) and (qpb4) are equivalent.

Notice that a classical right action of a group G on a manifold P is free if and only if the map $\alpha : P \times G \to P \times P$ given by $\alpha : (p, g) \mapsto (p, pg)$ is injective. Here the right action $\gamma : P \times G \to P$ is denoted by $(p, g) \mapsto pg$.

Exercise 12.1. *Prove this equivalence. You should be aware that we have not defined free in this context.*

Dualizing this situation gives the motivation for condition (qpb3). More precisely, we put $\mathcal{B} = \mathcal{F}(P)$, the functions on P, and $\mathcal{A} = \mathcal{F}(G)$, the functions on G. Therefore, the pullback $\alpha^* : \mathcal{B} \otimes \mathcal{B} \to \mathcal{B} \otimes \mathcal{A}$ is identified with β. And the right action $\gamma : P \times G \to P$ in turn induces the pullback $\gamma^* : \mathcal{B} \to \mathcal{B} \otimes \mathcal{A}$, which is identified with $_{\mathcal{B}}\Phi$.

Next, we consider $f \in \mathcal{B} \otimes \mathcal{B} \cong \mathcal{F}(P \times P)$. Under this last isomorphism, we have that $f : P \times P \to \mathbb{C}$. Now we take f of the form

$$f(p_1, p_2) = f_1(p_1) f_2(p_2)$$

for functions $f_1, f_2 : P \to \mathbb{C}$. So $f_1, f_2 \in \mathcal{F}(P) = \mathcal{B}$. In short, we are taking f of the form $f = f_1 \otimes f_2 \in \mathcal{B} \otimes \mathcal{B}$. Then for $p \in P$ and $g \in G$, we have that

$$\begin{aligned}
(\beta f)(p, g) &= (\alpha^* f)(p, g) \\
&= f(\alpha(p, g)) \\
&= f(p, pg) \\
&= f_1(p) f_2(\gamma(p, g)) \\
&= (f_1 \otimes 1)(p, g)\, \gamma^* f_2(p, g) \\
&= ((f_1 \otimes 1)\, _{\mathcal{B}}\Phi(f_2))(p, g),
\end{aligned}$$

which implies $\beta(f_1 \otimes f_2) = (f_1 \otimes 1)\, _{\mathcal{B}}\Phi(f_2)$. Since we identify f_1 (resp., f_2) with b_1 (resp., b_2) in condition (qpb3), this shows that the definition of β in (12.1) generalizes the classical case.

Example 12.1. As an elementary example of a quantum principal bundle, we consider the quantum correlate of the classical principal bundle

$$G \xrightarrow{id_G} G \to \{x\}$$

over a one-point space $\{x\}$. So we simply take $P = (\mathcal{A}, \mathcal{A}, \phi)$, where $\mathcal{A}$ is a $*$-Hopf algebra with 1 and $\phi : \mathcal{A} \to \mathcal{A} \otimes \mathcal{A}$ is its co-multiplication. Then the first three properties in the definition of quantum principal bundle are clearly satisfied. As for the fourth property, we note that in this case

$$\beta(b_1 \otimes b_2) = (b_1 \otimes 1)\phi(b_2) = r(b_1 \otimes b_2)$$

holds for all $b_1, b_2 \in \mathcal{A}$, where r was defined in (6.5). As we saw, the map r is actually an isomorphism, so in particular it is surjective. So $P = (\mathcal{A}, \mathcal{A}, \phi)$ is a quantum principal bundle.

It is instructive to identify $\mathcal{V}$ for this example. By definition, we have

$$\mathcal{V} = \{b \in \mathcal{A} \,|\, \phi(b) = b \otimes 1\}.$$

These are the "functions" on the base space in this quantum setting. Of course, the functions on the classical base space $\{x\}$ are a $*$-algebra isomorphic to $\mathbb{C}$, the complex numbers. And we always have $\mathbb{C} \cong \mathbb{C}1 \subset \mathcal{V}$. But is it possible for $\mathcal{V}$ to be larger than $\mathbb{C}1$ in this quantum setting? As we have seen before, by applying $\varepsilon \otimes id$ to both sides of $\phi(b) = b \otimes 1$ (where ε is the co-unit of $\mathcal{A}$), we obtain $b = \varepsilon(b)1 \in \mathbb{C}1$. So we have proved that $\mathcal{V} = \mathbb{C}1$.

Example 12.2. There is another elementary example of a quantum principal bundle. This is the quantum correlate of the classical principal bundle

$$\{e\} \to P \xrightarrow{\;id_P\;} P,$$

where $\{e\}$ is the group with exactly one element (the identity, e), the map $\{e\} \to P$ is arbitrary, and P is an arbitrary smooth manifold. So we take $\mathcal{A} = \mathbb{C}$ in this case, $\mathcal{B}$ to be an arbitrary $*$-algebra with unit, and

$$_\mathcal{B}\Phi : \mathcal{B} \to \mathcal{B} \otimes \mathcal{A} = \mathcal{B} \otimes \mathbb{C}$$

to be the trivial right co-action $_\mathcal{B}\Phi(b) := b \otimes 1$. We leave it to the reader to check the first three properties of a quantum principal bundle. For the fourth property, we see that

$$\beta(b_1 \otimes b_2) = (b_1 \otimes 1) \,_\mathcal{B}\Phi(b_2) = (b_1 \otimes 1)(b_2 \otimes 1) = b_1 b_2 \otimes 1 \cong b_1 b_2,$$

so that β corresponds to the multiplication in $\mathcal{B}$. But the multiplication map $\mathcal{B} \otimes \mathcal{B} \to \mathcal{B}$ is surjective, since $\mathcal{B}$ has a unit 1.

Exercise 12.2. *Identify $\mathcal{V}$ in Example 12.2. Does this compare well with the classical case?*

Example 12.3. This example includes the previous two examples as special cases, as the reader can check. Now we wish to consider the quantum correlate of the explicitly trivial classical principal bundle

$$pr_1 : M \times G \to M,$$

where pr_1 is the *projection* onto the first factor. So we take $P = (\mathcal{B}, \mathcal{A}, {}_\mathcal{B}\Phi)$ with $\mathcal{A}$ being a $*$-Hopf algebra with unit and $\mathcal{B} = \mathcal{W} \otimes \mathcal{A}$, where $\mathcal{W}$ is any $*$-algebra with unit. Obviously, we give $\mathcal{B}$ the usual product operations for its multiplication and its $*$-operation. Finally, we define

$$ {}_\mathcal{B}\Phi := id_\mathcal{W} \otimes \phi : \mathcal{W} \otimes \mathcal{A} \rightarrow \mathcal{W} \otimes \mathcal{A} \otimes \mathcal{A}. $$

Exercise 12.3. *Verify that the definitions in Example 12.3 give us a QPB. Also, identify the $*$-algebra $\mathcal{V}$ of its "functions" on the base space. Is this always equal to $\mathcal{W}$, as expected?*

These last three examples, of course, are trivial cases of quantum principal bundles. But they should shed a bit of light on the general case.

Exercise 12.4. *Any right action $P \times G \rightarrow P$ [denoted as $(p, g) \mapsto pg$] can be used to define a left action $G \times P \rightarrow P$ [denoted as $(g, p) \mapsto gp$] as $gp := pg^{-1}$.*

Similarly, the right co-action ${}_\mathcal{B}\Phi : \mathcal{B} \rightarrow \mathcal{B} \otimes \mathcal{A}$ can be used to define $\Phi_\mathcal{B} : \mathcal{B} \rightarrow \mathcal{A} \otimes \mathcal{B}$ as $\Phi_\mathcal{B} := \kappa^{-1}(b^{(1)}) \otimes b^{(0)}$, where ${}_\mathcal{B}\Phi(b) = b^{(0)} \otimes b^{(1)}$.

*Motivate the definition of $\Phi_\mathcal{B}$ and show that it is a left co-action. What happens if we use κ instead of κ^{-1} in the definition of $\Phi_\mathcal{B}$? **Hint:** Use (B.10). Even so, this is a laborious calculation. It seems that (B.10) is no longer an identity if one replaces κ with κ^{-1}. Of course, for a Hopf algebra with $\kappa^{-1} = \kappa$, things work out.*

12.2 The Translation Map

This is a rather technical section, but it concerns a basic structure in some formalisms of QPBs. However, in our approach we take the controversial position that the transition map is a side issue for an introductory book. Thus, this section is optional for reading the rest of this book though it is useful for preparing oneself for other approaches.

This story starts with the map $\beta : \mathcal{B} \otimes \mathcal{B} \rightarrow \mathcal{B} \otimes \mathcal{A}$ associated with a QPB $(\mathcal{B}, \mathcal{A}, {}_\mathcal{B}\Phi)$ given by $\beta(b_1 \otimes b_2) = (b_1 \otimes 1_\mathcal{A}) {}_\mathcal{B}\Phi(b_2)$ for $b_1, b_2 \in \mathcal{B}$ as defined above in Eq. (12.1). Recall that β is an essential part of the definition of a QPB. Being surjective, the map β factors through its quotient by the kernel of β and gives an isomorphism of that quotient with $\mathcal{B} \otimes \mathcal{A}$, namely,

$$ \mathcal{B} \otimes \mathcal{B} / \ker \beta \xrightarrow{\cong} \mathcal{B} \otimes \mathcal{A}. $$

A nontrivial theorem that holds for QPBs with a *compact quantum group* says that there exists a canonical isomorphism

$$ \mathcal{B} \otimes_\mathcal{V} \mathcal{B} \xrightarrow{\cong} \mathcal{B} \otimes \mathcal{B} / \ker \beta. $$

This is proved in [24]. However, for the present purposes in this section, we will merely assume that this is an isomorphism. In later discussions of the material of this section, we will implicitly assume this as a hypothesis. Then the importance of the result in [24] is that there is a large class of examples where this hypothesis holds.

Composing these isomorphisms, we get an isomorphism

$$X : \mathcal{B} \otimes_{\mathcal{V}} \mathcal{B} \xrightarrow{\cong} \mathcal{B} \otimes \mathcal{A}.$$

Definition 12.2. *We say that a QPB satisfies the* Hopf–Galois condition *(see [71] and [72]) if this holds:*

- *(qpb5) The map*

$$X : \mathcal{B} \otimes_{\mathcal{V}} \mathcal{B} \to \mathcal{B} \otimes \mathcal{A}$$

defined by $b_1 \otimes_{\mathcal{V}} b_2 \mapsto b_1 \, {}_{\mathcal{B}}\Phi(b_2)$ *is an isomorphism.*

This condition is often included in the definition of a QPB, but we have separated it since it will only be used in this section.

We define the associated *(quantum) translation map* $\tau : \mathcal{A} \to \mathcal{B} \otimes_{\mathcal{V}} \mathcal{B}$ by the formula

$$\tau(a) := X^{-1}(1 \otimes a) \qquad (12.2)$$

for all $a \in \mathcal{A}$. This gets us to the definition in seconds flat, but from where did this definition come?

The motivation for the definition of the translation map τ comes from the isomorphism $\delta : P \times_M P \to P \times G$ associated to a classical principal bundle $\pi : P \to M$ with group G. We now outline how this is done. First, the *fiber product* $P \times_M P$ is defined to be

$$P \times_M P := \{(p,q) \in P \times P \mid \pi(p) = \pi(q)\}.$$

And for any $(p,g) \in P \times G$, we define $\gamma(p,g) := (p, pg) \in P \times_M P$. Then γ is an isomorphism (for now, C^∞ diffeomorphism) in the appropriate category

$$\gamma : P \times G \to P \times_M P .$$

We then define $\delta := \gamma^{-1}$ so that

$$\delta : P \times_M P \to P \times G$$

is again an isomorphism (in the same category as γ). So δ^* corresponds to X^{-1} above. Actually, there is an explicit formula for δ given by

$$\delta(p_1, p_2) = (p_1, p_1^{-1} p_2),$$

where p_1 and p_2 are in the same fiber of P [i.e., $\pi(p_1) = \pi(p_2)$] and $g := p_1^{-1} p_2$ is the unique element in G such that $p_2 = p_1 g$.

Exercise 12.5. *Prove that $p_1^{-1} p_2$ exists and is unique. Then show that the map $P \times_M P \to G$ given by $(p_1, p_2) \mapsto p_1^{-1} p_2$ is smooth.*

Exercise 12.6. *Prove that γ and δ are inverse isomorphisms.*

We next define the *(classical) translation function* $t : P \times_M P \to G$ by

$$t(p_1, p_2) := p_1^{-1} p_2,$$

where $\pi(p_1) = \pi(p_2)$. This is also called the *affine operation* in [76]. It follows that

$$\delta(p_1, p_2) = (p_1, p_1^{-1} p_2) = (p_1, t(p_1, p_2))$$

so that the essential part of δ is simply the classical transition map. Then in this case we have that

$$X^{-1} = \delta^* : \mathcal{F}(P \times G) \to \mathcal{F}(P \times_M P).$$

Let $a \in \mathcal{F}(G)$, that is, $a : G \to \mathbb{R}$. Let $1 \times a : P \times G \to \mathbb{R}$ be given by $(p, g) \mapsto 1(p) a(g) = a(g)$. Then for $(p_1, p_2) \in P \times_M P$, we have

$$X^{-1}(1 \times a)(p_1, p_2) = \delta^*(1 \times a)(p_1, p_2)$$
$$= (1 \times a)\delta(p_1, p_2)$$
$$= (1 \times a)(p_1, t(p_1, p_2))$$
$$= a(t(p_1, p_2))$$
$$= (t^* a)(p_1, p_2).$$

Identifying $1 \times a$ with $1 \otimes a$ as usual, we then have

$$\tau(a) = X^{-1}(1 \otimes a) = t^*(a).$$

So the translation map τ on the quantum side corresponds to the translation function t on the classical side via pullback. This not only motivates the definition of τ but also gives a setting (namely, classical differential geometry) in which we can see what is going on.

Exercise 12.7. *Find formulas for the translation functions of the QPBs defined in Examples 12.1, 12.2, and 12.3.*

Next, we are going to present some of the basic properties of the quantum translation map τ. Obviously, these are essential ingredients in the theory. Unfortunately, these are somewhat painful just to state, let alone to prove! This seems to be a price that one has to pay (at least for now) to get an abstract theory. A reference for this material is [8].

First, we introduce a Sweedler-type notation for the quantum translation map: $\tau(a) = [a]_1 \otimes [a]_2$. Recall that the tensor product in this notation is actually $\otimes_V$. Also, we have $[a]_1, [a]_2 \in \mathcal{B}$.

Theorem 12.1. *Let $\Phi_\mathcal{B}$ be as in Exercise 12.4, let $f \in \mathcal{V}$, and let $a, c \in \mathcal{A}$. Then the following identities hold:*

1. $[a^{(1)}]_1 \otimes [a^{(1)}]_2 \otimes a^{(2)} = [a]_1 \otimes {}_\mathcal{B}\Phi[a]_2.3pt$

2. $a^{(1)} \otimes [a^{(2)}]_1 \otimes [a^{(2)}]_2 = \Phi_\mathcal{B}[a]_1 \otimes [a]_2.3pt$

3. $[a]_1 [a]_2 = \varepsilon(a)1.3pt$

4. $f[a]_1 \otimes [a]_2 = [a]_1 \otimes [a]_2 f.3pt$

5. $[a]_2^* \otimes [a]_1^* = [\kappa(a)^*]_1 \otimes [\kappa(a)^*]_2.3pt$

6. $[c]_1 [a]_1 \otimes [a]_2 [c]_2 = [ac]_1 \otimes [ac]_2.$

In the first two identities, we have used the usual Sweedler notation associated with the co-multiplication, namely, $\phi(a) = a^{(1)} \otimes a^{(2)}$.

Proof. To show part 1, we consider this diagram:

$$
\begin{array}{ccc}
\mathcal{B} \otimes \mathcal{B} & \xrightarrow{\ id_\mathcal{B} \otimes {}_\mathcal{B}\Phi\ } & \mathcal{B} \otimes \mathcal{B} \otimes \mathcal{A} \\
{\scriptstyle id_\mathcal{B} \otimes {}_\mathcal{B}\Phi}\Big\downarrow & & \Big\downarrow{\scriptstyle id_\mathcal{B} \otimes {}_\mathcal{B}\Phi \otimes id_\mathcal{A}} \\
\mathcal{B} \otimes \mathcal{B} \otimes \mathcal{A} & \xrightarrow{\ id_\mathcal{B} \otimes id_\mathcal{B} \otimes \phi\ } & \mathcal{B} \otimes \mathcal{B} \otimes \mathcal{A} \otimes \mathcal{A} \\
{\scriptstyle \mu_\mathcal{B} \otimes id_\mathcal{A}}\Big\downarrow & & \Big\downarrow{\scriptstyle \mu_\mathcal{B} \otimes id_\mathcal{A} \otimes id_\mathcal{A}} \\
\mathcal{B} \otimes \mathcal{A} & \xrightarrow{\ id_\mathcal{B} \otimes \phi\ } & \mathcal{B} \otimes \mathcal{A} \otimes \mathcal{A}.
\end{array}
$$

The upper square commutes because ${}_\mathcal{B}\Phi$ is a right co-action. The lower square trivially commutes. The composite of the vertical arrows on the left side of the diagram gives precisely β, while the vertical arrows on the right compose to give $\beta \otimes id_\mathcal{A}$. We let $a \in \mathcal{A}$ be an arbitrary element. We then consider $[a]_1 \otimes [a]_2$ in the upper left corner, namely, in $\mathcal{B} \otimes \mathcal{B}$. We chase this element all the way straight down (getting $1 \otimes a$) and then across to $1 \otimes a^{(1)} \otimes a^{(2)}$. This element in turn is the image of $[a^{(1)}]_1 \otimes [a^{(1)}]_2 \otimes a^{(2)}$ in $\mathcal{B} \otimes \mathcal{B} \otimes \mathcal{A}$ under the composed vertical arrow $\beta \otimes id_\mathcal{A}$. On the other hand, just pushing $[a]_1 \otimes [a]_2$ across gives $[a]_1 \otimes {}_\mathcal{B}\Phi[a]_2$. So these two elements in $\mathcal{B} \otimes \mathcal{B} \otimes \mathcal{A}$ are equal modulo an element in the kernel of the vertical arrow $\beta \otimes id_\mathcal{A}$; that is, they are equal in $\mathcal{B} \otimes_V \mathcal{B} \otimes \mathcal{A}$. And this proves part 1.

The proof of part 2 is dual to that of part 1 in some sense. We leave the details to the reader.

For the proof of part 3, we use the following commutative diagram:

$$
\begin{array}{ccccc}
\mathcal{B} \otimes \mathcal{B} & \xrightarrow{id \otimes {}_{\mathcal{B}}\Phi} & \mathcal{B} \otimes \mathcal{B} \otimes \mathcal{A} & \xrightarrow{id \otimes id \otimes \varepsilon} & \mathcal{B} \otimes \mathcal{B} \otimes \mathbb{C} \\
\beta \downarrow & & \downarrow \mu_{\mathcal{B}} \otimes id & & \downarrow \mu_{\mathcal{B}} \otimes id \\
\mathcal{B} \otimes \mathcal{A} & \xrightarrow{id} & \mathcal{B} \otimes \mathcal{A} & \xrightarrow{id \otimes \varepsilon} & \mathcal{B} \otimes \mathbb{C}.
\end{array}
$$

Note that the square on the left commutes by the definition of β, while the other square trivially commutes. We now chase an element through this diagram, but not in the standard way, since we start with an element of the form $1 \otimes a$ in $\mathcal{B} \otimes \mathcal{A}$ in the lower left corner. This goes straight across to the lower right corner: $1 \otimes a \mapsto 1 \otimes \varepsilon(a) \cong \varepsilon(a)1$. Now we use the hypothesis that β is surjective to chase $1 \otimes a$ up the vertical arrow on the left to get an element $[a]_1 \otimes [a]_2$ in $\mathcal{B} \otimes \mathcal{B}$. Next, we observe that the composition of the two horizontal arrows at the top of the diagram is (essentially) the identity, since ${}_{\mathcal{B}}\Phi$ is a right co-action. So chasing $[a]_1 \otimes [a]_2$ across to the upper right corner simply gives us $[a]_1 \otimes [a]_2 \otimes 1$. This element then gets mapped down to $[a]_1[a]_2 \otimes 1 \cong [a]_1[a]_2$ in the lower right corner. Equating the results of these two chases yields part 3.

Since this material is optional for this book, we do not prove parts 4–6. See [8] for more details. ∎

12.3 Extending Properties of FODCs

In order to discuss the differential calculus of a QPB, we will need more information about extending properties of first-order differential calculi (FODCs) to their HODCs. It is no surprise that FODCs and HODCs enter into the theory of differential calculi of QPBs. But it may come as a surprise that we still need to develop our understanding of them even further.

We let (Γ, d) be a bicovariant FODC over $\mathcal{A}$. (Actually, we can do some of this theory in the more general setting where Γ is only left covariant over $\mathcal{A}$. See [22] for the details.) We recall some definitions, notations, and properties already established in Chapter 3 and Section 6.4. So Γ is a left $\mathcal{A}$-module, and there exists the left co-action of $\mathcal{A}$ on Γ denoted as

$$
\Phi_\Gamma : \Gamma \to \mathcal{A} \otimes \Gamma,
$$

which is a left $\mathcal{A}$-module morphism. Also, the differential $d : \mathcal{A} \to \Gamma$ is a left $\mathcal{A}$-co-module morphism. The vector subspace of left invariant elements in Γ is defined by

$$
{}_{inv}\Gamma := \{\omega \in \Gamma \mid \Phi_\Gamma(\omega) = 1 \otimes \omega\}.
$$

Since (Γ, d) is a left covariant FODC over $\mathcal{A}$, it can be realized (up to an isomorphism) as

$$\Gamma = \mathcal{A} \otimes \left(\frac{\ker \varepsilon}{\mathcal{R}} \right),$$

where $\mathcal{R}$ can be any right ideal in $\mathcal{A}$ satisfying $\mathcal{R} \subset \ker \varepsilon$. Then the left co-action Φ_Γ of $\mathcal{A}$ on Γ is given by $\phi \otimes id$. Also, the subspace of left invariant elements is

$$_{\text{inv}}\Gamma = 1 \otimes \left(\frac{\ker \varepsilon}{\mathcal{R}} \right) \cong \frac{\ker \varepsilon}{\mathcal{R}}.$$

The quantum germs map $\pi : \mathcal{A} \to {_{\text{inv}}}\Gamma$ is given by

$$\pi(a) = \kappa(a^{(1)}) \, da^{(2)}.$$

It is surjective, and $\ker \pi = \mathcal{R} \oplus \mathbb{C}1$. There is a right action of $\mathcal{A}$ on $_{\text{inv}}\Gamma$ given by

$$\pi(a) \circ b = \pi(ab - \varepsilon(a)b)$$

for all $a, b \in \mathcal{A}$. This right action of $\mathcal{A}$ on $_{\text{inv}}\Gamma$ is uniquely determined by the criterion that the vector space isomorphism of $_{\text{inv}}\Gamma$ with $\ker \varepsilon / \mathcal{R}$ be an isomorphism of right $\mathcal{A}$-modules as well, where the latter space has the right action of $\mathcal{A}$ induced by the multiplication in $\mathcal{A}$.

Some identities for these structures are

$$da = a^{(1)} \pi(a^{(2)}),$$

$$\pi(ab) = \varepsilon(a)\pi(b) + \pi(a) \circ b,$$

$$\pi(a)^* = -\pi\big(\kappa(a)^*\big),$$

$$\pi_{\text{inv}}(da) = \pi(a).$$

The first identity is Proposition 6.4, part 6, the second identity is Eq. (6.29), and the third identity is Eq. (11.13). The last equality is (6.33).

We now take all these known structures and extend them and their properties to various enveloping graded vector spaces. The rest of this section contains an inordinate number of new definitions and their immediate consequences.

First, we recall the notations $\Gamma^\otimes$ for the tensor algebra on Γ and $\Gamma^\wedge$ for the universal enveloping algebra of Γ. The left co-actions ϕ (which is the co-multiplication of $\mathcal{A}$) and Φ_Γ of $\mathcal{A}$ on $\mathcal{A}$ in degree 0 and Γ in degree 1, respectively, have unique extensions to graded algebra, differential morphisms

$$\Phi_{\Gamma^\otimes} : \Gamma^\otimes \to \mathcal{A} \otimes \Gamma^\otimes,$$

$$\Phi_{\Gamma^\wedge} : \Gamma^\wedge \to \mathcal{A} \otimes \Gamma^\wedge,$$

which are left co-actions.

Exercise 12.8. *Write out and prove the assertions in the previous paragraph.*

Next, for $k \geq 1$, we define

$$_{\mathrm{inv}}\Gamma^{\otimes k} := {_{\mathrm{inv}}}\Gamma \otimes \cdots \otimes {_{\mathrm{inv}}}\Gamma$$

with k factors. The tricky bit is the "correct" definition of $_{\mathrm{inv}}\Gamma^{\otimes 0}$. To make things work out, this will not be $\mathcal{A}$ in general, but rather

$$_{\mathrm{inv}}\Gamma^{\otimes 0} := \{a \in \mathcal{A} \,|\, \phi(a) = 1 \otimes a\}.$$

This is the natural definition, because $\phi = \Phi^0_{\Gamma\otimes}$ is the left co-action of $\mathcal{A}$ in degree 0. But by applying $id \otimes \varepsilon$ to $\phi(a) = 1 \otimes a$, we see that $a = \varepsilon(a)1$ (as we have noted several times before). Therefore, $_{\mathrm{inv}}\Gamma^{\otimes 0} \subset \mathbb{C}1$. The opposite inclusion is trivial and so $_{\mathrm{inv}}\Gamma^{\otimes 0} = \mathbb{C}1 \cong \mathbb{C}$.

So given these definitions, we can add them up to define

$$({_{\mathrm{inv}}}\Gamma)^{\otimes} := {_{\mathrm{inv}}}\Gamma^{\otimes} := \oplus_{k\geq 0} {_{\mathrm{inv}}}\Gamma^{\otimes k}.$$

There exists a natural isomorphism

$$_{\mathrm{inv}}\Gamma^{\otimes} \cong {_{\mathrm{inv}}}(\Gamma^{\otimes}),$$

where the space on the right side consists of the left invariant elements of $\Gamma^{\otimes}$ with respect to the co-action $\Phi_{\Gamma\otimes}$, namely,

$$_{\mathrm{inv}}(\Gamma^{\otimes}) := \{\omega \in \Gamma^{\otimes} \,|\, \Phi_{\Gamma\otimes}(\omega) = 1 \otimes \omega\}.$$

Since $\Phi^{\otimes}_{\Gamma}$ is a multiplicative morphism, $_{\mathrm{inv}}(\Gamma^{\otimes})$ is a subalgebra of $\Gamma^{\otimes}$.

Much in the same vein we define

$$_{\mathrm{inv}}\Gamma^{\wedge} := \{\omega \in \Gamma^{\wedge} \,|\, \Phi_{\Gamma^{\wedge}}(\omega) = 1 \otimes \omega\},$$

the subalgebra of left invariant elements in $\Gamma^{\wedge}$. Following the pattern in (6.31), we have a canonical identification

$$\Gamma^{\wedge} = \mathcal{A} \otimes {_{\mathrm{inv}}}\Gamma^{\wedge}.$$

Using this identification, we define the *canonical projection map*

$$\pi_{\mathrm{inv}} : \Gamma^{\wedge} \to {_{\mathrm{inv}}}\Gamma^{\wedge}$$

by $\pi_{\mathrm{inv}} := \varepsilon \otimes id$. This is just the natural extension of the map π_{inv} referred to above in degree 0.

Now we extend the right action $\circ$ on $_{\text{inv}}\Gamma$ to $_{\text{inv}}(\Gamma^{\otimes})$. [And an exactly analogous procedure extends this action on $_{\text{inv}}\Gamma$ to $_{\text{inv}}(\Gamma^{\wedge})$.] We already have this action defined on the degree-1 elements (that is, on the elements of $_{\text{inv}}\Gamma$), so we have two tasks: First, define the action on degree-0 elements (that is, on $\mathbb{C}$) and second on elements of degree 2 and above. For $b \in \mathcal{A}$ and $\lambda \in \mathbb{C}$, we define

$$\lambda \circ a := \varepsilon(a)\lambda.$$

So here we are thinking of the left invariant element λ in degree 0 being acted on by a. Next, for $\theta_1, \ldots, \theta_k \in {}_{\text{inv}}\Gamma$ with $k \geq 2$, we define

$$(\theta_1 \otimes \cdots \otimes \theta_k) \circ a := (\theta_1 \circ a^{(1)}) \otimes (\theta_2 \circ a^{(2)}) \otimes \cdots \otimes (\theta_k \circ a^{(k)}), \qquad (12.3)$$

where the $(k-1)$-fold co-multiplication applied to $a \in \mathcal{A}$ is written in Sweedler's notation as $a^{(1)} \otimes a^{(2)} \otimes \cdots \otimes a^{(k)}$. Note that this action sends elements of degree k to elements of degree k for all $k \geq 0$.

Exercise 12.9. *Verify that this is a right action of $\mathcal{A}$ on $_{\text{inv}}(\Gamma^{\otimes})$ [or on $_{\text{inv}}(\Gamma^{\wedge})$ if one replaces $\otimes$ with $\wedge$ throughout].*

Exercise 12.10. *Check that this is the unique extension of $\circ$ to a right action of $\mathcal{A}$ on $_{\text{inv}}(\Gamma^{\otimes})$ [or analogously on $_{\text{inv}}(\Gamma^{\wedge})$] that satisfies*

$$1 \circ a = \varepsilon(a)1$$

$$(\theta \otimes \eta) \circ a = (\theta \circ a^{(1)}) \otimes (\eta \circ a^{(2)})$$

for all $a \in \mathcal{A}$ and all $\theta, \eta \in {}_{\text{inv}}(\Gamma^{\otimes})$, where as usual $\phi(a) = a^{(1)} \otimes a^{(2)}$ (which is what the onefold co-multiplication is).

Proposition 12.1. *The graded algebra $_{\text{inv}}(\Gamma^{\wedge})$ is invariant under the action of the differential d of $\Gamma^{\wedge}$. So restricting d to $_{\text{inv}}(\Gamma^{\wedge})$ makes it a graded differential algebra.*

Proof. First, we take a degree-0 element $\theta \in {}_{\text{inv}}(\Gamma^{\wedge 0}) \cong \mathbb{C}$. We must show that $d\theta$ is also left invariant. But $d\theta = 0$, which is trivially left invariant. We note that this argument partially motivates our definition of the degree-0 elements in $_{\text{inv}}(\Gamma^{\wedge})$.

The proof for $\deg(\theta) = |\theta| \geq 1$ is left to the reader. ∎

Exercise 12.11. *Complete the proof of the previous proposition.* **Hint:** *The differential d is left covariant.*

Theorem 12.2. *We have the following identity:*

$$d(\theta \circ a) = (d\theta) \circ a - \pi(a^{(1)}) \otimes (\theta \circ a^{(2)}) + (-1)^{|\theta|}(\theta \circ a^{(1)}) \otimes \pi(a^{(2)}) \qquad (12.4)$$

for all $a \in \mathcal{A}$ and all $\theta \in {}_{\text{inv}}(\Gamma^{\otimes})$, where again $\phi(a) = a^{(1)} \otimes a^{(2)}$ is Sweedler's notation for the co-multiplication.

Proof. For the case when θ is left invariant of degree 0, that is, $\theta = \lambda \in \mathbb{C}$, we have $d\lambda = 0$, implying $(d\lambda) \circ a = 0$. Also, $d(\lambda \circ a) = d(\varepsilon(a)\lambda) = 0$. The last two terms on the right side of (12.4) in this case are

$$-\pi(a^{(1)})(\lambda \circ a^{(2)}) + (\lambda \circ a^{(1)})\pi(a^{(2)}) = -\pi(a^{(1)})(\varepsilon(a^{(2)})\lambda) + (\varepsilon(a^{(1)})\lambda)\pi(a^{(2)})$$

$$= -\lambda\pi(\varepsilon(a^{(2)})a^{(1)}) + \lambda\pi(\varepsilon(a^{(1)})a^{(2)})$$

$$= -\lambda\pi(a) + \lambda\pi(a)$$

$$= 0.$$

As is customary, we have not written the symbol $\otimes$ when one of the factors is in $\mathbb{C}$. So this case of (12.4) is quickly verified.

For the case where θ has degree 1, that is, $\theta \in {}_{\mathrm{inv}}\Gamma$, we write $\theta = \pi(b)$ for some $b \in \mathcal{A}$ and then calculate as follows. (But be warned that this case is not as quickly settled as the previous one.) First, we have

$$d(\theta \circ a) = d(\pi(b) \circ a)$$

$$= d\left(\pi\left((ba) - \varepsilon(b)a\right)\right)$$

$$= d\left(\pi(ba) - \varepsilon(b)\,\pi(a)\right)$$

$$= d\left(\pi(ba)\right) - \varepsilon(b)\,d\pi(a)$$

$$= -\pi((ba)^{(1)}) \otimes \pi((ba)^{(2)}) + \varepsilon(b)\,\pi(a^{(1)}) \otimes \pi(a^{(2)})$$

$$= -\pi(b^{(1)}a^{(1)}) \otimes \pi(b^{(2)}a^{(2)}) + \varepsilon(b)\,\pi(a^{(1)}) \otimes \pi(a^{(2)}). \quad (12.5)$$

Next, we see that

$$d(\theta) \circ a = d(\pi(b)) \circ a$$

$$= -\left(\pi(b^{(1)}) \otimes \pi(b^{(2)})\right) \circ a$$

$$= -\left(\pi(b^{(1)}) \circ a^{(1)}\right) \otimes \left(\pi(b^{(2)}) \circ a^{(2)}\right)$$

$$= -\left(\pi(b^{(1)}a^{(1)}) - \varepsilon(b^{(1)})\pi(a^{(1)})\right) \otimes \left(\pi(b^{(2)}a^{(2)}) - \varepsilon(b^{(2)})\pi(a^{(2)})\right)$$

$$= -\pi(b^{(1)}a^{(1)}) \otimes \pi(b^{(2)}a^{(2)}) - \varepsilon(b^{(1)})\varepsilon(b^{(2)})\pi(a^{(1)}) \otimes \pi(a^{(2)})$$

$$\quad + \varepsilon(b^{(1)})\pi(a^{(1)}) \otimes \pi(b^{(2)}a^{(2)}) + \pi(b^{(1)}a^{(1)}) \otimes \varepsilon(b^{(2)})\pi(a^{(2)})$$

$$= -\pi(b^{(1)}a^{(1)}) \otimes \pi(b^{(2)}a^{(2)}) - \varepsilon(b^{(1)})\varepsilon(b^{(2)})\pi(a^{(1)}) \otimes \pi(a^{(2)})$$

$$\quad + \pi(a^{(1)}) \otimes \pi\left(\varepsilon(b^{(1)})b^{(2)}a^{(2)}\right) + \pi\left(\varepsilon(b^{(2)})b^{(1)}a^{(1)}\right) \otimes \pi(a^{(2)})$$

$$= -\pi(b^{(1)}a^{(1)}) \otimes \pi(b^{(2)}a^{(2)}) - \varepsilon(b)\pi(a^{(1)}) \otimes \pi(a^{(2)})$$

$$\quad + \pi(a^{(1)}) \otimes \pi\left(ba^{(2)}\right) + \pi\left(ba^{(1)}\right) \otimes \pi(a^{(2)}), \quad (12.6)$$

where in the last equality we used $b = \varepsilon(b^{(1)})b^{(2)} = \varepsilon(b^{(2)})b^{(1)}$, which also implies that $\varepsilon(b) = \varepsilon(b^{(1)})\varepsilon(b^{(2)})$.

We also have

$$\pi(a^{(1)}) \otimes \big(\pi(b) \circ a^{(2)}\big) = \pi(a^{(1)}) \otimes \big(\pi(ba^{(2)}) - \varepsilon(b)\pi(a^{(2)})\big)$$
$$= \pi(a^{(1)}) \otimes \pi(ba^{(2)}) - \varepsilon(b)\pi(a^{(1)}) \otimes \pi(a^{(2)}) \qquad (12.7)$$

as well as

$$\big(\pi(b) \circ a^{(1)}\big) \otimes \pi(a^{(2)}) = \big(\pi(ba^{(1)}) - \varepsilon(b)\pi(a^{(1)})\big) \otimes \pi(a^{(2)})$$
$$= \pi(ba^{(1)}) \otimes \pi(a^{(2)}) - \varepsilon(b)\pi(a^{(1)}) \otimes \pi(a^{(2)}). \qquad (12.8)$$

The three identities (12.5), (12.7), and (12.8) sum to give

$$d(\theta \circ a) + \pi(a^{(1)})\big(\pi(b) \circ a^{(2)}\big) + \big(\pi(b) \circ a^{(1)}\big)\pi(a^{(2)})$$
$$= -\pi(b^{(1)}a^{(1)}) \otimes \pi(b^{(2)}a^{(2)}) - \varepsilon(b)\,\pi(a^{(1)}) \otimes \pi(a^{(2)})$$
$$+ \pi(a^{(1)}) \otimes \pi(ba^{(2)}) + \pi(ba^{(1)}) \otimes \pi(a^{(2)})$$
$$= d(\theta) \circ a. \qquad (12.9)$$

Here we used (12.6) in the last equality. And (12.9) is equivalent to (12.4) in the case when θ has degree 1, that is, $|\theta| = 1$.

The remainder of proof is an induction argument on the degree of θ, using the definition of the right action for degrees ≥ 2 given in (12.3). ∎

Exercise 12.12. *Write out that induction argument.*

Theorem 12.3. *Let Γ be a $*$-covariant FODC over a $*$-Hopf algebra $\mathcal{A}$. Then*

$$(\omega \circ a)^* = \omega^* \circ \kappa(a)^* \qquad (12.10)$$

for all $a \in \mathcal{A}$ and all $\omega \in {}_{\mathrm{inv}}(\Gamma^{\otimes})$.

Remark. As we have emphasized before, the expression $\kappa(a)^*$ is not to be confused with $\kappa(a^*)$, since κ need not be a $*$-morphism.

Proof. As should be expected by now, this is a proof by cases of the degree of ω. First, for $\omega = \lambda \in \mathbb{C}$ of degree 0, we obtain

$$(\omega \circ a)^* = (\lambda \circ a)^*$$
$$= (\varepsilon(a)\lambda)^*$$
$$= (\varepsilon(a))^*\lambda^*$$
$$= (\varepsilon\kappa(a))^*\lambda^*$$

$$= \varepsilon(\kappa(a)^*)\lambda^*$$

$$= \lambda^* \circ \kappa(a)^*$$

$$= \omega^* \circ \kappa(a)^*,$$

where we used the Hopf algebra identity $\varepsilon\kappa = \varepsilon$, the $*$-Hopf algebra identity $(\varepsilon(a))^* = \varepsilon(a^*)$, and definitions.

Next, for $\omega \in {}_{\mathrm{inv}}\Gamma$ of degree 1, we can write $\omega = \pi(b)$ for some $b \in \mathcal{A}$. Recall that $(\pi(a))^* = -\pi(\kappa(a)^*)$. Then we calculate on the one hand

$$(\omega \circ a)^* = (\pi(b) \circ a)^*$$

$$= \big(\pi(ba - \varepsilon(b)a)\big)^*$$

$$= -\pi\Big(\big(\kappa(ba - \varepsilon(b)a)\big)^*\Big)$$

$$= -\pi\Big(\big(\kappa(a)\kappa(b) - \varepsilon(b)\kappa(a)\big)^*\Big)$$

$$= -\pi\Big(\kappa(b)^*\kappa(a)^* - \varepsilon(b)^*\kappa(a)^*\Big).$$

And on the other hand, we have

$$\omega^* \circ \kappa(a)^* = \pi(b)^* \circ \kappa(a)^*$$

$$= -\pi(\kappa(b)^*) \circ \kappa(a)^*$$

$$= -\pi\Big(\kappa(b)^*\kappa(a)^* - \varepsilon(\kappa(b)^*)\kappa(a)^*\Big)$$

$$= -\pi\Big(\kappa(b)^*\kappa(a)^* - [\varepsilon(\kappa(b))]^*\kappa(a)^*\Big)$$

$$= -\pi\Big(\kappa(b)^*\kappa(a)^* - \varepsilon(b)^*\kappa(a)^*\Big),$$

which proves the identity (12.10) in this case.

Finally, for $|\omega| = k \geq 2$, we consider the decomposable subcase

$$\omega = \omega_1 \otimes \omega_2 \otimes \cdots \otimes \omega_k,$$

with each ω_j having degree 1. We use the previous case in the following:

$$(\omega \circ a)^* = \big((\omega_1 \otimes \cdots \otimes \omega_k) \circ a\big)^*$$

$$= \big((\omega_1 \circ a^{(1)}) \otimes \cdots \otimes (\omega_k \circ a^{(k)})\big)^*$$

$$= \big((\omega_k \circ a^{(k)})^* \otimes \cdots \otimes (\omega_1 \circ a^{(1)})^*\big)$$

$$= \big((\omega_k^* \circ \kappa(a^{(k)})^*) \otimes \cdots \otimes (\omega_1^* \circ \kappa(a^{(1)})^*)\big)$$

$$= (\omega_k^* \otimes \cdots \otimes \omega_1^*) \circ \kappa(a)^*$$
$$= (\omega_1 \otimes \cdots \otimes \omega_k)^* \circ \kappa(a)^*$$
$$= \omega^* \circ \kappa(a)^*.$$

In the fifth equality, we used the fact that the $(k-1)$-fold co-product ϕ^{k-1} of ϕ is a $*$-morphism and that the co-inverse κ is anti-co-multiplicative to get

$$\phi^{k-1}(\kappa(a)^*) = (\phi^{k-1}(\kappa(a)))^*$$
$$= (\kappa(a)^{(1)} \otimes \cdots \otimes \kappa(a)^{(k)})^*$$
$$= (\kappa(a^{(k)}) \otimes \cdots \otimes \kappa(a^{(1)}))^*$$
$$= \kappa(a^{(k)})^* \otimes \cdots \otimes \kappa(a^{(1)})^*.$$

That proves (12.10) in this last case.   ∎

Exercise 12.13. *Show that the $(k-1)$-fold co-product of $\kappa(a)$ is as indicated at the end of the last proof.* **Hint:** *Induction on k.*

12.4   Differential Calculus of QPBs

Having now defined a quantum principal bundle, we next want to define what a differential calculus is on a quantum principal bundle. It is almost impossible to overemphasize the importance of this section! The point is that in the classical theory, once we have a principal bundle in hand, we also immediately have the functorially given differential calculi of the three spaces involved: the Lie group, the total space, and the base space. So the differential geometry of the principal bundle is mainly given by the spaces themselves, though connections appear later as an added feature. Not so for the quantum case! For QPBs, we still have to impose the differential calculus structure via axioms to be satisfied, not by functors. Again, the nonuniqueness of the differential structure gives us a much richer theory than the classical case.

As an intuitive starting point, this theory should have *three* differential calculi on the algebras $\mathcal{A}$, $\mathcal{B}$, and $\mathcal{V}$, which correspond respectively to the Lie group G, the total space E, and the base manifold M of a classical principal bundle $E \to M$ with structure group G. Then there must also be some compatibility conditions among these differential calculi. Minimally, the properties of these three differential calculi should extend to all degrees the properties we already have for the given degree-0 forms (namely, the algebras $\mathcal{A}$, $\mathcal{B}$, and $\mathcal{V}$ themselves) since these properties for zero forms correspond to pullbacks of classical properties and the classical k-forms also have the analogous properties via pullback.

Definition 12.3. *A (graded) differential calculus on a QPB* $P = (\mathcal{B}, \mathcal{A}, {}_\mathcal{B}\Phi)$ *is given as these objects and these properties:*

- *(diff1a) A graded differential $*$-algebra $(\Omega(P), d_P)$ over $\mathcal{B}$ such that $\Omega(P)$ is generated as a graded differential calculus by $\Omega^0(P) = \mathcal{B}$, the elements of degree 0. [In particular, $d_P : \Omega^0(P) = \mathcal{B} \to \Omega^1(P)$ is an FODC and $d_P : \Omega(P) \to \Omega(P)$ has degree 1.]*
- *(diff1b) A bicovariant and $*$-covariant FODC (Γ, d) over $\mathcal{A}$.*
- *(diff2) An extension of the right co-action ${}_\mathcal{B}\Phi : \mathcal{B} \to \mathcal{B} \otimes \mathcal{A}$ of the QPB to a necessarily unique right co-action of $\Gamma^\wedge$ on $\Omega(P)$, denoted by*

$$\Omega_{(P)}\Psi : \Omega(P) \to \Omega(P) \otimes \Gamma^\wedge,$$

where $\Gamma^\wedge$ is the universal differential calculus of the FODC (Γ, d) and ${}_{\Omega(P)}\Psi$ is a differential, unital $*$-morphism of graded algebras. Saying that ${}_{\Omega(P)}\Psi$ is a right co-action of $\Gamma^\wedge$ on $\Omega(P)$ means that

$$({}_{\Omega(P)}\Psi \otimes id)\, {}_{\Omega(P)}\Psi = (id \otimes \hat{\phi})\, {}_{\Omega(P)}\Psi$$

and $\qquad (id \otimes \varepsilon)\, {}_{\Omega(P)}\Psi = id,$

where $\hat{\phi} : \Gamma^\wedge \to \Gamma^\wedge \otimes \Gamma^\wedge$ is the unique extension of the co-multiplication $\phi : \mathcal{A} \to \mathcal{A} \otimes \mathcal{A}$ as a morphism of graded differential algebras (as introduced in Theorem 10.6).

Whenever we say that P is a QPB with differential calculus, we mean that all of the notation of this definition applies unless otherwise stated. The labeling of these properties was chosen to make them correspond more or less to the labeling in the definition in [22]. The notation for ${}_{\Omega(P)}\Psi$ in Đurđevich's papers is $\hat{F}$.

Just as the algebra $\mathcal{V}$ corresponding to the base space, formally denoted as M, of a QPB was defined as a secondary object, here in the case of the differential calculus on a QPB the base space differential calculus, denoted as $\Omega(M)$, is also defined as a secondary object, namely, as the right invariant elements in the differential calculus $\Omega(P)$ of the total space:

$$\Omega(M) := \{\omega \in \Omega(P) \mid {}_{\Omega(P)}\Psi(\omega) = \omega \otimes 1_\mathcal{A}\} = \Omega(P)_{\text{inv}}. \qquad (12.11)$$

Here the notation $\otimes$ in the co-domains of ${}_{\Omega(P)}\Psi(\omega)$ and $\hat{\phi}$ denotes the tensor product of graded differential $*$-algebras. This is in agreement with our convention that the notation $\otimes$ always means the tensor product in the category under consideration. This terminology itself deserves careful definitions, which we now give.

Definition 12.4. *For $\mathbb{N}$-graded vector spaces $V = \oplus_{i \geq 0} V^i$ and $W = \oplus_{j \geq 0} W^j$ (where $\mathbb{N}$ is the set of nonnegative integers), their tensor product $V \otimes W$ is the $\mathbb{N}$-graded vector space $\oplus_{k \geq 0} (V \otimes W)^k$ with*

$$(V \otimes W)^k := \oplus_{i+j=k} (V^i \otimes W^j),$$

where the $\otimes$ on the left side is the tensor product of ordinary (ungraded) vector spaces.

As a general principle, the tensor product of structures based on graded vector spaces has the same definition of tensor product on the underlying graded vector space plus whatever other structures that are appropriate. We will now see this principle applied in several of the following definitions. First, recall from Chapter 8 that $G = \oplus_{i \geq 0} G^i$ is said to be a *graded algebra* if G is an algebra with identity 1 satisfying $1 \in G^0$ and $G^i G^j \subset G^{i+j}$. It seems that the next definition should have been given much earlier on, but here it is at last.

Definition 12.5. *Let $G = \oplus_{k \geq 0} G^k$ be a graded algebra. We say that G is a* graded differential algebra *if there exists a linear map $d : G \to G$ that is a differential, that is, $d^2 = 0$, and has degree $+1$, that is, $d : G^k \to G^{k+1}$ for all $k \geq 0$ and satisfies the graded Leibniz rule.*

We say that G is a graded differential *-algebra *if it is a graded differential algebra and a *-algebra satisfying $(G^k)^* \subset G^k$ for all $k \geq 0$ and the differential d is a *-morphism.*

If $H = \oplus_{k \geq 0} H^k$ is also a graded algebra, then we say that a linear map $\alpha : G \to H$ is a morphism of graded differential algebras *provided that α is a morphism of algebras (i.e., preserves multiplication and identity) that satisfies $\alpha : G^k \to H^k$ for all $k \geq 0$ and also intertwines the differentials, that is, $d_H \alpha = \alpha d_G$, where d_G (resp., d_H) is the differential in G (resp., in H).*

Furthermore, we say that α is a morphism of graded differential *-algebras *if it is also a *-morphism.*

Definition 12.6. *Let $V = \oplus_{i \geq 0} V^i$ and $W = \oplus_{j \geq 0} W^j$ be graded algebras. Then we define their* tensor product $V \otimes W$ *to be the graded algebra with multiplication defined by*

$$(v_1 \otimes w_1)(v_2 \otimes w_2) := (-1)^{jk} v_1 v_2 \otimes w_1 w_2,$$

where $v_2 \in V^j$ and $w_1 \in W^k$ are homogeneous elements.

Exercise 12.14. *Prove that this is actually a graded algebra and that $1 \otimes 1$ is its identity. In particular, the multiplication must be shown to be associative.*

Definition 12.7. *Let $V = \oplus_{i \geq 0} V^i$ and $W = \oplus_{j \geq 0} W^j$ be graded differential algebras with differentials d_V and d_W, respectively. Then we define their* tensor product $V \otimes W$ *to be the graded differential algebra with differential d defined by*

$$d(v \otimes w) := d_V(v) \otimes w + (-1)^j v \otimes d_W(w),$$

where $v \in V^j$ and $w \in W^k$.

The crucial point of this definition is that d is indeed a differential, that is, $d^2 = 0$. This depends on the presence of the sign $(-1)^j$ in the definition.

Exercise 12.15. *Prove that $d^2 = 0$.*

Given this extensive list of definitions, we now know exactly what it means when we say that $\Omega(P)\Gamma$ and $\hat{\phi}$ are morphisms of differential algebras. The reader is advised to review the definition of QPB with differential calculus at this point to make sure that this is all understood well.

Also, a right co-action of $\Gamma^\wedge$ or any other enveloping graded algebra of $\mathcal{A}$ on a graded vector space W will be denoted as $_W\Psi$. Similarly, a left co-action of $\Gamma^\wedge$ on W will be denoted as Ψ_W. Recall that the right (resp., left) co-action of $\mathcal{A}$ on a graded vector space W is denoted by $_W\Phi$ (resp., Φ_W).

Notice that there need not be classical objects P and M in this quantum setting even though the notation could mislead one into thinking that they do exist. The notation $\Gamma^\wedge$ for the universal differential calculus is used to maintain agreement with the notation in Đurđevich's papers. Perhaps a more consistent notation for this object would have been $\Omega(G)$, granted that one understands that the classical object G need not exist.

The choice of the universal differential calculus $\Gamma^\wedge$ instead of some other differential calculus enveloping the FODC (Γ, d) is just that: a *choice*. Any other choice works to give a similar theory. See Appendix A of [22] for more details about this when one uses the braided exterior calculus of Woronowicz from [86], which was discussed in Section 10.1.

Proposition 12.2. $\Omega(M)$ *is a graded differential *-subalgebra of* $\Omega(P)$.

Proof. Recall that $\Omega(M)$ was defined in Eq. (12.11). For all $\omega \in \Omega(M)$, we have

$$_{\Omega(P)}\Psi(\omega^*) = (_{\Omega(P)}\Psi(\omega))^* = (\omega \otimes 1)^* = \omega^* \otimes 1,$$

which says that $\omega^* \in \Omega(M)$. So $\Omega(M)$ is closed under the *-operation. As for the differential d_P, we have for all $\omega \in \Omega^k(M)$ that

$$\begin{aligned}
_{\Omega(P)}\Psi(d_P\,\omega) &= d(_{\Omega(P)}\Psi(\omega)) \\
&= d(\omega \otimes 1) \\
&= (d_P\omega) \otimes 1 + (-1)^k \omega \otimes d\,1 \\
&= (d_P\omega) \otimes 1.
\end{aligned}$$

This means that $d_P\omega \in \Omega(M)$ and so $\Omega(M)$ is invariant under the action of d_P. We denote by d_M the restriction of d_P to $\Omega(M)$. So d_M is a graded differential on $\Omega(M)$.

The fact that $_{\Omega(P)}\Psi$ is unital, that is, $_{\Omega(P)}\Psi(1) = 1 \otimes 1$, implies that $1 \in \Omega(M)$.

Finally, suppose that $\omega_1, \omega_2 \in \Omega(M)$. So we have $_{\Omega(P)}\Psi(\omega_1) = \omega_1 \otimes 1$ and $_{\Omega(P)}\Psi(\omega_2) = \omega_2 \otimes 1$. Then

$$_{\Omega(P)}\Psi(\omega_1\omega_2) = {}_{\Omega(P)}\Psi(\omega_1)\,_{\Omega(P)}\Psi(\omega_2) = (\omega_1 \otimes 1)(\omega_2 \otimes 1) = \omega_1\omega_2 \otimes 1,$$

using that $_{\Omega(P)}\Psi$ is a multiplicative homomorphism and $\deg 1 = 0$. Therefore, $\omega_1\omega_2 \in \Omega(M)$, which shows that $\Omega(M)$ is closed under multiplication. ∎

We next define a map [which will turn out to be a right co-action of $\mathcal{A}$ on $\Omega(P)$] by composing with $_{\Omega(P)}\Psi$ as follows:

$$_{\Omega(P)}\Phi := (id \otimes p_0)\, _{\Omega(P)}\Psi : \Omega(P) \to \Omega(P) \otimes \mathcal{A}.$$

Here $p_0 : \Gamma^\wedge \to \mathcal{A}$ is the projection map that is zero on all positive-degree elements and is the identity on zero-degree elements.

Proposition 12.3. *The map* $_{\Omega(P)}\Phi$ *is a right co-action of* $\mathcal{A}$ *on* $\Omega(P)$. *The differential* d_P *of* $\Omega(P)$ *is right covariant with respect to this co-action.*

Example 12.4. *We amplify on Example 12.3 of the explicit product QPB,* $P = (\mathcal{W} \otimes \mathcal{A}, \mathcal{A}, id_{\mathcal{W}} \otimes \phi)$. *We let* Γ *be a bicovariant and* $*$-*covariant FODC over* $\mathcal{A}$ *and* $\Gamma^\wedge$ *be its universal differential calculus. We also let* Υ *be any graded, differential* $*$-*algebra generated as a graded, differential algebra by* $\Upsilon^0 = \mathcal{W}$. *Then we define the graded, differential* $*$-*algebra*

$$\Omega(P) := \Upsilon \otimes \Gamma^\wedge.$$

We also define

$$_{\Omega(P)}\Psi := id_\Upsilon \otimes \hat{\phi} : \Upsilon \otimes \Gamma^\wedge \to \Upsilon \otimes \Gamma^\wedge \otimes \Gamma^\wedge.$$

This is called the explicit product differential calculus *on* P *associated to the factors* Υ *and* $\Gamma^\wedge$.

It is important to note that in the setting of classical differential geometry, this example corresponds to the product of the differential calculi of the base and fiber spaces. However, in this noncommutative setting, there are other possible differential calculi on an explicit product bundle; that is, even though the bundle is trivial (that is, factorizable), it is not necessarily the case that the differential calculus on the bundle is trivial. We will see an example of this strictly quantum phenomenon in Section 14.3, where we construct a "twisted" differential calculus on an explicit product bundle. Moreover, that example is itself just a special case of a very general "twisted" construction given in [23].

Exercise 12.16. *Prove that the construction in Example 12.4 is a differential calculus for the explicit product bundle* P. *Identify* $\Omega(M)$ *in this case.*

12.5   Horizontal Forms

Note that for a classical principal bundle $G \xrightarrow{\iota_p} P \xrightarrow{\pi} M$ we always have the concept of vertical tangent vectors to points in P. Here $\iota_p : G \to P$ is any map of the form $g \mapsto p \cdot g$, with $p \in P$ being arbitrary. The image of each ι_p forms a fiber that is an orbit of the action of G. These fibers are called the *vertical submanifolds* of P. The tangents to these vertical submanifolds form a vector bundle over P defined by

$$\mathrm{Vert}(T(P)) := \mathrm{Ker}\, T(\pi).$$

The relevant commutative diagram here with exact upper row is

$$
\begin{array}{ccccccccc}
0 & \to & T(G) & \xrightarrow{T(\iota)} & T(P) & \xrightarrow{T(\pi)} & T(M) & \to & 0 \\
 & & \tau_G \downarrow & & \downarrow \tau_P & & \downarrow \tau_M & & \\
 & & G & \xrightarrow{\iota} & P & \xrightarrow{\pi_M} & M. & &
\end{array}
$$

So $\mathrm{Vert}(T(P))$ is a vector subbundle of $T(P)$, which is defined without any reference to a connection. And the subbundle $\mathrm{Vert}(T(P))$ is readily seen to be invariant under the right action of G on $T(P)$ induced by taking the derivative of the right action of G on P. Moreover, $\mathrm{Vert}(T(P))$ is invariant under the exterior derivative of $T(P)$. Then one way to define a connection is as an "adequate" way of selecting a subbundle complementary to $\mathrm{Vert}(T(P))$, this being called the subbundle of *horizontal vectors* in $T(P)$. (For example, see [76].) The meaning of "adequate" in this setting is that this complementary subbundle should also be invariant under the right co-action of G and under the exterior derivative. But we can consider the quotient bundle $T(P)/\mathrm{Vert}(T(P))$ as a model for that complementary subbundle. So we always have this short exact sequence, even before introducing a connection:

$$0 \to \mathrm{Vert}(T(P)) \to T(P) \to T(P)/\mathrm{Vert}(T(P)) \to 0.$$

A connection then is a G-invariant, exterior derivative invariant splitting of this short exact sequence. All this is basic classical differential geometry.

But going over to the dual structures (co-vectors instead of vectors), we can work with 1-forms instead. Then the appropriate intrinsic concept dual to vertical vector is *horizontal co-vector*, which is nothing other than a co-vector that annihilates all vertical vectors. We will consider *vertical forms* in the quantum setting in the next section. For now we start off with the concept of a horizontal form, which we emphasize again is defined without any reference to a connection. But we warn the reader that the classical case does not fully motivate the next definition!

Definition 12.8. *Suppose we are given a differential calculus on a QPB. We define*

$$\mathfrak{h} \equiv \mathfrak{hor}(P) := \{\omega \in \Omega(P) \,|\, _{\Omega(P)}\Psi(\omega) \in \Omega(P) \otimes \mathcal{A}\}.$$

We say that the elements of $\mathfrak{hor}(P)$ *are* horizontal forms.

Let us first of all note that $\mathfrak{hor}^0(P) = \Omega^0(P) = \mathcal{B}$ since $_{\Omega(P)}\Psi = {}_{\mathcal{B}}\Phi$ in degree 0. So all the "functions" on the total space are considered to be horizontal, even those that vary along the "vertical" fibers! Notice that the HODC associated to the FODC (Γ, d) over $\mathcal{A}$ only enters in this definition indirectly via the map $_{\Omega(P)}\Psi$. So we want a definition that makes good intuitive sense for any HODC for the "fiber space." More on this in a moment.

Clearly, more motivation is needed for this definition, which does not completely follow from the prior discussion in any intuitive sense. Let's consider a form $\omega \in \Omega^k(P)$ of the form

$$\omega = b_0 \, db_1 \cdots db_k,$$

where each $b_j \in \mathcal{B}$. By definition, such forms span $\Omega^k(P)$. To see whether this form is horizontal, we calculate

$$\begin{aligned}
_{\Omega(P)}\Psi(\omega) &= {}_{\Omega(P)}\Psi(b_0 \, db_1 \cdots db_k) \\
&= {}_{\Omega(P)}\Psi(b_0) \, d({}_{\Omega(P)}\Psi(b_1)) \cdots d({}_{\Omega(P)}\Psi(b_k)) \\
&= {}_{\mathcal{B}}\Phi(b_0) \, d({}_{\mathcal{B}}\Phi(b_1)) \cdots d({}_{\mathcal{B}}\Phi(b_k)),
\end{aligned} \tag{12.12}$$

where in the second equality we used that $_{\Omega(P)}\Psi$ is a morphism of differential algebras and in the last equality that $_{\Omega(P)}\Psi$ extends $_{\mathcal{B}}\Phi$. [By the way, this little argument shows that $_{\Omega(P)}\Psi$ is uniquely determined by $_{\mathcal{B}}\Phi$.] The first factor $_{\mathcal{B}}\Phi(b_0) \in \mathcal{B}$ does not really matter, as we shall see. But the following k factors determine the outcome of this calculation. For simplicity, we take $b \in \mathcal{B}$ and consider the generic expression

$$d({}_{\mathcal{B}}\Phi(b)) = d(b^{(0)} \otimes b^{(1)}) = (d_P b^{(0)}) \otimes b^{(1)} + b^{(0)} \otimes db^{(1)},$$

where $_{\mathcal{B}}\Phi(b)) = b^{(0)} \otimes b^{(1)} \in \mathcal{B} \otimes \mathcal{A}$. Here d on the far left side is the differential in $\mathcal{B} \otimes \mathcal{A}$, while the d on the far right is the differential in the FODC (Γ, d) over $\mathcal{A}$. It is important to note where these two terms live. We have

$$(d_P b^{(0)}) \otimes b^{(1)} \in \Omega^1(P) \otimes \mathcal{A} \qquad \text{and} \qquad b^{(0)} \otimes db^{(1)} \in \mathcal{B} \otimes \Gamma.$$

We see that the terms in (12.12) of the second form are going to cause us trouble, but the terms of the first form, together with the coefficient $_{\Omega(P)}\Phi(b_0)$, will give us an element in $\Omega(P) \otimes \mathcal{A}$, which is to say a horizontal form according to our definition. So to get horizontal forms, it is sufficient to require $db_j^{(1)} = 0$ for $j = 1, \ldots, k$.

And this works for any HODC extending the FODC (Γ, d). Now what does $db_j^{(1)} = 0$ mean? For the case of a classical Lie group with its de Rham exterior calculus, this means that the function $b_j^{(1)}$ on G is locally constant or, in other words, its does not vary locally. Since the function $b^{(0)} \otimes b^{(1)}$ does not vary in the "directions" of the vertical part G, we consider its differential to be a horizontal form. The same interpretation applies in the quantum case because then $db_j^{(1)} = 0$ is taken to mean that the quantum calculus does not detect "variation" of the vertical part $b_j^{(1)} \in \mathcal{A}$. So the intuition is that the form ω is horizontal if its differential is zero along the vertical directions. It seems at first glance that the definition of the horizontal subspace has nothing to do with the differential structure of $\Omega(P)$, but this is not so, as the argument in this paragraph shows. So where does the differential structure enter into Definition 12.8? Precisely in the criterion that $_{\Omega(P)}\Psi$ is a morphism of differential algebras.

It is a well-established tradition in noncommutative geometry to work with a noncommutative version of forms rather than a noncommutative version of tangent vectors. This even has been converted into a sort of folk wisdom of the discipline, namely, that "quantum tangent vectors don't work." (As mentioned earlier, the works in [16–18] form an exception to this "rule" though only in special cases.) However, for purely conventional reasons the co-actions in a QPB are taken to be right co-actions. In the classical (resp., quantum) theory, the choice of the "side" of the action (resp., co-action) is simply a convention.

So the definition of horizontal form given in Definition 12.8 is a step in the right direction when viewed in the light of the classical situation because it says the right co-action of ω, which a priori lies in $\Omega(P) \otimes \Gamma^{\wedge}$, actually lies in $\Omega(P) \otimes \mathcal{A}$. So there is the possibility that the subspace $\mathfrak{hor}(P)$ is invariant under the co-action $_{\Omega}\Psi$ of $\mathcal{A}$. This is true and will be proved ahead in Proposition 12.5. This fact is an example of the covariance imperative and serves as further justification for Definition 12.8.

Proposition 12.4. $\mathfrak{hor}(P)$ *is a graded differential $*$-subalgebra of $\Omega(P)$.*

Exercise 12.17. *Prove this proposition.* **Hint:** *Think about the proof of Proposition 12.2.*

Proposition 12.5. *The subalgebra $\mathfrak{hor}(P)$ of $\Omega(P)$ is invariant under the right co-action $_{\Omega(P)}\Psi$ when restricted to $\mathfrak{hor}(P)$; that is,*

$$_{\Omega(P)}\Psi : \mathfrak{hor}(P) \to \mathfrak{hor}(P) \otimes \Gamma^{\wedge}.$$

Actually, something even stronger is true, namely, that

$$_{\Omega(P)}\Psi : \mathfrak{hor}(P) \to \mathfrak{hor}(P) \otimes \mathcal{A}.$$

Proof. We will prove the second assertion, which clearly implies the first assertion. This is a question of showing that $_{\Omega(P)}\Psi(\omega) \in \mathfrak{hor}(P) \otimes \mathcal{A}$ for all $\omega \in \mathfrak{hor}(P)$, that is, for all $\omega \in \Omega(P)$ satisfying $_{\Omega(P)}\Psi(\omega) \in \Omega(P) \otimes \mathcal{A}$.

But $\Omega_{(P)}\Psi(\omega) \in \mathfrak{hor}(P) \otimes \mathcal{A}$ simply means that

$$(\Omega_{(P)}\Psi \otimes id)(\Omega_{(P)}\Psi(\omega)) \in \Omega(P) \otimes \mathcal{A} \otimes \mathcal{A}.$$

Using that $\Omega_{(P)}\Psi$ is a co-action of $\Gamma^\wedge$ gives

$$(\Omega_{(P)}\Psi \otimes id)\,\Omega_{(P)}\Psi(\omega) = (id \otimes \hat{\phi})\,\Omega_{(P)}\Psi(\omega) \in (id \otimes \hat{\phi})(\Omega(P) \otimes \mathcal{A}).$$

However,

$$(id \otimes \hat{\phi})(\Omega(P) \otimes \mathcal{A}) = (id \otimes \phi)(\Omega(P) \otimes \mathcal{A}) \subset \Omega(P) \otimes \mathcal{A} \otimes \mathcal{A},$$

since $\hat{\phi}$ extends the co-multiplication $\phi : \mathcal{A} \to \mathcal{A} \otimes \mathcal{A}$. ∎

Notation. $\mathfrak{h}(P)\Phi := \Omega_{(P)}\Psi|_{\mathfrak{hor}(P)}$, the restriction of $\Omega_{(P)}\Psi$ to $\mathfrak{hor}(P)$.

So $\mathfrak{h}(P)\Phi$ is a right co-action of $\mathcal{A}$ on $\mathfrak{hor}(P)$. This is why Φ rather than Ψ is used in the notation for it. As noted in detail above, this is dual to the classical situation where we have a natural right action of the structure group of a principal bundle on the vertical (tangent) vectors.

Now we are in a position to give another characterization of $\Omega(M)$. This is an immediate consequence of the preceding discussion.

Proposition 12.6. $\Omega(M) = \{\omega \in \mathfrak{hor}(P) \mid \mathfrak{h}(P)\Phi(\omega) = \omega \otimes 1\}$; *that is,* $\Omega(M)$ *is the subspace of those horizontal forms on the total space that are invariant under the right co-action* $\mathfrak{h}(P)\Phi$ *of* $\mathcal{A}$ *on* $\mathfrak{hor}(P)$.

Be careful here! The differential calculus on the base space is not the horizontal forms on the total space, but rather the right invariant horizontal forms on the total space.

Exercise 12.18. *Identify* $\mathfrak{hor}(P)$ *for the differential calculus introduced in Example 12.4. Does this make sense intuitively?*

Exercise 12.19. *Prove that* $\Omega^0(M) = \mathcal{V}$ *by recalling the definition of these two spaces.*

Notice that we are *not* requiring that $\Omega^0(M)$ generates $\Omega(M)$ even though this is so for many interesting examples. However, it is also false for other interesting examples! This is a good moment to point out that the algebra $\mathfrak{hor}(P)$ need not be generated by $\mathfrak{hor}^0(P) = \mathcal{B}$ and $\mathfrak{hor}^1(P)$.

This next lemma is quite useful, though for now it looks like a rather technical result.

Lemma 12.1. *Let* $\omega \in \Omega^n(P)$ *be a given homogeneous element and let* k *be a given integer satisfying* $0 \leq k \leq n$. *Suppose that for every integer* $l > k$ *we have that*

$$(id \otimes p_l)\,\Omega_{(P)}\Psi(\omega) = 0.$$

Then there are two finite families of elements (each having m elements), namely, $\varphi_1, \ldots, \varphi_m \in \mathfrak{hor}^{n-k}(P)$ *and* $\theta_1, \ldots, \theta_m \in {}_{inv}\Gamma^{\wedge k}$, *satisfying*

$$(id \otimes p_k)\, \Omega(P)\Psi(\omega) = \sum_{j=1}^{m} \Omega(P)\Phi(\varphi_j)\,\theta_j = \sum_{j=1}^{m} \mathfrak{h}(P)\Phi(\varphi_j)\,\theta_j. \qquad (12.13)$$

Remark. Recall that p_j is the projection of $\Gamma^\wedge$ onto $\Gamma^{\wedge j}$. So one hypothesis says that $\Omega(P)\Psi(\omega)$ has no nonzero "piece" with second factor in $\Gamma^{\wedge l}$ for all $l > k$. The conclusion says that $\Omega(P)\Psi(\omega)$ has a very special structure. Let's see what that is. The second equality in (12.13) follows from the definition of $\mathfrak{h}(P)\Phi$. Since $\varphi_j \in \mathfrak{hor}^{n-k}(P)$, we have

$$\mathfrak{h}(P)\Phi(\varphi_j) \in \mathfrak{Hor}^{n-k}(P) \otimes \mathcal{A}$$

for each j, and therefore, since $\theta_j \in {}_{inv}\Gamma^{\wedge k}$, we also have

$$\mathfrak{h}(P)\Phi(\varphi_j)\,\theta_j \in \mathfrak{Hor}^{n-k}(P) \otimes {}_{inv}\Gamma^{\wedge k} \subset \Omega^{n-k}(P) \otimes \Gamma^{\wedge k} \subset (\Omega(P) \otimes \Gamma^\wedge)^n.$$

So (12.13) implies that the part of $\Omega(P)\Psi(\omega)$ that lies in $\Omega^{n-k}(P) \otimes \Gamma^{\wedge k}$ (which by hypothesis is the highest degree in $\Gamma^\wedge$ where this element can have a nonzero piece) actually lies in the smaller subspace $\mathfrak{Hor}^{n-k}(P) \otimes {}_{inv}\Gamma^{\wedge k}$.

Proof. We use Đurđevich's notation $\hat{F} = \Omega(P)\Psi$. We also let *id* denote the identity map of any object, letting context dictate the exact meaning of this symbol. By the above remark, we see that

$$(id \otimes p_k)\hat{F}(\omega) \in \Omega^{n-k}(P) \otimes \Gamma^{\wedge k} = \Omega^{n-k}(P) \otimes \mathcal{A} \otimes {}_{inv}\Gamma^{\wedge k},$$

using the identification $\Gamma^{\wedge k} = \mathcal{A} \otimes {}_{inv}\Gamma^{\wedge k}$. Let $\{\theta_i \mid i \in I\}$ be a (Hamel) basis of ${}_{inv}\Gamma^{\wedge k}$. So we expand this as *finite* sums over i and j, getting

$$(id \otimes p_k)\hat{F}(\omega) = \sum_i \eta_i \otimes \theta_i = \sum_{ij} \xi_{ij} \otimes a_{ij} \otimes \theta_i, \qquad (12.14)$$

where $\eta_i = \sum_j \xi_{ij} \otimes a_{ij} \in \Omega^{n-k}(P) \otimes \mathcal{A}$ for each i. Thus, $\xi_{ij} \in \Omega^{n-k}(P)$ and $a_{ij} \in \mathcal{A}$ for each pair i, j. Applying $\hat{F} \otimes id$ to the previous and using a defining property of $\hat{F}$, we obtain

$$\sum \hat{F}(\xi_{ij}) \otimes a_{ij} \otimes \theta_i = (\hat{F} \otimes p_k)\hat{F}(\omega)$$

$$= (id \otimes p_k)(\hat{F} \otimes id)\hat{F}(\omega)$$

$$= (id \otimes p_k)(id \otimes \hat{\phi})\hat{F}(\omega)$$

$$= (id \otimes p_k)(id \otimes \hat{\phi})\Big(\sum_{ij} \xi_{ij} \otimes a_{ij} \otimes \theta_i + R\Big),$$

where the terms denoted by R lie in $\bigoplus_{j<k}(\Omega^{n-j}(P)\otimes\Gamma^{\wedge j})$ by hypothesis. But the combined action of $\hat{\phi}$ followed by p_k maps all such terms to zero. So we now have

$$\sum\hat{F}(\xi_{ij})\otimes a_{ij}\otimes\theta_i = (id\otimes p_k)(id\otimes\hat{\phi})\Big(\sum\xi_{ij}\otimes a_{ij}\otimes\theta_i\Big)$$

$$= (id\otimes p_k)\sum\xi_{ij}\otimes\hat{\phi}(a_{ij}\otimes\theta_i)$$

$$= (id\otimes p_k)\sum\xi_{ij}\otimes\phi(a_{ij})\hat{\phi}(\theta_i)$$

$$= (id\otimes p_k)\sum\xi_{ij}\otimes\phi(a_{ij})\big(1_{\mathcal{A}}\otimes\theta_i+\mathrm{ad}(\theta_i)\big).$$

Now the term with $\mathrm{ad}(\theta_i)$ gets annihilated by p_k, provided we also assume that $k>0$. (We will return later to the special case $k=0$.) So, using Sweedler's notation $\phi(a_{ij})=a_{ij}^{(1)}\otimes a_{ij}^{(2)}$, we get in that case

$$\sum\hat{F}(\xi_{ij})\otimes a_{ij}\otimes\theta_i = (id\otimes p_k)\sum\xi_{ij}\otimes\phi(a_{ij})\big(1_{\mathcal{A}}\otimes\theta_i\big)$$

$$= (id\otimes p_k)\sum\xi_{ij}\otimes(a_{ij}^{(1)}\otimes a_{ij}^{(2)})\big(1_{\mathcal{A}}\otimes\theta_i\big)$$

$$= (id\otimes p_k)\sum\xi_{ij}\otimes a_{ij}^{(1)}\otimes a_{ij}^{(2)}\theta_i$$

$$= \sum\xi_{ij}\otimes a_{ij}^{(1)}\otimes a_{ij}^{(2)}\theta_i.$$

Recalling that $a_{ij}\otimes\theta_i\in\mathcal{A}\otimes_{\mathrm{inv}}\Gamma^{\wedge k}\cong\Gamma^{\wedge k}$, we apply $id\otimes\pi_{\mathrm{inv}}$ to the previous (recalling that $\pi_{\mathrm{inv}}=\varepsilon\otimes id$) and get

$$\sum\varepsilon(a_{ij})\hat{F}(\xi_{ij})\otimes\theta_i = \sum\xi_{ij}\otimes a_{ij}^{(1)}\otimes\varepsilon(a_{ij}^{(2)})\theta_i$$

$$= \sum\xi_{ij}\otimes\varepsilon(a_{ij}^{(2)})a_{ij}^{(1)}\otimes\theta_i$$

$$= \sum\xi_{ij}\otimes a_{ij}\otimes\theta_i.$$

Next, we define $\varphi_i := \sum_j\varepsilon(a_{ij})\xi_{ij}\in\Omega^{n-k}(P)$, so that the previous identity becomes

$$\sum_i\hat{F}(\varphi_i)\otimes\theta_i = \sum_{ij}\xi_{ij}\otimes a_{ij}\otimes\theta_i.$$

Since the set $\{\theta_i\}$ is linearly independent, this implies for every i that

$$\hat{F}(\varphi_i) = \sum_j\xi_{ij}\otimes a_{ij}\in\Omega^{n-k}(P)\otimes\mathcal{A},$$

which in turn says by the very definition that each φ_i is horizontal, that is, $\varphi_i \in \mathfrak{hor}^{n-k}(P)$. Substituting this last result back into (12.14), we see that

$$(id \otimes p_k)\hat{F}(\omega) = \sum_{ij} \xi_{ij} \otimes a_{ij} \otimes \theta_i = \sum_i \hat{F}(\varphi_i) \otimes \theta_i, \qquad (12.15)$$

which is the first equality in (12.13). As remarked earlier, the second equality in (12.13) follows immediately from the first equality.

We still have to prove the special case $k = 0$. But then the hypothesis says that $(id \otimes p_l)\hat{F}(\omega) = 0$ for all $l > 0$, so that

$$\hat{F}(\omega) = (id \otimes p_0)\hat{F}(\omega) \in \Omega^n(P) \otimes \mathcal{A},$$

which means that ω is a horizontal form. So then we have the trivial identity

$$(id \otimes p_0)\hat{F}(\omega) = \hat{F}(\varphi)1_{\mathbb{C}},$$

where $\varphi := \omega$ is horizontal and $1_{\mathbb{C}} \in \mathbb{C} \cong {}_{inv}\Gamma^{\wedge 0}$. And this is the desired form in this special case. ∎

The following corollary of Lemma 12.1 is also rather handy.

Corollary 12.1. *Let $\omega \in \Omega^n(P)$ be a given homogeneous element. Then there are two finite families of elements (each having m elements), namely, $b_1, \ldots, b_m \in \mathcal{B}$ and $\theta_1, \ldots, \theta_m \in {}_{inv}\Gamma^{\wedge n}$, satisfying*

$$(id \otimes p_n)_{\Omega(P)}\Psi(\omega) = \sum_{j=1}^{m} {}_{\mathcal{B}}\Phi(b_j)\,\theta_j. \qquad (12.16)$$

Proof. We want to apply the previous lemma in the case when $k = n$. To justify this, we note that $(id \otimes p_l)_{\Omega(P)}\Psi(\omega) = 0$ for all $l > n$ since $_{\Omega(P)}\Psi$ is a degree-preserving map. Then, by the previous lemma, there exist m elements $\varphi_1, \ldots, \varphi_m \in \mathfrak{hor}^0(P) = \mathcal{B}$ (which we denote as $b_1, \ldots, b_m$) and $\theta_1, \ldots, \theta_m \in {}_{inv}\Gamma^{\wedge n}$ that satisfy (12.13). But (12.13) in this special case becomes (12.16). ∎

12.6 The Vertical Algebra

We continue in this section with a definition and discussion of the vertical algebra. This will correspond to vertical forms in the classical case. In this noncommutative setting, we will define a model covariant vertical differential algebra.

We continue studying a QPB $P = (\mathcal{B}, \mathcal{A}, {}_{\mathcal{B}}\Phi)$ with a bicovariant $*$-FODC Γ over a $*$-Hopf algebra $\mathcal{A}$. We consider the ungraded algebra $\mathcal{B}$ as the graded algebra, which is zero in all gradings except in grading 0, where it is equal to $\mathcal{B}$. Then we define the *vertical algebra*

$$\mathfrak{v}(P) \equiv \mathfrak{ver}(P) := \mathcal{B} \otimes {}_{inv}\Gamma^{\wedge}$$

as a graded vector space for the time being. Notice that this definition does not
depend on the choice of a differential calculus on the QPB. Of course, $_{inv}\Gamma^\wedge$ is the
graded subalgebra of $\Gamma^\wedge$ of left invariant elements, namely,

$$_{inv}\Gamma^\wedge := \{\omega \in \Gamma^\wedge \mid \Phi_{\Gamma^\wedge}(\omega) = 1 \otimes \omega\}.$$

The grading on $\mathfrak{ver}(P)$ is the usual grading on the tensor product of two graded
vector spaces, which in this case is

$$\mathfrak{ver}^j(P) := \oplus_{k+l=j}\left(\mathcal{B}^k \otimes {}_{inv}\Gamma^{\wedge l}\right) = \mathcal{B} \otimes {}_{inv}\Gamma^{\wedge j}.$$

In particular, in degree 0 we have

$$\mathfrak{ver}^0(P) = \mathcal{B} \otimes {}_{inv}\Gamma^{\wedge 0} = \mathcal{B} \otimes \mathbb{C} \cong \mathcal{B}. \tag{12.17}$$

As the notation suggests, these correspond to vertical differential forms in the
classical case and will be thought of as such in the noncommutative case as
well. However, in some ways this construction runs against "common" intuition.
For example, both $\mathfrak{hor}^0(P)$ and $\mathfrak{ver}^0(P)$ are identified with $\mathcal{B}$. However, this is
exactly what happens in the classical case. In other ways, it seems a quite natural
construction, which is actually a very special case of a construction that is defined
and studied in [23]. We will have more comments about this in the Notes section at
the end of this chapter.

Definition 12.9. *Suppose* $P = (\mathcal{B}, \mathcal{A}, {}_{\mathcal{B}}\Phi)$ *is a QPB and* Γ *is a bicovariant
$*$-FODC over* $\mathcal{A}$. *Suppose that* $a, b \in \mathcal{B}$ *and* $\omega, \eta \in {}_{inv}\Gamma^\wedge$. *Using Sweedler's
notation, put* $_{\mathcal{B}}\Phi(b) = b^{(0)} \otimes b^{(1)} \in \mathcal{B} \otimes \mathcal{A}$ *for* $b \in \mathcal{B}$. *We define a multiplication in*
$\mathfrak{ver}(P)$ *by*

$$(a \otimes \omega)(b \otimes \eta) := ab^{(0)} \otimes (\omega \circ b^{(1)})\eta. \tag{12.18}$$

We next define a $*$-operation on $\mathfrak{ver}(P)$ *by*

$$(b \otimes \eta)^* := b^{(0)*} \otimes (\eta^* \circ b^{(1)*}) \tag{12.19}$$

on decomposable elements and extend antilinearly to $\mathfrak{ver}(P)$. *Finally, we define a
differential on* $\mathfrak{ver}(P)$ *by*

$$d_v(b \otimes \eta) := b \otimes d\eta + b^{(0)} \otimes \pi(b^{(1)})\eta. \tag{12.20}$$

It is important to remark that the definitions of these operations, just like the
definition of $\mathfrak{ver}(P)$ itself, do not depend on the choice of a differential calculus
on the QPB. Notice that the right co-action $_{\mathcal{B}}\Phi$ of the QPB is used in all three

operations to "twist" the element $b \in \mathcal{B}$. If the right co-action is trivial, namely, $_{\mathcal{B}}\Phi(b) = b \otimes 1$ for all $b \in \mathcal{B}$, then we have the special case of operations without twisting or, more strictly speaking, with trivial twisting.

Proposition 12.7. *With the operations of Definition 12.9, $\mathfrak{ver}(P)$ becomes a graded differential $*$-algebra, known as the* vertical algebra. *Moreover, as a differential algebra, $\mathfrak{ver}(P)$ is generated by $\mathfrak{ver}^0(P) \cong \mathcal{B}$.*

In order not to interrupt the flow of the current discussion, we will give the proof of this proposition at the end of this section.

Exercise 12.20. *Identify $\mathfrak{ver}(P)$ for the differential calculus introduced in Example 12.4. Identify the multiplication, the $*$-operation, and the differential d_v as well for this case. Does all this make sense intuitively?*

In the next result we will use the alternative notation $\mathfrak{v}(P)$ instead of the lengthier $\mathfrak{ver}(P)$, but only in subscripts.

Theorem 12.4. *There exists a unique degree-preserving morphism of graded differential $*$-algebras*

$$_{\mathfrak{v}(P)}\Psi : \mathfrak{ver}(P) \to \mathfrak{ver}(P) \otimes \Gamma^{\wedge},$$

which is equal to $_{\mathcal{B}}\Phi : \mathcal{B} \to \mathcal{B} \otimes \mathcal{A}$ in degree 0. Moreover, $_{\mathfrak{v}(P)}\Psi$ satisfies

$$\left(_{\mathfrak{v}(P)}\Psi \otimes id\right) {}_{\mathfrak{v}(P)}\Psi = (id \otimes \hat{\phi}) {}_{\mathfrak{v}(P)}\Psi,$$

which says that the following diagram is commutative:

$$
\begin{array}{ccc}
\mathfrak{ver}(P) & \xrightarrow{\;\;_{\mathfrak{v}(P)}\Psi\;\;} & \mathfrak{ver}(P) \otimes \Gamma^{\wedge} \\
{\scriptstyle _{\mathfrak{v}(P)}\Psi} \downarrow & & \downarrow {\scriptstyle id \otimes \hat{\phi}} \\
\mathfrak{ver}(P) \otimes \Gamma^{\wedge} & \xrightarrow{\;\;_{\mathfrak{v}(P)}\Psi \otimes id\;\;} & \mathfrak{ver}(P) \otimes \Gamma^{\wedge} \otimes \Gamma^{\wedge}.
\end{array}
\tag{12.21}
$$

Recall that $\hat{\phi} : \Gamma^{\wedge} \to \Gamma^{\wedge} \otimes \Gamma^{\wedge}$ was defined and analyzed in Theorem 10.6.

Proof. The uniqueness of $_{\mathfrak{v}(P)}\Psi$ follows from its being equal to the given map $_{\mathcal{B}}\Phi$ in degree 0 and that the degree-0 elements (namely, $\mathcal{B}$) generate $\mathfrak{ver}(P)$ as a differential algebra.

To define $_{\mathfrak{v}(P)}\Psi$, we take $b \in \mathcal{B}$ and $\theta \in {}_{inv}\Gamma$. We use Sweedler's notation to write

$$_{\mathcal{B}}\Phi(b) = b^{(0)} \otimes b^{(1)} \in \mathcal{B} \otimes \mathcal{A}$$

and

$$\hat{\phi}(\theta) = \theta^{(0)} \otimes \theta^{(1)} \in {}_{inv}\Gamma^{\wedge} \otimes \Gamma^{\wedge}.$$

Then define

$$\mathfrak{v}(P)\Psi(b \otimes \theta) := b^{(0)} \otimes \theta^{(0)} \otimes b^{(1)}\theta^{(1)} \in \mathcal{B} \otimes {}_{\mathrm{inv}}\Gamma^{\wedge} \otimes \Gamma^{\wedge}.$$

We can think of this definition as a way of "pasting together" the maps ${}_{\mathcal{B}}\Phi$ and $\hat{\phi}$.

It now is an extended exercise left to the reader to show that ${}_{\mathfrak{v}(P)}\Psi$ satisfies all the properties stated in the theorem. ∎

Exercise 12.21. *Do that extended exercise.* **Hint:** *Read the proof of the next theorem and adapt those arguments to the present context.*

Now we can use the projection $p_0 : \Gamma \to \mathcal{A}$ (which is the identity map in degree 0 and the zero map in all other degrees) in order to define a right co-action of $\mathcal{A}$ on $\mathfrak{ver}(P)$ as follows:

$$\mathfrak{v}(P)\Phi := (id \otimes p_0) \, {}_{\mathfrak{v}(P)}\Psi : \mathfrak{ver}(P) \to \mathfrak{ver}(P) \otimes \mathcal{A}.$$

In Sweedler's notation as established above, this can be written as

$$\mathfrak{v}(P)\Phi(b \otimes \theta) = b^{(0)} \otimes \theta^{(0)} \otimes p_0(b^{(1)}\theta^{(1)}).$$

The next result is related to, but different from, the previous theorem. But first here is a little result that we will use to prove the next theorem.

Exercise 12.22. *Prove that $\dot{\phi}p_0 = (p_0 \otimes p_0)\hat{\phi}$.* **Hint:** *Write this identity as a commutative diagram. The content of the problem will remain unchanged, but this might help you think about what is happening.*

Theorem 12.5. *The map ${}_{\mathfrak{v}(P)}\Phi$ is a right co-action of $\mathcal{A}$ on $\mathfrak{ver}(P)$. This means that these two diagrams commute:*

$$
\begin{array}{ccc}
\mathfrak{ver}(P) & \xrightarrow{\;{}_{\mathfrak{v}(P)}\Phi\;} & \mathfrak{ver}(P) \otimes \mathcal{A} \\
{\scriptstyle {}_{\mathfrak{v}(P)}\Phi} \downarrow & & \downarrow {\scriptstyle {}_{\mathfrak{v}(P)}\Phi \otimes id} \\
\mathfrak{ver}(P) \otimes \mathcal{A} & \xrightarrow{\;id \otimes \phi\;} & \mathfrak{ver}(P) \otimes \mathcal{A} \otimes \mathcal{A}
\end{array}
\qquad (12.22)
$$

and

$$
\begin{array}{ccc}
\mathfrak{ver}(P) & \xrightarrow{\;{}_{\mathfrak{v}(P)}\Phi\;} & \mathfrak{ver}(P) \otimes \mathcal{A} \\
{\scriptstyle id} \downarrow & & \downarrow {\scriptstyle id \otimes \varepsilon} \\
\mathfrak{ver}(P) & \xrightarrow{\;\cong\;} & \mathfrak{ver}(P) \otimes \mathbb{C}.
\end{array}
\qquad (12.23)
$$

Also, the differential d_v is right covariant with respect to $_{\mathfrak{v}(P)}\Phi$, which means that this diagram commutes:

$$\begin{array}{ccc} \mathfrak{ver}(P) & \xrightarrow{\;_{\mathfrak{v}(P)}\Phi\;} & \mathfrak{ver}(P) \otimes \mathcal{A} \\ {\scriptstyle d_v}\downarrow & & \downarrow{\scriptstyle d_v \otimes id} \\ \mathfrak{ver}(P) & \xrightarrow{\;_{\mathfrak{v}(P)}\Phi\;} & \mathfrak{ver}(P) \otimes \mathcal{A}. \end{array} \tag{12.24}$$

Proof. We start with a diagram chase through (12.22). So we pick an element

$$b \otimes \theta \in \mathfrak{ver}(P) = \mathcal{B} \otimes_{\mathrm{inv}} \Gamma^{\wedge}.$$

First, we go down and across, getting

$$(id \otimes \phi)_{\mathfrak{v}(P)}\Phi(b \otimes \theta) = (id \otimes \phi)(b^{(0)} \otimes \theta^{(0)} \otimes p_0(b^{(1)}\theta^{(1)}))$$

$$= b^{(0)} \otimes \theta^{(0)} \otimes \phi(p_0(b^{(1)}\theta^{(1)})).$$

Next, we chase this element across and then down, obtaining

$$(_{\mathfrak{v}(P)}\Phi \otimes id)_{\mathfrak{v}(P)}\Phi(b \otimes \theta) = (_{\mathfrak{v}(P)}\Phi \otimes id)(b^{(0)} \otimes \theta^{(0)} \otimes p_0(b^{(1)}\theta^{(1)}))$$

$$= b^{(00)} \otimes \theta^{(00)} \otimes p_0(b^{(01)}\theta^{(01)}) \otimes p_0(b^{(1)}\theta^{(1)}))$$

$$= b^{(0)} \otimes \theta^{(0)} \otimes p_0(b^{(11)}\theta^{(11)}) \otimes p_0(b^{(12)}\theta^{(12)}))$$

$$= b^{(0)} \otimes \theta^{(0)} \otimes (p_0 \otimes p_0)(b^{(11)}\theta^{(11)} \otimes b^{(12)}\theta^{(12)})$$

$$= b^{(0)} \otimes \theta^{(0)} \otimes (p_0 \otimes p_0)((b^{(11)} \otimes b^{(12)})(\theta^{(11)} \otimes \theta^{(12)}))$$

$$= b^{(0)} \otimes \theta^{(0)} \otimes (p_0 \otimes p_0)(\phi(b^{(1)})\hat{\phi}(\theta^{(1)}))$$

$$= b^{(0)} \otimes \theta^{(0)} \otimes (p_0 \otimes p_0)(\hat{\phi}(b^{(1)}\theta^{(1)}))$$

$$= b^{(0)} \otimes \theta^{(0)} \otimes (\phi p_0(b^{(1)}\theta^{(1)})),$$

where we used the previous exercise in the last step plus the usual number of identities and definitions in the other steps. In short, the two chases give the same result, and so (12.22) commutes.

Next, we consider (12.23). As in the previous chase, we start with $b \otimes \theta$ in the upper right corner. Then going across and down, we get

$$(id \otimes \varepsilon)_{\mathfrak{v}(P)}\Phi(b \otimes \theta) = (id \otimes \varepsilon)(b^{(0)} \otimes \theta^{(0)} \otimes p_0(b^{(1)}\theta^{(1)}))$$

$$= b^{(0)} \otimes \theta^{(0)} \otimes \varepsilon(p_0(b^{(1)}\theta^{(1)}))$$

$$= b^{(0)} \otimes \theta^{(0)} \otimes \varepsilon(b^{(1)}p_0(\theta^{(1)}))$$

$$\cong \varepsilon(b^{(1)})\varepsilon(p_0(\theta^{(1)}))(b^{(0)} \otimes \theta^{(0)})$$

$$= \varepsilon(b^{(1)})b^{(0)} \otimes \varepsilon(p_0(\theta^{(1)}))\theta^{(0)}$$

$$= b \otimes \varepsilon(p_0(\theta^{(1)}))\theta^{(0)}$$

$$= b \otimes (id \otimes \varepsilon)(id \otimes p_0)(\theta^{(0)} \otimes \theta^{(1)}))$$

$$= b \otimes (id \otimes \varepsilon)(id \otimes p_0)(1_A \otimes \theta + \mathrm{ad}(\theta)))$$

$$= b \otimes (id \otimes \varepsilon)(\mathrm{ad}(\theta)))$$

$$\cong b \otimes \theta,$$

where we used $b^{(1)} \in \mathcal{A}$ in the third equality and in the third line from the bottom the identity $\hat{\phi}(\theta) = 1_A \otimes \theta + \mathrm{ad}(\theta)$. Finally, in the last line we used that ad is a right co-action. We also used $(id \otimes p_0)(1_A \otimes \theta) = 0$, which is true if $\deg(\theta) \geq 1$. We leave the case $\deg(\theta) = 0$ to the reader.

Since the down-and-across chase of (12.23) trivially gives the same result, we have that the diagram (12.23) is commutative.

To establish that (12.24) is a commutative diagram, we again do a chase and again starting with $b \otimes \theta$ in the upper right corner. We recall that the definition of the derivation d_v in the usual Sweedler's notation is

$$d_v(b \otimes \theta) = b \otimes d\theta + b^{(0)} \otimes \pi(b^{(1)})\theta.$$

Starting the chase by going down and then across, we find that

$$\mathfrak{v}(P)\Phi(d_v(b \otimes \theta)) = \mathfrak{v}(P)\Phi\big(b \otimes d\theta + b^{(0)} \otimes \pi(b^{(1)})\theta\big)$$

$$= b^{(0)} \otimes (d\theta)^{(0)} \otimes p_0(b^{(1)}(d\theta)^{(1)})$$

$$+ b^{(00)} \otimes (\pi(b^{(1)})\theta)^{(0)} \otimes p_0(b^{(01)}(\pi(b^{(1)})\theta)^{(1)})$$

$$= b^{(0)} \otimes d\theta^{(0)} \otimes p_0(b^{(1)}\theta^{(1)})$$

$$+ b^{(0)} \otimes (\pi(b^{(2)})\theta)^{(0)} \otimes p_0(b^{(1)}(\pi(b^{(2)})\theta)^{(1)}), \qquad (12.25)$$

where in the first term we used

$$(d\theta)^{(0)} \otimes (d\theta)^{(0)} = {}_\Gamma\Phi(d\theta)$$

$$= (d \otimes id)\,{}_\Gamma\Phi(\theta)$$

$$= (d \otimes id)\,(\theta^{(0)} \otimes \theta^{(1)})$$

$$= d\theta^{(0)} \otimes \theta^{(1)},$$

while in the second term we changed to the single-digit Sweedler's notation.
To get a better expression for the second term, we have to consider

$$(\pi(b^{(2)})\theta)^{(0)} \otimes (\pi(b^{(2)})\theta)^{(1)} = \hat{\phi}(\pi(b^{(2)})\theta) = \hat{\phi}(\pi(b^{(2)})\hat{\phi}(\theta)$$

$$= \big(1 \otimes \pi(b^{(2)}) + \mathrm{ad}(\pi(b^{(2)}))\big)\hat{\phi}(\theta)$$

$$= \left(1 \otimes \pi(b^{(2)}) + (\pi \otimes id)\mathrm{ad}(b^{(2)})\right)\hat{\phi}(\theta)$$

$$= \left(1 \otimes \pi(b^{(2)}) + (\pi \otimes id)\mathrm{ad}(b^{(2)})\right)\hat{\phi}(\theta)$$

$$= \left(1 \otimes \pi(b^{(2)}) + \pi(b^{(3)}) \otimes \kappa(b^{(2)})\, b^{(4)}\right)\hat{\phi}(\theta)$$

$$= \left(1 \otimes \pi(b^{(2)}) + \pi(b^{(3)}) \otimes \kappa(b^{(2)})\, b^{(4)}\right)\left(\theta^{(0)} \otimes \theta^{(1)}\right)$$

$$= \theta^{(0)} \otimes \pi(b^{(2)})\theta^{(1)} + \pi(b^{(3)})\theta^{(0)} \otimes \kappa(b^{(2)})\, b^{(4)}\theta^{(1)}.$$

Notice that this calculation rewrites a simple Sweedler's expression, which is in general a sum, as the sum of two simple Sweedler's expressions, each of which is itself in general a sum. Keeping this subtlety in mind, we substitute this calculation back into (12.25). This yields

$$v(P)\Phi(d_v(b \otimes \theta))$$

$$= b^{(0)} \otimes d\theta^{(0)} \otimes p_0(b^{(1)}\theta^{(1)}) + b^{(0)} \otimes (\pi(b^{(2)})\theta)^{(0)} \otimes p_0(b^{(1)}(\pi(b^{(2)})\theta)^{(1)})$$

$$= b^{(0)} \otimes d\theta^{(0)} \otimes p_0(b^{(1)}\theta^{(1)})$$

$$+ b^{(0)} \otimes \theta^{(0)} \otimes p_0(b^{(1)}\pi(b^{(2)})\theta^{(1)}) + b^{(0)} \otimes \pi(b^{(3)})\theta^{(0)} \otimes p_0(b^{(1)}\kappa(b^{(2)})\, b^{(4)}\theta^{(1)})$$

$$= b^{(0)} \otimes d\theta^{(0)} \otimes p_0(b^{(1)}\theta^{(1)}) + b^{(0)} \otimes \pi(b^{(3)})\theta^{(0)} \otimes p_0(b^{(1)}\kappa(b^{(2)})\, b^{(4)}\theta^{(1)})$$

$$= b^{(0)} \otimes d\theta^{(0)} \otimes p_0(b^{(1)}\theta^{(1)}) + b^{(0)} \otimes \pi(b^{(2)})\theta^{(0)} \otimes p_0(\varepsilon(b^{(1)})\, b^{(3)}\theta^{(1)})$$

$$= b^{(0)} \otimes d\theta^{(0)} \otimes p_0(b^{(1)}\theta^{(1)}) + b^{(0)} \otimes \pi(\varepsilon(b^{(1)})\, b^{(2)})\theta^{(0)} \otimes p_0(b^{(3)}\theta^{(1)})$$

$$= b^{(0)} \otimes d\theta^{(0)} \otimes p_0(b^{(1)}\theta^{(1)}) + b^{(0)} \otimes \pi(b^{(1)})\theta^{(0)} \otimes p_0(b^{(2)}\theta^{(1)}).$$

Here we used $\deg \pi(b^{(2)}) = 1$ to conclude that $p_0(b^{(1)}\pi(b^{(2)})\theta^{(1)}) = 0$, plus the usual identities. On the other hand, going first across and then down yields

$$(d_v \otimes id)\, v(P)\Phi(b \otimes \theta) = (d_v \otimes id)(b^{(0)} \otimes \theta^{(0)} \otimes p_0(b^{(1)}\theta^{(1)}))$$

$$= d_v(b^{(0)} \otimes \theta^{(0)}) \otimes p_0(b^{(1)}\theta^{(1)})$$

$$= (b^{(0)} \otimes d\theta^{(0)} + b^{(00)} \otimes \pi(b^{(01)})\theta^{(0)}) \otimes p_0(b^{(1)}\theta^{(1)})$$

$$= b^{(0)} \otimes d\theta^{(0)} \otimes p_0(b^{(1)}\theta^{(1)}) + b^{(00)} \otimes \pi(b^{(01)})\theta^{(0)} \otimes p_0(b^{(1)}\theta^{(1)})$$

$$= b^{(0)} \otimes d\theta^{(0)} \otimes p_0(b^{(1)}\theta^{(1)}) + b^{(0)} \otimes \pi(b^{(1)})\theta^{(0)} \otimes p_0(b^{(2)}\theta^{(1)}),$$

where we again changed to the single-digit Sweedler's notation in the second term. Since the two chases coincide, the diagram does commute. ∎

Now we have a crucial result relating the vertical forms to the space of all the forms. Recall that in classical differential geometry, the vertical tangent vectors form a subbundle of the bundle of all the tangent vectors. And the inclusion map of the subbundle is a map that preserves all of the relevant properties. Here the dual relation holds; namely, the vertical forms are a quotient of all of the forms, and the quotient map preserves all of the relevant properties.

Theorem 12.6. *There exists a unique degree-preserving, differential, unital algebra morphism*

$$\pi_\mathfrak{v} : \Omega(P) \to \mathfrak{ver}(P),$$

which in degree 0 corresponds to the identity map $id_\mathcal{B} : \mathcal{B} \to \mathcal{B}$ under the isomorphism $\mathfrak{ver}^0(P) \cong \mathcal{B}$ given in (12.17). Also, the following two diagrams are commutative:

$$
\begin{array}{ccc}
\Omega(P) & \xrightarrow{\ \pi_\mathfrak{v}\ } & \mathfrak{ver}(P) \\
{\scriptstyle \Omega(P)}\Phi \downarrow & & \downarrow {\scriptstyle \mathfrak{v}(P)}\Phi \\
\Omega(P) \otimes \mathcal{A} & \xrightarrow{\ \pi_\mathfrak{v} \otimes id_\mathcal{A}\ } & \mathfrak{ver}(P) \otimes \mathcal{A}
\end{array}
\qquad (12.26)
$$

and

$$
\begin{array}{ccc}
\Omega(P) & \xrightarrow{\ \pi_\mathfrak{v}\ } & \mathfrak{ver}(P) \\
{\scriptstyle \Omega(P)}\Psi \downarrow & & \downarrow {\scriptstyle \mathfrak{v}(P)}\Psi \\
\Omega(P) \otimes \Gamma^\wedge & \xrightarrow{\ \pi_\mathfrak{v} \otimes id_{\Gamma^\wedge}\ } & \mathfrak{ver}(P) \otimes \Gamma^\wedge.
\end{array}
\qquad (12.27)
$$

Moreover, this map $\pi_\mathfrak{v}$ is also a $\mathcal{B}$-bimodule morphism, a $*$-morphism, and a surjection.

Proof. Since by definition the elements of degree-0 $\Omega^0(P) = \mathcal{B}$ generate $\Omega(P)$ as a graded, differential algebra, the differential algebra morphism $\pi_\mathfrak{v}$ is uniquely determined by the condition that it is equal to the identity of $\mathcal{B}$ in degree 0. Also, being the identity in degree 0 implies it is unital. So the argument reduces to showing its existence and its properties.

Let $\omega \in \Omega^k(P)$ be a homogeneous element of degree k. Recall that $p_k : \Omega(P) \to \Omega^k(P)$ is the projection onto the degree-k elements, that is, the linear map that is the identity on $\Omega^k(P)$ and is zero on every $\Omega^l(P)$ for $l \neq k$. Then we define

$$\pi_\mathfrak{v}(\omega) := (id \otimes \pi_{\text{inv}})(id \otimes p_k) {\scriptstyle \Omega(P)}\Psi(\omega). \qquad (12.28)$$

This definition merits the most careful attention. First, we have

$${\scriptstyle \Omega(P)}\Psi : \Omega^k(P) \to \left(\Omega(P) \otimes \Gamma^\wedge \right)^k = \oplus_{i+j=k} \left(\Omega^i(P) \otimes \Gamma^{\wedge j} \right).$$

Next, we apply $(id \otimes p_k)$ to the image of ω in the last space. Because of the definition of p_k, we have that its image lies in

$$\Omega^0(P) \otimes \Gamma^{\wedge k} = \mathcal{B} \otimes \Gamma^{\wedge k}.$$

Finally, we recall that $\pi_{\text{inv}} : \Gamma^{\wedge k} \to {}_{\text{inv}}\Gamma^{\wedge k}$, so that when all is said and done,

$$\pi_{\mathfrak{v}}(\omega) \in \mathcal{B} \otimes {}_{\text{inv}}\Gamma^{\wedge k} = \mathfrak{ver}^k(P).$$

In short, $\pi_{\mathfrak{v}} : \Omega^k(P) \to \mathfrak{ver}^k(P)$, as desired.

If we take $\omega \in \Omega^0(P) = \mathcal{B}$, then we have

$$
\begin{aligned}
\pi_{\mathfrak{v}}(\omega) &= (id \otimes \pi_{\text{inv}})(id \otimes p_0)\, {}_{\Omega(P)}\Psi(\omega) \\
&= (id \otimes \pi_{\text{inv}})(id \otimes p_0)\, {}_{\mathcal{B}}\Phi(\omega) \\
&= (id \otimes \pi_{\text{inv}})\, {}_{\mathcal{B}}\Phi(\omega) \\
&= (id \otimes \varepsilon \otimes id_{\mathcal{C}})\, {}_{\mathcal{B}}\Phi(\omega) \\
&\cong (id \otimes \varepsilon)\, {}_{\mathcal{B}}\Phi(\omega) \\
&= \omega.
\end{aligned}
$$

Here we used the fact that ${}_{\Omega(P)}\Psi$ equals ${}_{\mathcal{B}}\Phi$ in degree 0, the fact that p_0 is the identity in degree 0, the definition of π_{inv}, and one of the properties that follows from ${}_{\mathcal{B}}\Phi(\omega)$ being a right co-action of $\mathcal{A}$. Thus, $\pi_{\mathfrak{v}}$ is the identity map in degree 0.

To show that $\pi_{\mathfrak{v}}$ is surjective, it suffices to note that it is the identity on $\mathcal{B}$ in degree 0 and that both $\Omega(P)$ and $\mathfrak{ver}(P)$ are generated as differential graded algebras by their degree-0 elements. So the element $b_0 d_{\mathfrak{v}} b_1 \cdots d_{\mathfrak{v}} b_k$ in $\mathfrak{ver}^k(P)$ (where all the b_js are in $\mathcal{B}$) is the image of $b_0 d_P b_1 \cdots d_P b_k$ in $\Omega^k(P)$ under $\pi_{\mathfrak{v}}$, a morphism of differential algebras. This shows the surjectivity.

Also, this mapping property $\pi_{\mathfrak{v}} : b_0 d_P b_1 \cdots d_P b_k \mapsto b_0 d_{\mathfrak{v}} b_1 \cdots d_{\mathfrak{v}} b_k$ can be used to immediately show that $\pi_{\mathfrak{v}}$ is a $*$-morphism, since both d_P and $d_{\mathfrak{v}}$ are $*$-morphisms.

Now we will prove that $\pi_{\mathfrak{v}}$ is a multiplicative morphism. So let $\omega \in \Omega^k(P)$ and $\upsilon \in \Omega^l(P)$ be given. Then we use Corollary 12.1 to write

$$(id \otimes p_k)\, {}_{\Omega(P)}\Psi(\omega) = \sum_{j=1}^{r} {}_{\mathcal{B}}\Phi(b_j)\, \theta_j$$

and

$$(id \otimes p_k)\, {}_{\Omega(P)}\Psi(\upsilon) = \sum_{i=1}^{s} {}_{\mathcal{B}}\Phi(c_i)\, \eta_i,$$

where r and s are integers, $b_j, c_i \in \mathcal{B}$, $\theta_j \in {}_{\text{inv}}\Gamma^{\wedge k}$, and $\eta_i \in {}_{\text{inv}}\Gamma^{\wedge l}$.

By the definitions of $\pi_\mathfrak{v}$ and π_{inv}, we have that

$$\pi_\mathfrak{v}(\omega) = (id \otimes \pi_{\mathrm{inv}})(id \otimes p_k)_{\Omega(P)}\Psi(\omega)$$

$$= \sum_j (id \otimes \pi_{\mathrm{inv}})_B\Phi(b_j)\,\theta_j$$

$$= \sum_j (id \otimes \pi_{\mathrm{inv}})(b_j^{(0)} \otimes b_j^{(1)})\theta_j$$

$$= \sum_j (id \otimes \pi_{\mathrm{inv}})(b_j^{(0)} \otimes (b_j^{(1)}\theta_j))$$

$$= \sum_j b_j^{(0)} \otimes (\varepsilon(b_j^{(1)})\theta_j)$$

$$= \sum_j \varepsilon(b_j^{(1)})b_j^{(0)} \otimes \theta_j$$

$$= \sum_j b_j \otimes \theta_j.$$

Similarly, $\pi_\mathfrak{v}(\upsilon) = \sum_{i=1}^{s} c_i \otimes \eta_i$.
   Then we calculate

$$\pi_\mathfrak{v}(\omega\upsilon) = (id \otimes \pi_{\mathrm{inv}})(id \otimes p_{k+l})_{\Omega(P)}\Psi(\omega\upsilon)$$

$$= (id \otimes \pi_{\mathrm{inv}})(id \otimes p_k)_{\Omega(P)}\Psi(\omega)(id \otimes p_l)_{\Omega(P)}\Psi(\upsilon)$$

$$= \sum_{ij}(id \otimes \pi_{\mathrm{inv}})(_B\Phi(b_j)\,\theta_j)(_B\Phi(c_i)\,\eta_i)$$

$$= \sum_{ij}(id \otimes \pi_{\mathrm{inv}})(b_j^{(0)} \otimes (b_j^{(1)}\theta_j))(c_j^{(0)} \otimes (c_j^{(1)}\eta_i))$$

$$= \sum_{ij}(id \otimes \pi_{\mathrm{inv}})\left(b_j^{(0)}c_i^{(00)} \otimes ((b_j^{(1)}\theta_j) \circ c_i^{(01)})c_i^{(1)}\eta_i\right)$$

$$= \sum_{ij}(id \otimes \pi_{\mathrm{inv}})\left(b_j^{(0)}c_i^{(00)} \otimes (b_j^{(1)}(\theta_j \circ c_i^{(01)}))c_i^{(1)}\eta_i\right)$$

$$= \sum_{ij}(id \otimes \pi_{\mathrm{inv}})\left(b_j^{(0)}c_i^{(00)} \otimes b_j^{(1)}c_i^{(1)}((\theta_j \circ c_i^{(01)})\eta_i)\right)$$

$$= \sum_{ij}b_j^{(0)}c_i^{(00)} \otimes \varepsilon(b_j^{(1)}c_i^{(1)})((\theta_j \circ c_i^{(01)})\eta_i)$$

$$= \sum_{ij}\varepsilon(b_j^{(1)})b_j^{(0)}c_i^{(0)} \otimes ((\theta_j \circ \varepsilon(c_i^{(12)})c_i^{(11)})\eta_i)$$

$$= \sum_{ij} b_j c_i^{(0)} \otimes \left((\theta_j \circ c_i^{(1)}) \eta_i \right)$$

$$= \left(\sum_j b_j \otimes \theta_j \right) \left(\sum_i c_i \otimes \eta_i \right)$$

$$= \pi_\mathfrak{v}(\omega) \pi_\mathfrak{v}(\upsilon).$$

All of this is standard definitions and identities, though the second equality uses a bit more than the fact that $_{\Omega(P)}\Psi$ is a multiplicative morphism, and so the reader should reflect on that step of the calculation. Thus, we have shown that $\pi_\mathfrak{v}$ is a multiplicative morphism.

Next, we will check that $\pi_\mathfrak{v}$ intertwines the derivations. Before doing so, we wish to comment again that by property (diff2) of Definition 12.3,

$$_{\Omega(P)}\Psi : \Omega(P) \to \Omega(P) \otimes \Gamma^\wedge$$

is a map that intertwines the differential d_P on $\Omega(P)$ with the tensor product differential on $\Omega(P) \otimes \Gamma^\wedge$, which we will simply denote as D here. Since we will be applying $id \otimes p_{k+1}$ after applying D, the only term in D that will survive is that which comes from

$$id \otimes d^\wedge : \mathcal{B} \otimes \Gamma^{\wedge k} \to \mathcal{B} \otimes \Gamma^{\wedge k+1}.$$

We also continue with the notations that were established above, including

$$(id \otimes p_k)\,_{\Omega(P)}\Psi(\omega) = \sum_j {}_\mathcal{B}\Phi(b_j)\,\theta_j.$$

To establish the intertwining, we calculate

$$\pi_\mathfrak{v} d(\omega) = (id \otimes \pi_{\text{inv}})(id \otimes p_{k+1})\,_{\Omega(P)}\Psi(d(\omega))$$

$$= (id \otimes \pi_{\text{inv}})(id \otimes p_{k+1}) D(_{\Omega(P)}\Psi(\omega))$$

$$= (id \otimes \pi_{\text{inv}})(id \otimes d^\wedge)(id \otimes p_k)_{\Omega(P)}\Psi(\omega)$$

$$= \sum_j (id \otimes \pi_{\text{inv}})(id \otimes d^\wedge)\,_\mathcal{B}\Phi(b_j)\,\theta_j$$

$$= \sum_j (id \otimes \pi_{\text{inv}})(id \otimes d^\wedge)(b_j^{(0)} \otimes b_j^{(1)}\theta_j)$$

$$= \sum_j (id \otimes \pi_{\text{inv}})(b_j^{(0)} \otimes db_j^{(1)}\theta_j + b_j^{(0)} \otimes b_j^{(1)} d^\wedge \theta_j)$$

$$= \sum_j (b_j^{(0)} \otimes \pi_{\text{inv}}(db_j^{(1)}\theta_j) + b_j^{(0)} \otimes \pi_{\text{inv}}(b_j^{(1)} d^\wedge \theta_j))$$

$$= \sum_j (b_j^{(0)} \otimes \pi_{\text{inv}}(db_j^{(1)})\theta_j + b_j^{(0)} \otimes \varepsilon(b_j^{(1)}) d^\wedge \theta_j)$$

$$= \sum_j (b_j^{(0)} \otimes \pi(b_j^{(1)})\theta_j + b_j \otimes d^\wedge \theta_j)$$

$$= \sum_j d_{\mathfrak{v}}(b_j \otimes \theta_j)$$

$$= d_{\mathfrak{v}}\pi_{\mathfrak{v}}(\omega).$$

And this establishes the desired intertwining.

Next, we will show that the diagram (12.27) is commutative. Notice that both of the compositions from the upper left to the lower right corner are morphisms of differential algebras. Next, we restrict these two compositions to degree-0 elements in the upper right corner; that is, we restrict to B. Since all the maps in the diagram preserve degree, this means that we are restricting to the degree-0 elements in all four spaces. With this restriction, each of these two compositions is equal to $\Omega(P)\Phi$. But the space $\Omega(P)$ in the upper left corner is generated as a differential algebra by its zero-degree elements. And so the two compositions are equal for all degrees; that is, the diagram is commutative.

The commutativity of the diagram (12.26) follows from the commutativity of diagram (12.27) as an exercise, which we leave for the reader. ∎

Exercise 12.23. *Prove that the diagram (12.26) commutes.*

The next result will be used in the theorem following it. However, it gives some nice formulas and so merits its own lemma.

Lemma 12.2. *Suppose that we have a QPB with differential calculus and that $\omega \in \Omega^1(P)$, which we write as $\omega = \sum_k b_k \, d_P c_k$, a finite sum, where $b_k, c_k \in B$. Then*

$$\hat{F}(\omega) = \sum_k \left(b_k^{(0)} d_P c_k^{(0)} \otimes b_k^{(1)} c_k^{(1)} + b_k^{(0)} c_k^{(0)} \otimes b_k^{(1)} dc_k^{(1)} \right)$$

$$\text{and} \quad \pi_v(\omega) = \sum_k b_k \, c_k^{(0)} \otimes \pi(c_k^{(1)}),$$

where $F(b_k) = b_k^{(0)} \otimes b_k^{(1)}$, $F(c_k) = c_k^{(0)} \otimes c_k^{(1)} \in B \otimes A$. Here we are using Đurđevich's notations: $\hat{F} = {}_{\Omega(P)}\Psi$ and $F = {}_B\Phi$.

Proof. We start with the rather direct proof of the first formula:

$$\hat{F}(\omega) = \sum_k \hat{F}(b_k \, d_P c_k)$$

$$= \sum_k \hat{F}(b_k)\hat{F}(d_P c_k)$$

$$= \sum_k F(b_k)\hat{F}(d_P c_k)$$

$$= \sum_k (b_k^{(0)} \otimes b_k^{(1)}) \hat{F}(d_P c_k)$$

$$= \sum_k (b_k^{(0)} \otimes b_k^{(1)})(d_P c_k^{(0)} \otimes c_k^{(1)} + c_k^{(0)} \otimes dc_k^{(1)})$$

$$= \sum_k (b_k^{(0)} d_P c_k^{(0)} \otimes b_k^{(1)} c_k^{(1)} + b_k^{(0)} c_k^{(0)} \otimes b_k^{(1)} dc_k^{(1)}).$$

The second formula requires more work. We first note that

$$(id \otimes p_1)\hat{F}(\omega) = b_k^{(0)} c_k^{(0)} \otimes b_k^{(1)} dc_k^{(1)}.$$

Then we have that

$$\pi_v(\omega) = (id \otimes \pi_{inv})(id \otimes p_1)\hat{F}(\omega) = \sum_k b_k^{(0)} c_k^{(0)} \otimes \pi_{inv}(b_k^{(1)} dc_k^{(1)}).$$

Next, we use that $\pi_{inv}(\phi) = \kappa(\phi^{(0)})\phi^{(1)}$ for all $\phi \in \Gamma$, where $\Phi_\Gamma(\phi) = \phi^{(0)} \otimes \phi^{(1)}$ is the left co-action of $\mathcal{A}$ on Γ. Recall that

$$\Phi_\Gamma(a\, dc) = \phi(a)(id \otimes d)\phi(c)$$

$$= (a^{(1)} \otimes a^{(2)})(c^{(1)} \otimes dc^{(2)})$$

$$= a^{(1)} c^{(1)} \otimes a^{(2)} dc^{(2)}$$

for $a \in \mathcal{A}$ and $c \in \mathcal{B}$, so that

$$\pi_{inv}(a\, dc) = \kappa(a^{(1)} c^{(1)})a^{(2)} dc^{(2)}.$$

Using the last identity, we then get

$$\pi_v(\omega) = \sum_k b_k^{(0)} c_k^{(0)} \otimes \pi_{inv}(b_k^{(1)} dc_k^{(1)})$$

$$= \sum_k b_k^{(0)} c_k^{(0)} \otimes \kappa(b_k^{(11)} c_k^{(11)})b_k^{(12)} dc_k^{(12)}$$

$$= \sum_k b_k^{(0)} c_k^{(0)} \otimes \kappa(c_k^{(11)})\kappa(b_k^{(11)})b_k^{(12)} dc_k^{(12)}$$

$$= \sum_k b_k^{(0)} c_k^{(0)} \otimes \kappa(c_k^{(11)})\varepsilon(b_k^{(1)})dc_k^{(12)}$$

$$= \sum_k \varepsilon(b_k^{(1)})b_k^{(0)} c_k^{(0)} \otimes \kappa(c_k^{(11)})dc_k^{(12)}$$

$$= \sum_k b_k \, c_k^{(0)} \otimes \kappa(c_k^{(11)}) dc_k^{(12)}$$

$$= \sum_k b_k \, c_k^{(0)} \otimes \pi(c_k^{(1)}),$$

where we used a lot of standard stuff plus $\pi(a) = \kappa(a^{(1)}) da^{(2)}$ for $a \in \mathcal{A}$ in the final step. ∎

Theorem 12.7. *This sequence of vector spaces is exact:*

$$0 \to \mathfrak{hor}^1(P) \overset{\iota}{\longrightarrow} \Omega^1(P) \overset{\pi_\mathfrak{v}}{\longrightarrow} \mathfrak{ver}^1(P) \to 0.$$

Remarks. The second map ι here is the inclusion of the subspace into its ambient space. Notice that these are spaces of degree-1 elements. And all of these are maps of $\mathcal{B}$-bimodules and are $*$-morphisms. This theorem gives some justification for the definition of the vertical forms $\mathfrak{ver}(P)$ of degree 1. This result is rather important, so we present two proofs of it.

First proof. We showed that the map $\pi_\mathfrak{v}$ is surjective in the previous theorem. The inclusion map ι is trivially injective. So it only remains to prove the exactness at $\Omega^1(P)$. First, let $\omega \in \mathfrak{hor}^1(P)$ be given. Then

$$\iota\pi_\mathfrak{v}(\omega) = \pi_\mathfrak{v}(\omega) = (id \otimes \pi_{\mathrm{inv}})(id \otimes p_1) \, _{\Omega(P)}\Psi(\omega).$$

Now $_{\Omega(P)}\Psi(\omega) \in \Omega^1 \otimes \mathcal{A}$ since ω is horizontal, and so, using that $p_1(\mathcal{A}) = 0$, we see that $\iota\pi_\mathfrak{v} = 0$. This shows that $\mathrm{Ran}\, \iota \subset \ker \pi_\mathfrak{v}$.

To prove the opposite inclusion, we take an arbitrary element

$$\omega \in \ker \pi_\mathfrak{v} \cap \Omega^1(P).$$

We have to show that ω lies in the image of the inclusion ι; that is, we have to show that ω is horizontal.

Now using the previous lemma, we have that

$$\sum_k b_k c_k^{(0)} \otimes \pi(c_k^{(1)}) = 0.$$

Applying $_\mathcal{B}\Phi \otimes id$ to this and calculating, we obtain

$$0 = \sum_k {}_\mathcal{B}\Phi(b_k c_k^{(0)}) \otimes \pi(c_k^{(1)})$$

$$= \sum_k {}_\mathcal{B}\Phi(b_k) \, _\mathcal{B}\Phi(c_k^{(0)}) \otimes \pi(c_k^{(1)})$$

$$= \sum_k (b_k^{(0)} \otimes b_k^{(1)})(c_k^{(00)} \otimes c_k^{(01)}) \otimes \pi(c_k^{(1)})$$

$$= \sum_k b_k^{(0)} c_k^{(00)} \otimes b_k^{(1)} c_k^{(01)} \otimes \pi(c_k^{(1)})$$

$$= \sum_k b_k^{(0)} c_k^{(0)} \otimes b_k^{(1)} c_k^{(01)} \otimes \pi(c_k^{(12)}),$$

where we used various standard identities as well as the co-action property

$$c_k^{(00)} \otimes c_k^{(01)} \otimes c_k^{(1)} = c_k^{(0)} \otimes c_k^{(01)} \otimes c_k^{(12)}$$

in the final step.

Next, we apply $id \otimes L$, where $L : \mathcal{A} \otimes \Gamma \to \Gamma$ is the left action of $\mathcal{A}$ on the $\mathcal{A}$-bimodule Γ, thereby getting

$$0 = \sum_k b_k^{(0)} c_k^{(0)} \otimes b_k^{(1)} c_k^{(01)} \pi(c_k^{(12)})$$

$$= \sum_k b_k^{(0)} c_k^{(0)} \otimes b_k^{(1)} d\, c_k^{(1)},$$

where we used the identity $da = a^{(1)} \pi(a^{(2)})$ for all $a \in \mathcal{A}$. Using the identity proved in the previous lemma in the first equation, we see that

$$\Omega(P) \Psi(\omega) = \sum_k \left(b_k^{(0)} d_P c_k^{(0)} \otimes b_k^{(1)} c_k^{(1)} + b_k^{(0)} c_k^{(0)} \otimes b_k^{(1)} dc_k^{(1)} \right)$$

$$= \sum_k (b_k^{(0)} d_P c_k^{(0)} \otimes b_k^{(1)} c_k^{(1)}) \in \mathcal{B} \otimes \mathcal{A}.$$

But $\Omega(P) \Psi(\omega) \in \mathcal{B} \otimes \mathcal{A}$ means by definition that ω is horizontal. And this was the last thing we needed to prove. ∎

The second proof exhibits the power of Lemma 12.1.

Second proof. As in the first proof, the only nontrivial step is showing that $\ker \pi_\upsilon \cap \Omega^1(P) \subset \mathfrak{hor}^1(P)$. But this time we start with an arbitrary element $\omega \in \Omega^1(P)$. Applying Lemma 12.1 for $n = k = 1$, we have that

$$(id \otimes p_1)_{\Omega(P)} \Psi(\omega) = \sum_j {}_{\Omega(P)}\Psi(b_j)\theta_j = \sum_j {}_\mathcal{B}\Phi(b_j)\theta_j, \tag{12.29}$$

where $b_j \in \mathfrak{hor}^0(P) = \mathcal{B}$ and $\theta_j \in {}_{inv}\Gamma$. Then by the definition of π_υ, we get

$$\pi_\upsilon(\omega) = \sum_j b_j \otimes \theta_j.$$

Now we also assume that $\omega \in \ker \pi_{\mathfrak{v}}$, which then implies $\sum b_j \otimes \theta_j = 0$. So by (12.29), this in turn implies $(id \otimes p_1)_{\Omega(P)} \Psi(\omega) = 0$, and then that gives us

$$\Omega(P) \Psi(\omega) = (id \otimes p_0)_{\Omega(P)} \Psi(\omega) \in \Omega^1(P) \otimes \mathcal{A}.$$

But this says by definition that ω is horizontal. ■

Some comments on our notation are in order since we are not using the notation of Ðurđevich's articles. Where we have used either Φ or Ψ with various subscripts, he uses variations based on the letter F. Here is a list detailing how Ðurđevich's notations and ours correspond:

$$_{\mathcal{B}}\Phi = F : \mathcal{B} \to \mathcal{B} \otimes \mathcal{A},$$

$$_{\Omega(P)}\Psi = \hat{F} : \Omega(P) \to \Omega(P) \otimes \Gamma^{\wedge},$$

$$_{\Omega(P)}\Phi = F^{\wedge} : \Omega(P) \to \Omega(P) \otimes \mathcal{A},$$

$$_{\mathfrak{v}(P)}\Psi = \hat{F}_v : \mathfrak{ver}(P) \to \mathfrak{ver}(P) \otimes \Gamma^{\wedge},$$

$$_{\mathfrak{v}(P)}\Phi = F_v : \mathfrak{ver}(P) \to \mathfrak{ver}(P) \otimes \mathcal{A}.$$

We now return to some unfinished business.

Proof of Proposition 12.7:. Throughout this proof we let a, b, c denote elements of $\mathcal{B}$ while ω, η, θ will be elements of $_{\mathrm{inv}}\Gamma^{\wedge}$.

First, we check that $\mathfrak{ver}(P)$ is an algebra with unit element. The obvious candidate for the zero-degree unit element is $1 \otimes 1$. Let's see if the obvious is indeed correct! So we calculate

$$(1 \otimes 1)(b \otimes \eta) = b^{(0)} \otimes (1 \circ b^{(1)})\eta$$

$$= b^{(0)} \otimes (\varepsilon(b^{(1)})1)\eta$$

$$= \varepsilon(b^{(1)})b^{(0)} \otimes \eta$$

$$= b \otimes \eta,$$

where $b = \varepsilon(b^{(1)})b^{(0)}$ since $_{\mathcal{B}}\Phi$ is a right co-action. We also have

$$(a \otimes \omega)(1 \otimes 1) = a1 \otimes (\omega \circ 1)1 = a \otimes \omega,$$

since $_{\mathcal{B}}\Phi(1) = 1 \otimes 1$. So multiplication by $1 \otimes 1$ on either side acts as the identity, which means it is the unit element.

Next, we address the issue of whether this multiplication is associative, because, if that is not so, then this argument will come to a grinding halt. In the following calculations, we will repeatedly use Sweedler's notation for the co-action $_B\Phi$ as established above in the statement of the proposition. So we first associate the product of three factors in one way,

$$(a \otimes \omega)\big((b \otimes \eta)(c \otimes \theta)\big) = (a \otimes \omega)\big(bc^{(0)} \otimes (\eta \circ c^{(1)})\theta\big)$$
$$= a(bc^{(0)})^{(0)} \otimes \big(\omega \circ (bc^{(0)})^{(1)}\big)(\eta \circ c^{(1)})\theta$$
$$= a(b^{(0)}c^{(00)}) \otimes \big(\omega \circ (b^{(1)}c^{(01)})\big)(\eta \circ c^{(1)})\theta,$$

where the first two equalities use the definition of the multiplication in $\mathfrak{ver}(P)$ and the third equality comes from the multiplicativity of $_B\Phi$, which implies that $_B\Phi(bc^{(0)}) = {}_B\Phi(b)\,{}_B\Phi(c^{(0)})$. And then we associate these factors in the other way to get

$$\big((a \otimes \omega)(b \otimes \eta)\big)(c \otimes \theta) = \big(ab^{(0)} \otimes (\omega \circ b^{(1)})\eta\big)(c \otimes \theta)$$
$$= ab^{(0)}c^{(0)} \otimes \big([(\omega \circ b^{(1)})\eta] \circ c^{(1)}\big)\theta$$
$$= ab^{(0)}c^{(0)} \otimes \big((\omega \circ b^{(1)}) \circ c^{(10)}\big)(\eta \circ c^{(11)})\theta$$
$$= ab^{(0)}c^{(0)} \otimes \big(\omega \circ (b^{(1)}c^{(10)})\big)(\eta \circ c^{(11)})\theta,$$

where in the third equality we used the identity for the action of $\circ$ on a product and in the last equality we used that $\circ$ is an action. Now the equality of these two ways of associating comes down to verifying that

$$c^{(00)} \otimes c^{(01)} \otimes c^{(1)} = c^{(0)} \otimes c^{(10)} \otimes c^{(11)}.$$

But this in turn is an immediate consequence of the co-action identity of $_B\Phi$, namely, $(_B\Phi \otimes id_A)\,_B\Phi = (id_B \otimes \phi)_B\Phi$, when it acts on the element $c \in B$. This proves the associativity of the multiplication.

Next, we will consider the $*$-operation. First, we remark that this is well defined since $(\lambda b \otimes \eta)^* = \lambda^*(b \otimes \eta)^* = (b \otimes \lambda \eta)^*$ for all $\lambda \in \mathbb{C}$. Explicitly, we have

$$(b \otimes \lambda \eta)^* = b^{(0)*} \otimes (\lambda\eta)^*b^{(1)*} = \lambda^*(b^{(0)*} \otimes \eta^*b^{(1)*}) = \lambda^*(b \otimes \eta)^*.$$

Also, using $_B\Phi(\lambda b) = \lambda\,_B\Phi(b) = \lambda(b^{(0)} \otimes b^{(1)}) = (\lambda b^{(0)}) \otimes b^{(1)}$, we have

$$(\lambda b \otimes \eta)^* = \lambda^* b^{(0)*} \otimes \eta^* b^{(1)*} = \lambda^*(b \otimes \eta)^*.$$

Second, we show that this is an involution. We recall that $\phi(a^*) = (\phi(a))^*$ for all $a \in \mathcal{A}$. In particular, $a^{(0)*} \otimes a^{(1)*} = a^{*(0)} \otimes a^{*(1)}$. So Sweedler indices commute with the $*$-operation in $\mathcal{A}$. Then, using the definitions and standard identities, we have that

$$(b \otimes \eta)^{**} = \left(b^{(0)*} \otimes (\eta^* \circ b^{(1)*})\right)^*$$

$$= b^{(0)*(0)*} \otimes \left((\eta^* \circ b^{(1)*})^* \circ b^{(0)*(1)*}\right)$$

$$= b^{(00)} \otimes \left((\eta^{**} \circ \kappa(b^{(1)*})^*) \circ b^{(01)}\right)$$

$$= b^{(00)} \otimes \left((\eta \circ \kappa^{-1}(b^{(1)})) \circ b^{(01)}\right)$$

$$= b^{(00)} \otimes \left((\eta \circ (\kappa^{-1}(b^{(1)}))b^{(01)})\right)$$

$$= b^{(0)} \otimes \left((\eta \circ (\kappa^{-1}(b^{(12)}))b^{(11)})\right)$$

$$= b^{(0)} \otimes \left(\eta \circ \varepsilon(b^{(1)})1\right)$$

$$= \varepsilon(b^{(1)})b^{(0)} \otimes (\eta \circ 1)$$

$$= b \otimes \eta.$$

Here we used (11.2) in the form $\kappa(a^*)^* = \kappa^{-1}(a)$ and the fact that κ^{-1} is the co-inverse for the Hopf algebra $\mathcal{A}^{\mathrm{cop}}$ with the reverse co-multiplication ϕ^{cop} instead of ϕ (see [45]).

Therefore, the $*$-operation is an involution; that is, when applied twice, it is the identity operation. Next, we will see that the $*$-operation is actually a graded $*$-operation. This means that the expressions $((a \otimes \omega)(b \otimes \eta))^*$ and $(b \otimes \eta)^*(a \otimes \omega)^*$ should be equal up to a sign that depends on the degrees of the factors. But we will also need this little fact about Sweedler's notation that we might not have used before:

$$_B\Phi(ab^{(0)}) = {}_B\Phi(a)\,{}_B\Phi(b^{(0)})$$

$$= (a^{(0)} \otimes a^{(1)})(b^{(00)} \otimes b^{(01)})$$

$$= a^{(0)}b^{(00)} \otimes a^{(1)}b^{(01)},$$

which says that $(ab^{(0)})^{(0)} = a^{(0)}b^{(00)}$ and $(ab^{(0)})^{(1)} = a^{(1)}b^{(01)}$. We use this in the third equality below, but with single-digit Sweedler's notation.

Now we first compute with homogeneous elements ω and η. For simplicity, we set $\delta = \deg(\omega)\deg(\eta)$ and use single-digit Sweedler's notation. We start with one side of the desired identity:

$$((a \otimes \omega)(b \otimes \eta))^* = (ab^{(0)} \otimes (\omega \circ b^{(1)})\eta)^*$$

$$= ((ab^{(0)})^{(0)})^* \otimes (((\omega \circ b^{(1)})\eta)^* \circ ((ab^{(0)})^{(1)})^*)$$

$$= (-1)^\delta (a^{(0)}b^{(0)})^* \otimes ((\eta^*(\omega \circ b^{(2)})^*) \circ (a^{(1)}b^{(1)})^*)$$

$$= (-1)^\delta (a^{(0)}b^{(0)})^* \otimes ((\eta^*(\omega^* \circ \kappa(b^{(2)})^*)) \circ (b^{(1)*}a^{(1)*}))$$

$$= (-1)^\delta (a^{(0)}b^{(0)})^* \otimes (\eta^* \circ (b^{(1)*}a^{(1)*}))((\omega^* \circ \kappa(b^{(3)})^*) \circ (b^{(2)*}a^{(2)*}))$$

$$= (-1)^\delta (a^{(0)} b^{(0)})^* \otimes (\eta^* \circ (b^{(1)*} a^{(1)*}))(\omega^* \circ (\kappa(b^{(3)})^* b^{(2)*} a^{(2)*}))$$

$$= (-1)^\delta (a^{(0)} b^{(0)})^* \otimes (\eta^* \circ (b^{(1)*} a^{(1)*}))(\omega^* \circ (\varepsilon(b^{(2)})^* a^{(2)*}))$$

$$= (-1)^\delta (a^{(0)} b^{(0)})^* \otimes (\eta^* \circ (\varepsilon(b^{(2)})^* b^{(1)*} a^{(1)*}))(\omega^* \circ a^{(2)*})$$

$$= (-1)^\delta (a^{(0)} b^{(0)})^* \otimes (\eta^* \circ (b^{(1)*} a^{(1)*}))(\omega^* \circ a^{(2)*}).$$

On the other hand, we have

$$(b \otimes \eta)^* (a \otimes \omega)^* = (b^{(0)*} \otimes (\eta^* \circ b^{(1)*}))(a^{(0)*} \otimes (\omega^* \circ a^{(1)*}))$$

$$= b^{(0)*} a^{(0)*} \otimes (\eta^* \circ b^{(1)*}) \circ a^{(1)*})(\omega^* \circ a^{(2)*})$$

$$= (a^{(0)} b^{(0)})^* \otimes (\eta^* \circ (b^{(1)*} a^{(1)*}))(\omega^* \circ a^{(2)*}).$$

Recall that a and b have degree 0, so that

$$\deg(a \otimes \omega) = \deg(\omega) \quad \text{and} \quad \deg(b \otimes \eta) = \deg(\eta).$$

Putting all the pieces together, we have shown that

$$((a \otimes \omega)(b \otimes \eta))^* = (-1)^{\deg(a \otimes \omega) \deg(b \otimes \eta)} (b \otimes \eta)^* (a \otimes \omega)^*,$$

which is the required relation between the $*$-operation and the multiplication in a graded algebra.

Now we turn our attention to d_v. First, we show that it is a differential, namely, that $d_v^2 = 0$:

$$d_v^2(b \otimes \eta) = d_v(b \otimes d\eta + b^{(0)} \otimes \pi(b^{(1)})\eta)$$

$$= d_v(b \otimes d\eta) + d_v(b^{(0)} \otimes \pi(b^{(1)})\eta)$$

$$= b \otimes d^2\eta + b^{(0)} \otimes \pi(b^{(1)})d\eta + b^{(0)} \otimes d(\pi(b^{(1)})\eta) + b^{(00)} \otimes \pi(b^{(01)})\pi(b^{(1)})\eta$$

$$= b^{(0)} \otimes \pi(b^{(1)})d\eta + b^{(0)} \otimes d(\pi(b^{(1)}))\eta - b^{(0)} \otimes (\pi(b^{(1)})d\eta)$$

$$+ b^{(00)} \otimes \pi(b^{(01)})\pi(b^{(1)})\eta$$

$$= b^{(0)} \otimes (-\pi(b^{(10)})\pi(b^{(11)}))\eta + b^{(00)} \otimes \pi(b^{(01)})\pi(b^{(1)})\eta$$

$$= 0.$$

We used the co-action property $b^{(0)} \otimes b^{(10)} \otimes b^{(11)} = b^{(00)} \otimes b^{(01)} \otimes b^{(1)}$. We also used $\deg(\pi(b^{(1)})) = 1$ and the Maurer–Cartan formula (10.14) for $d(\pi(a))$.

Now we will show that d_v satisfies the graded Leibniz rule:

$$d_v\big((a \otimes \omega)(b \otimes \eta)\big) = d_v\big(ab^{(0)} \otimes (\omega \circ b^{(1)})\eta\big)$$

$$= ab^{(0)} \otimes d((\omega \circ b^{(1)})\eta) + a^{(0)}b^{(00)} \otimes \pi(a^{(1)}b^{(01)})(\omega \circ b^{(1)})\eta$$

$$= ab^{(0)} \otimes (d(\omega \circ b^{(1)})\eta + (-1)^{|\omega|}ab^{(0)} \otimes (\omega \circ b^{(1)})d\eta)$$

$$+ a^{(0)}b^{(00)} \otimes \big(\varepsilon(a^{(1)})\pi(b^{(01)})\big)(\omega \circ b^{(1)})\eta$$

$$+ a^{(0)}b^{(00)} \otimes \big(\pi(a^{(1)}) \circ b^{(01)}\big)(\omega \circ b^{(1)})\eta$$

$$= ab^{(0)} \otimes (d(\omega \circ b^{(1)})\eta + (-1)^{|\omega|}ab^{(0)} \otimes (\omega \circ b^{(1)})d\eta)$$

$$+ \varepsilon(a^{(1)})a^{(0)}b^{(00)} \otimes \pi(b^{(01)})(\omega \circ b^{(1)})\eta$$

$$+ a^{(0)}b^{(00)} \otimes \big(\pi(a^{(1)}) \circ b^{(01)}\big)(\omega \circ b^{(1)})\eta$$

$$= ab^{(0)} \otimes \big(d\omega \circ b^{(1)} - \pi(b^{(10)})(\omega \circ b^{(11)}) + (-1)^{|\omega|}(\omega \circ b^{(10)})\pi(b^{(11)})\big)\eta$$

$$+ (-1)^{|\omega|}ab^{(0)} \otimes (\omega \circ b^{(1)})d\eta$$

$$+ ab^{(00)} \otimes \pi(b^{(01)})(\omega \circ b^{(1)})\eta + a^{(0)}b^{(00)} \otimes \big(\pi(a^{(1)}) \circ b^{(01)}\big)(\omega \circ b^{(1)})\eta$$

$$= ab^{(0)} \otimes (d\omega \circ b^{(1)})\eta + (-1)^{|\omega|}ab^{(0)} \otimes \big((\omega \circ b^{(10)})\pi(b^{(11)})\big)\eta$$

$$+ (-1)^{|\omega|}ab^{(0)} \otimes (\omega \circ b^{(1)})d\eta + a^{(0)}b^{(00)} \otimes \big(\pi(a^{(1)}) \circ b^{(01)}\big)(\omega \circ b^{(1)})\eta.$$

On the other hand,

$$\big(d_v(a \otimes \omega)\big)(b \otimes \eta) = \big(a \otimes d\omega + a^{(0)} \otimes \pi(a^{(1)})\omega\big)(b \otimes \eta)$$

$$= ab^{(0)} \otimes (d\omega \circ b^{(1)})\eta + a^{(0)}b^{(0)} \otimes \big((\pi(a^{(1)})\omega) \circ b^{(1)}\big)\eta$$

$$= ab^{(0)} \otimes (d\omega \circ b^{(1)})\eta + a^{(0)}b^{(0)} \otimes \big((\pi(a^{(1)}) \circ b^{(10)}))(\omega \circ b^{(11)})\big)\eta$$

and

$$(a \otimes \omega)\big(d_v(b \otimes \eta)\big) = (a \otimes \omega)\big(b \otimes d\eta + b^{(0)} \otimes \pi(b^{(1)})\eta\big)$$

$$= ab^{(0)} \otimes (\omega \circ b^{(1)})d\eta + ab^{(00)} \otimes (\omega \circ b^{(01)})(\pi(b^{(1)})\eta).$$

Now putting this all together, using the co-action identity

$$b^{(00)} \otimes b^{(01)} \otimes b^{(1)} = b^{(0)} \otimes b^{(10)} \otimes b^{(11)}$$

as well as $|\omega| = \deg(\omega) = \deg(a \otimes \omega)$ yields

$$d_v\big((a \otimes \omega)(b \otimes \eta)\big) = \big(d_v(a \otimes \omega)\big)(b \otimes \eta) + (-1)^{\deg(a \otimes \omega)}(a \otimes \omega)\big(d_v(b \otimes \eta)\big),$$

which is the graded Leibniz equality, as desired.

Next, we will show that d_v is a *-morphism. First, using definitions and identities, we calculate that

$$d_v((b \otimes \eta)^*) = d_v(b^{(0)*} \otimes (\eta^* \circ b^{(1)*}))$$

$$= b^{(0)*} \otimes d(\eta^* \circ b^{(1)*}) + b^{(00)*} \otimes \pi(b^{(01)*})(\eta^* \circ b^{(1)*})$$

$$= b^{(0)*} \otimes \left(d(\eta^*) \circ b^{(1)*} - \pi(b^{(10)*})(\eta^* \circ b^{(11)*}) + (-1)^{|\eta|}(\eta^* \circ b^{(10)*})\pi(b^{(11)*}) \right)$$

$$+ b^{(00)*} \otimes \pi(b^{(01)*})(\eta^* \circ b^{(1)*})$$

$$= b^{(0)*} \otimes \left(d(\eta^*) \circ b^{(1)*} + (-1)^{|\eta|}(\eta^* \circ b^{(10)*})\pi(b^{(11)*}) \right)$$

$$= b^{(0)*} \otimes (d\eta)^* \circ b^{(1)*} + (-1)^{|\eta|} b^{(0)*} \otimes (\eta^* \circ b^{(10)*})\pi(b^{(11)*}), \tag{12.30}$$

where in the third line we canceled the second and fourth terms using the co-action property.

But we also have

$$\left(d_v(b \otimes \eta) \right)^* = \left(b \otimes d\eta + b^{(0)} \otimes \pi(b^{(1)})\eta \right)^*$$

$$= b^{(0)*} \otimes ((d\eta)^* \circ b^{(1)*})) + b^{(00)*} \otimes ((\pi(b^{(1)})\eta)^* \circ b^{(01)*})$$

$$= b^{(0)*} \otimes ((d\eta)^* \circ b^{(1)*})) + (-1)^{|\eta|} b^{(00)*} \otimes ((\eta^* \pi(b^{(1)})^*) \circ b^{(01)*})$$

$$= b^{(0)*} \otimes ((d\eta)^* \circ b^{(1)*})) + (-1)^{|\eta|} b^{(00)*} \otimes ((\eta^* \circ b^{(010)*})(\pi(b^{(1)})^* \circ b^{(011)*}))$$

$$= b^{(0)*} \otimes ((d\eta)^* \circ b^{(1)*})) + (-1)^{|\eta|} b^{(00)*} \otimes ((\eta^* \circ b^{(010)*})(\varepsilon(b^{(011)*})\pi(b^{(1)})^*))$$

$$= b^{(0)*} \otimes ((d\eta)^* \circ b^{(1)*})) + (-1)^{|\eta|} b^{(00)*} \otimes ((\eta^* \circ \varepsilon(b^{(011)*})b^{(010)*})(\pi(b^{(1)})^*))$$

$$= b^{(0)*} \otimes ((d\eta)^* \circ b^{(1)*})) + (-1)^{|\eta|} b^{(00)*} \otimes ((\eta^* \circ b^{(01)*})(\pi(b^{(1)})^*)). \tag{12.31}$$

Now we see that (12.30) and (12.31) are equal using the co-action property. And so $d_v((b \otimes \eta)^*) = (d_v(b \otimes \eta))^*$ is proved.

Finally, we now will show that $\mathfrak{ver}(P)$ is generated as a differential algebra by $\mathfrak{ver}^0(P) = \mathcal{B}$. Since $_\mathcal{B}\Phi(1_\mathcal{B}) = 1_\mathcal{B} \otimes 1_\mathcal{A}$ and $\pi(1_\mathcal{A}) = 0$, we have that $d_v(1_\mathcal{B} \otimes \eta) = 1_\mathcal{B} \otimes d\eta$ for all $\eta \in {}_{\text{inv}}\Gamma^\wedge$. And this shows that $\mathfrak{ver}^1(P)$ generates $\oplus_{k \geq 1} \mathfrak{ver}^k(P)$ as a differential $\mathcal{B}$-module. So it remains to show that $\mathfrak{ver}^0(P)$ generates $\mathfrak{ver}^1(P)$. But since $_\mathcal{B}\Phi : \mathcal{B} \to \mathcal{B} \otimes \mathcal{A}$ is surjective, we know that for every $a \in \mathcal{A}$ there exists some $b \in \mathcal{B}$ such that

$$_\mathcal{B}\Phi(b) = 1_\mathcal{B} \otimes a.$$

Then $d_v(b \otimes 1) = b^{(0)} \otimes \pi(b^{(1)})1 = 1 \otimes \pi(a)$. Since $a \in \mathcal{A}$ was arbitrary and π is *onto* $_{\text{inv}}\Gamma$, we conclude that $1 \otimes {}_{\text{inv}}\Gamma$ is contained in the image of d_v. But $1 \otimes {}_{\text{inv}}\Gamma$ generates $\mathcal{B} \otimes {}_{\text{inv}}\Gamma = \mathfrak{ver}^1(P)$ as a $\mathcal{B}$-module. And that finishes the proof of this last part of the theorem. ∎

12.7   Quantum Connections

Connections are an essential part of the modern theory of classical differential geometry. For example, the standard reference [49] is centered on this topic. It has also become an important part of contemporary physics, where it was developed independently as something called *gauge theory*. This concept has been introduced in noncommutative geometry as well.

Let P be a QPB with differential calculus throughout this section. Here is the definition of a quantum connection.

Definition 12.10. *A linear map* $\omega : {}_{inv}\Gamma \to \Omega^1(P)$ *is said to be a* quantum connection *if, for every* $\theta \in {}_{inv}\Gamma$, *we have*

$$\omega(\theta^*) = \omega(\theta)^* \tag{12.32}$$

$$\Omega(P)\Psi\big(\omega(\theta)\big) = (\omega \otimes id_\mathcal{A})\mathrm{ad}(\theta) + 1_\mathcal{B} \otimes \theta, \tag{12.33}$$

where $\mathrm{ad} : {}_{inv}\Gamma \to {}_{inv}\Gamma \otimes \mathcal{A}$ *is the right adjoint co-action of* $\mathcal{A}$ *on* ${}_{inv}\Gamma$ *and* $1_\mathcal{B}$ *is the identity element in* $\mathcal{B}$.

An underlying idea here is that each $\omega(\theta)$ is a vertical 1-form. We will say more on this later on.

Notice that we do not use the differential d_P of $\Omega(P)$ in this definition. Nor will d_P make an appearance in the rest of this section. However, d_P plays a role in the theory of curvature and of covariant derivatives.

The two conditions on ω merit motivation. The condition (12.32), which simply says that ω is a $*$-morphism, is not really all that essential. The point is that in the classical theory all (well, almost all) vector spaces are over $\mathbb{R}$, the real numbers, and so the conjugation reduces to the identity (a linear map!). Here if $\theta^* = \theta$, which is usually called a "self-adjoint" element though sometimes a "real" element, we have according to (12.32) that $\omega(\theta) = \omega(\theta)^*$. So ω maps real elements to real elements, thereby imitating the classical case.

However, in this theory where the underlying field is $\mathbb{C}$, the complex numbers, it is also reasonable to impose a condition such as $\omega(\theta^*) = \sqrt{-1}\,\omega(\theta)^*$ instead of (12.32). Moreover, if one wishes to develop a theory of quantum connections in algebras that do not necessarily have a $*$-operation, then one simply drops the condition (12.32). In that case, of course, one can use any field instead of $\mathbb{C}$. We will not do this explicitly, but it is rather straightforward to see how one would modify the subsequent theory.

On the other hand, (12.33) dualizes the algebraic condition for a classical connection and so is an essential condition. First, let's see that this equation makes sense in terms of the sets in which the three expressions lie. Starting with the left side, we note that $\omega(\theta) \in \Omega^1(P)$, so that

$$\Omega(P)\Psi\big(\omega(\theta)\big) \in \big(\Omega(P) \otimes \Gamma^\wedge\big)^1 = \big(\Omega^1(P) \otimes \mathcal{A}\big) \oplus \big(\mathcal{B} \otimes \Gamma\big), \tag{12.34}$$

using $\Gamma^{\wedge 0} = \mathcal{A}$, $\Gamma^{\wedge 1} = \Gamma$, and $\Omega^0(P) = \mathcal{B}$. For the first expression on the right side, we first note that $\mathrm{ad}(\theta) \in {}_{\mathrm{inv}}\Gamma \otimes \mathcal{A}$, which implies that

$$(\omega \otimes id_{\mathcal{A}})\mathrm{ad}(\theta) \in \Omega^1(P) \otimes \mathcal{A}.$$

Finally, for the second expression on the right side, we have immediately that

$$1_{\mathcal{B}} \otimes \theta \in \mathcal{B} \otimes {}_{\mathrm{inv}}\Gamma \subset \mathcal{B} \otimes \Gamma.$$

So we see that (12.33) is giving us the decomposition of the expression on its left side in terms of the direct sum in (12.34). Now we think about how this corresponds to a classical principal bundle $E \to M$ with structure Lie group G. More explicitly, the right co-action on $\omega(\theta)$ breaks into one piece in $\Omega^1(P) \otimes \mathcal{A}$ that corresponds classically to the right action of G on the tangent bundle $T(E)$ and into a second piece that corresponds to γ in the diagram (12.38) ahead.

The following discussion is meant to motivate the definition of a quantum connection, so do not expect the proof of anything! We will "translate" the conditions for an Ehresmann connection from the classical theory, but the rules of this translation are not intended to be rigorous for the time being. We will use the notation established in [76] and will not define everything in detail here. See [76] or your favorite differential geometry text for all the definitions and basic properties.

The definition of an *Ehresmann connection* can be formulated as a smooth map $w : T(E) \to \mathfrak{g}$, the Lie algebra of G, which is linear on each fiber of $T(E)$ and makes two diagrams commute. The first diagram is

$$
\begin{array}{ccc}
T(E) \times G & \xrightarrow{\alpha} & T(E) \\
{\scriptstyle w \times id_G}\Big\downarrow & & \Big\downarrow{\scriptstyle w} \\
\mathfrak{g} \times G & \xrightarrow{\beta} & \mathfrak{g},
\end{array}
\qquad (12.35)
$$

where α is the right action of G on $T(E)$ and $\beta(A, g) := \mathrm{Ad}_{g^{-1}}(A)$ is the right adjoint action of G on its Lie algebra. This diagram expresses the covariance of the Ehresmann connection w.

But we are still a way from forms. In fact, we do not even have vector fields on E [i.e., sections of $T(E)$] but just $T(E)$ in the diagram (12.35). However, each element of $\mathfrak{g}$, which is the tangent space of G at the identity element $e \in G$, does correspond to a unique left invariant vector field on G. And there is an isomorphism between the real vector space $\mathfrak{g}$ and the real vector space of left invariant vector fields on G. If we dualize this to the noncommutative setting, the corresponding object is ${}_{\mathrm{inv}}\Gamma$, the left invariant 1-forms on the noncommutative space. So formally dualizing the previous diagram, using our standard notation, yields this diagram of complex vector spaces:

$$
\begin{array}{ccc}
\Omega^1(P) \otimes \mathcal{A} & \xleftarrow{\alpha^*} & \Omega^1(P) \\
{\scriptstyle \omega \otimes id_A}\Big\uparrow & & \Big\uparrow{\scriptstyle \omega} \\
{}_{\mathrm{inv}}\Gamma \otimes \mathcal{A} & \xleftarrow{\mathrm{ad}} & {}_{\mathrm{inv}}\Gamma.
\end{array}
\qquad (12.36)
$$

The horizontal upper map α^* is identified with $(id \otimes p_0)_{\Omega(P)}\Psi$, which is the right co-action of $\mathcal{A}$ on $\Omega^1(P)$. So this diagram expresses the right covariance of the quantum connection ω.

Taking θ in the lower right corner in the diagram (12.36) and chasing it through this diagram to the upper left corner gives (but does not prove) the formula

$$(id \otimes p_0)_{\Omega(P)}\Psi(\omega(\theta)) = (\omega \otimes id_{\mathcal{A}})\mathrm{ad}(\theta). \tag{12.37}$$

The second, defining property of an Ehresmann connection is equivalent to the commutative diagram

$$
\begin{array}{ccc}
E \times \mathfrak{g} & \overset{\gamma}{\longrightarrow} & T(E) \\
\pi_2 \downarrow & & \downarrow w \\
\mathfrak{g} & \overset{id}{\longrightarrow} & \mathfrak{g},
\end{array}
\tag{12.38}
$$

where the horizontal upper map is $\gamma(x, A) := A^\sharp(x)$ and π_2 is the projection on the second factor. For each $A \in \mathfrak{g}$, we denote the corresponding vertical vector field on E by $A^\sharp$. The diagram (12.38) is equivalent to the equation

$$w(A^\sharp(x)) = A,$$

which is how this condition on w is usually expressed in textbooks. But we prefer diagrams, since we know how to dualize them immediately with no effort to speak of. This condition requires that the Lie group element, which an Ehresmann connection w associates to the standard vertical vector field generated by the Lie group element A, must be A itself. One often says colloquially that an Ehresmann connection maps standard vertical vector fields to their generators.

Formally dualizing this gives

$$
\begin{array}{ccc}
\mathcal{B} \otimes_{\mathrm{inv}} \Gamma & \overset{\gamma^*}{\longleftarrow} & \Omega^1(P) \\
\pi_2^* \uparrow & & \uparrow \omega \\
{}_{\mathrm{inv}}\Gamma & \overset{id}{\longleftarrow} & {}_{\mathrm{inv}}\Gamma.
\end{array}
\tag{12.39}
$$

The horizontal upper map γ^* is identified with $(id \otimes p_1)_{\Omega(P)}\Psi$, another "piece" of the co-action of $\Gamma^\wedge$ on $\Omega^1(P)$. As we have remarked before, $\pi_2^*(\theta) = 1_{\mathcal{B}} \otimes \theta$. Taking θ in the lower right corner and chasing it through this diagram to the upper left corner gives (but does not prove) the formula

$$(id \otimes p_1)_{\Omega(P)}\Psi(\omega(\theta)) = 1_{\mathcal{B}} \otimes \theta. \tag{12.40}$$

Summing (12.37) and (12.40) gives the condition (12.33). This ends our purely motivational discussion for (12.33).

Since (12.33) is *not* linear in ω, the set of quantum connections on P, denoted $\operatorname{qconn}(P)$, is an affine space if it is nonempty. More explicitly, the difference of two connections $\lambda = \omega_1 - \omega_2$ satisfies these two conditions linear in λ:

$$\lambda(\theta^*) = \lambda(\theta)^*, \tag{12.41}$$

$$\Omega(P)\Psi(\lambda(\theta)) = (\lambda \otimes id_A)\operatorname{ad}(\theta). \tag{12.42}$$

Any $\lambda : {}_{\operatorname{inv}}\Gamma \to \Omega^1(P)$ satisfying (12.41) and (12.42) is called a *quantum connection displacement, or QCD*. Since these are linear conditions on λ, the set of all quantum connection displacements forms a vector space, which is then the vector space associated to the affine space $\operatorname{qconn}(P)$, provided that $\operatorname{qconn}(P)$ is nonempty. We introduce the notation

$$\operatorname{qcd}(P) := \{\lambda : {}_{\operatorname{inv}}\Gamma \to \Omega^1(P) \,|\, \lambda \text{ is a QCD}\}$$

for this vector space. Then we have this well-known property of affine spaces; namely, if $\operatorname{qconn}(P) \neq \emptyset$, the empty set, then for any $\omega_0 \in \operatorname{qconn}(P)$, we have

$$\operatorname{qconn}(P) = \omega_0 + \operatorname{qcd}(P). \tag{12.43}$$

Exercise 12.24. *First, prove (12.43). Then think back to days gone by when you studied solutions to homogeneous and inhomogeneous linear differential equations and realize the common underlying structure of that and of (12.43).*

So the study of all quantum connections on P reduces to the study of just one of them plus the study of all the QCDs on P.

Exercise 12.25. *Prove that the second property (12.42) of a QCD implies that $\lambda(\theta)$ is a horizontal element in $\Omega^1(P)$ for all $\theta \in {}_{\operatorname{inv}}\Gamma$. In other words, $\lambda(\theta) \in \mathfrak{hor}^1(P)$.*

We now present a property of quantum connections that could have been used as a definition for them.

Theorem 12.8. *Suppose that ω is a quantum connection on a QPB. Define $\mu_\omega : \mathfrak{ver}^1(P) \to \Omega^1(P)$ by*

$$\mu_\omega(b \otimes \theta) := b\,\omega(\theta).$$

Then $\pi_\mathfrak{v} \circ \mu_\omega = id$, the identity on $\mathfrak{ver}^1(P)$. In particular, μ_ω is injective. Moreover, $\mu \equiv \mu_\omega$ satisfies these properties:

1. *μ is a left B-module morphism.*
2. *μ intertwines the right co-actions on $\mathfrak{ver}^1(P)$ and on $\Omega^1(P)$.*
3. *$\mu(1_B \otimes \theta^*) = (\mu(1_B \otimes \theta))^*$ for all $\theta \in {}_{\operatorname{inv}}\Gamma$.*

Conversely, if $\mu : \mathfrak{vet}^1(P) \to \Omega^1(P)$ is a linear map satisfying the above three properties as well as $\pi_{\mathfrak{v}} \circ \mu = id$, then there exists a unique quantum connection ω such that $\mu = \mu_\omega$.

The bijection between the set $\mathrm{qconn}(P)$ *and the set of all such maps μ is an isomorphism of affine spaces.*

Remarks. All of the conditions on μ are linear in μ except for the inhomogeneous condition $\pi_{\mathfrak{v}} \circ \mu = id$, which is commonly expressed by saying that μ splits the exact sequence in Theorem 12.7. Also, condition 3 on μ is imposed only to ensure that ω is a $*$-morphism. If the first property in the definition of a quantum connection is changed, then a corresponding change would have to be made here in condition 3. The upshot is the quantum connection ω gives us a way (by using the map μ_ω) of identifying $\mathfrak{vet}^1(P)$ with a subspace of $\Omega^1(P)$. This subspace itself is commonly referred to as the vertical subspace of $\Omega^1(P)$, but the reader should bear in mind that saying this means that one has a specific quantum connection ω under consideration. In particular, $\omega(\theta) = \mu_\omega(1 \otimes \theta) \in \Omega^1(P)$ is a vertical vector (with respect to the quantum connection ω itself) for every $\theta \in {}_{\mathrm{inv}}\Gamma$. We have already mentioned that this property is a basic idea behind the theory of quantum connections.

Proof. We start with an element $b \otimes \theta \in \mathfrak{vet}^1(P) = \mathcal{B} \otimes {}_{\mathrm{inv}}\Gamma$. Then we have $\mu_\omega(b \otimes \theta) = b\,\omega(\theta) \in \Omega^1(P)$. Now we want to project this element back to $\mathfrak{vet}^1(P)$ using $\pi_{\mathfrak{v}}$. And if everything works as claimed, then we will get $b \otimes \theta$. Let's see what happens. So we consider $\pi_{\mathfrak{v}}(\mu_\omega(b \otimes \theta)) = \pi_{\mathfrak{v}}(b\,\omega(\theta))$. Now $\pi_{\mathfrak{v}} = (id \otimes \pi_{\mathrm{inv}})(id \otimes p_1)\,{}_{\Omega(P)}\Psi$. So we first consider

$$
\begin{aligned}
{}_{\Omega(P)}\Psi(b\,\omega(\theta)) &= {}_{\Omega(P)}\Psi(b)\,{}_{\Omega(P)}\Psi(\omega(\theta)) \\
&= {}_{\mathcal{B}}\Phi(b)\,{}_{\Omega(P)}\Psi(\omega(\theta)) \\
&= (b^{(0)} \otimes b^{(1)})\big((\omega \otimes id_A)\mathrm{ad}(\theta) + 1_{\mathcal{B}} \otimes \theta\big).
\end{aligned}
$$

Next, we apply $(id \otimes p_1)$ to both sides, giving

$$
(id \otimes p_1)({}_{\Omega(P)}\Psi(b\,\omega(\theta))) = b^{(0)} \otimes b^{(1)}\theta.
$$

Finally, we apply $(id \otimes \pi_{\mathrm{inv}})$ to both sides of this last equation, thereby getting

$$
\begin{aligned}
\pi_{\mathfrak{v}}(\mu_\omega(b \otimes \theta)) &= b^{(0)} \otimes \pi_{\mathrm{inv}}(b^{(1)}\theta) \\
&= b^{(0)} \otimes \varepsilon(b^{(1)})\theta \\
&= \varepsilon(b^{(1)})b^{(0)} \otimes \theta \\
&= b \otimes \theta.
\end{aligned}
$$

To see that μ_ω is a left $\mathcal{B}$-module morphism, we take $c \in \mathcal{B}$ and calculate

$$\mu_\omega(c(b \otimes \theta)) = \mu_\omega((cb) \otimes \theta)$$
$$= (cb)\omega(\theta)$$
$$= c(b\omega(\theta))$$
$$= c\mu_\omega(b \otimes \theta).$$

To see that μ_ω intertwines the right co-actions on $\mathfrak{ver}^1(P)$ and on $\Omega^1(P)$, we first note that we have already shown above

$$_{\Omega(P)}\Psi(\mu_\omega(b \otimes \theta)) = (b^{(0)} \otimes b^{(1)})\big((\omega \otimes id_{\mathcal{A}})\mathrm{ad}(\theta) + 1_{\mathcal{B}} \otimes \theta\big).$$

If we write $\mathrm{ad}(\theta) = \theta^{(0)} \otimes \theta^{(1)} \in {}_{inv}\Gamma \otimes \mathcal{A}$, this becomes

$$_{\Omega(P)}\Psi(\mu_\omega(b \otimes \theta)) = (b^{(0)} \otimes b^{(1)})\big((\omega \otimes id_{\mathcal{A}})(\theta^{(0)} \otimes \theta^{(1)}) + 1_{\mathcal{B}} \otimes \theta\big)$$
$$= (b^{(0)} \otimes b^{(1)})\big(\omega(\theta^{(0)}) \otimes \theta^{(1)} + 1_{\mathcal{B}} \otimes \theta\big)$$
$$= b^{(0)}\omega(\theta^{(0)}) \otimes b^{(1)}\theta^{(1)} + b^{(0)} \otimes b^{(1)}\theta.$$

On the other hand, using

$$\hat{\phi}(\theta) = 1_{\mathcal{A}} \otimes \theta + \mathrm{ad}(\theta) = 1_{\mathcal{A}} \otimes \theta + \theta^{(0)} \otimes \theta^{(1)},$$

we have by the definition of $_v\Psi$ that

$$_v\Psi(b \otimes \theta) = b^{(0)} \otimes 1_{\mathcal{A}} \otimes b^{(1)}\theta + b^{(0)} \otimes \theta^{(0)} \otimes b^{(1)}\theta^{(1)}.$$

Next, we have to apply $\mu_\omega \otimes id$ to this last expression. But so far that is only defined for the second term. For the first term we define

$$(\mu_\omega \otimes id)(b \otimes a \otimes \theta) := b \otimes \theta.$$

Then we have

$$(\mu_\omega \otimes id)\,_v\Psi(b \otimes \theta) = b^{(0)} \otimes b^{(1)}\theta + b^{(0)}\omega(\theta^{(0)}) \otimes b^{(1)}\theta^{(1)}.$$

To see that μ_ω satisfies condition 3, we get immediately

$$(\mu_\omega(1_{\mathcal{B}} \otimes \theta))^* = (\omega(\theta))^* = \omega(\theta^*) = \mu_\omega(1_{\mathcal{B}} \otimes \theta^*).$$

Notice that we recover the quantum connection ω from the map μ_ω from the formula

$$\mu_\omega(1_{\mathcal{B}} \otimes \theta) = \omega(\theta),$$

which holds for every $\theta \in {}_{inv}\Gamma$. So for the converse we take any map μ with the assumed properties, and we define the associated map

$$\omega(\theta) := \mu(1_B \otimes \theta)$$

for all $\theta \in {}_{inv}\Gamma$. So we have $\omega : {}_{inv}\Gamma \to \Omega^1(P)$. It remains to show that ω satisfies the two characteristic properties of a quantum connection. First, we use condition 3 to obtain

$$\omega(\theta^*) := \mu(1_B \otimes \theta^*) = (\mu(1_B \otimes \theta))^* = (\omega(\theta))^*.$$

Next, we use a lot of properties, but in particular condition 2 in the second equality and ${}_B\Phi(1_B) = 1_B \otimes 1_A$ in the third equality, to obtain

$$
\begin{aligned}
{}_{\Omega(P)}\Psi(\omega(\theta)) &= {}_{\Omega(P)}\Psi(\mu(1_B \otimes \theta)) \\
&= (\mu \otimes id)\,{}_v\Psi(1_B \otimes \theta) \\
&= (\mu \otimes id)(1_B \otimes 1_A \otimes \theta + 1_A \otimes \theta^{(0)} \otimes \theta^{(1)}) \\
&= 1_B \otimes \theta + \mu(1_B \otimes \theta^{(0)}) \otimes \theta^{(1)} \\
&= 1_B \otimes \theta + \omega(\theta^{(0)}) \otimes \theta^{(1)} \\
&= 1_B \otimes \theta + (\omega \otimes id)(\theta^{(0)} \otimes \theta^{(1)}) \\
&= 1_B \otimes \theta + (\omega \otimes id)\,\mathrm{ad}(\theta).
\end{aligned}
$$

And this is the second property of a quantum connection.

We have not yet used condition 1. But we also have not yet proved that the quantum connection ω is uniquely determined by μ nor that $\mu = \mu_\omega$. (We are still arguing the converse and using the notation of that part of the proof.) First, we note that by using condition 1 we get

$$\mu_\omega(b \otimes \theta) = b\omega(\theta) = b\mu(1 \otimes \theta) = \mu(b(1 \otimes \theta)) = \mu(b \otimes \theta).$$

This proves $\mu_\omega = \mu$ and that the right side, namely μ, is uniquely determined by ω on the left side via the construction μ_ω.

The final statement of the theorem we leave to the reader. ∎

One of the consequences of having $*$-operations on the various spaces under consideration is that then one can define a $*$-operation on connections. We first do this for any linear map between such vector spaces.

Theorem 12.9. *Let V and W be vector spaces, each with a $*$-operation. Let $T : V \to W$ be linear. Define $T^* : V \to W$ by*

$$T^*(v) := (T(v^*))^* \quad \text{for all } v \in V.$$

Then T^* is linear. Moreover, if we let $\mathrm{Mor}(V, W)$ denote the vector space of all linear maps $T : V \to W$, the mapping $T \mapsto T^*$ is a $*$-operation on $\mathrm{Mor}(V, W)$.

The proof is left to the reader.

So for any quantum connection $\omega : {}_{inv}\Gamma \to \Omega^1(P)$, we have the linear map $\omega^* : {}_{inv}\Gamma \to \Omega^1(P)$. Then we have the next result in this setting.

Proposition 12.8. *The map ω^* is a quantum connection.*

Again, the proof is not very complicated and will be left to the reader.

Definition 12.11. *Let $\mathcal{R} \subset \ker \varepsilon$ be the right ideal in $\mathcal{A}$ such that*

$$\Gamma \cong \mathcal{A} \otimes (\ker \varepsilon / \mathcal{R}).$$

Then a quantum connection ω is said to be multiplicative *if for all $a \in \mathcal{R}$, we have that*

$$\omega(\pi(a^{(1)}))\, \omega(\pi(a^{(1)})) = 0,$$

where $\phi(a) = a^{(1)} \otimes a^{(2)} \in \mathcal{A} \otimes \mathcal{A}$ is the co-multiplication in Sweedler's notation.

Remark. We have $\pi(a^{(j)}) \in {}_{inv}\Gamma$, so that $\omega(\pi(a^{(j)})) \in \Omega^1(P)$ for $j = 1, 2$. So we have in general that

$$\omega(\pi(a^{(1)}))\, \omega(\pi(a^{(1)})) \in \Omega^2(P).$$

For a multiplicative quantum connection, all of these elements for $a \in \mathcal{R}$ are zero. These products lie in $\Omega^2(P)$, and so the conditions that we are imposing on a quantum connection for it to be multiplicative are *quadratic conditions*.

It is also important to emphasize what should be obvious (just in case it is not!), namely, that this definition uses the existence of the HODC on P, not just the FODC.

We now will sketch the motivation behind employing the terminology "multiplicative" for such a quantum connection. The maps $\omega : {}_{inv}\Gamma \to \Omega^1(P)$ in degree 1 and $\mathbb{C} \to \Omega^0(P) = \mathcal{B}$ in degree 0 (the latter is $\lambda \to \lambda 1_\mathcal{B}$) induce a unital, multiplicative map $\omega^\otimes : {}_{inv}\Gamma^\otimes \to \Omega(P)$. Then ω is multiplicative if and only if ${}_{inv}S^\wedge \subset \ker \omega^\otimes$ since ${}_{inv}S^\wedge$ is by definition the two-sided ideal in ${}_{inv}\Gamma^\otimes$ generated by the *quadratic elements*

$$\pi(a^{(1)}) \otimes \pi(a^{(2)}) \quad \text{with} \quad a \in \mathcal{R}.$$

Therefore, if ω is multiplicative, then $\omega^\otimes$ factors through the quotient

$${}_{inv}\Gamma^\wedge = {}_{inv}\Gamma^\otimes / {}_{inv}S^\wedge$$

to give a unital, multiplicative map $\omega^\wedge : {}_{inv}\Gamma^\wedge \to \Omega(P)$. Since we are requiring that ω be a $*$-morphism, it follows that $\omega^\otimes$ is a $*$-morphism as well. Moreover, if ω is multiplicative, then $\omega^\wedge$ is also a $*$-morphism.

Definition 12.12. *A quantum connection ω is said to be* regular *if for all $\varphi \in \mathfrak{hor}^k(P)$ and all $\theta \in {}_{inv}\Gamma$, we have this identity in $\Omega^{k+1}(P)$:*

$$\omega(\theta)\,\varphi = (-1)^k \varphi^{(0)}\, \omega(\theta \circ \varphi^{(1)}),$$

where $\,_{\Omega(P)}\Psi(\varphi) = \varphi^{(0)} \otimes \varphi^{(1)} \in \mathfrak{hor}^k(P) \otimes \mathcal{A}$.

Remark. This definition uses the existence of the full HODC $\Omega(P)$ as well as the existence of the horizontal forms $\mathfrak{hor}(P)$. Unfortunately, there seems to be no good motivation for employing the overused terminology "regular" for such a quantum connection. On the other hand, I have not been able to find a reasonable replacement.

Warning: We will continue to use the notation $\,_{\Omega(P)}\Psi(\varphi) = \varphi^{(0)} \otimes \varphi^{(1)}$ for the rest of this section.

Corollary 12.2. *Let ω be a regular quantum connection. Then every form $\varphi \in \Omega^k(M)$ grade commutes with every element in $\mathrm{Ran}\,\omega \subset \Omega^1(P)$.*

Proof. By definition, $\varphi \in \Omega^k(M)$ means that $\,_{\Omega(P)}\Psi(\varphi) = \varphi \otimes 1_{\mathcal{A}}$. Thus, since ω is regular, we have for all $\theta \in {}_{inv}\Gamma$ that

$$\omega(\theta)\,\varphi = (-1)^k \varphi^{(0)}\, \omega(\theta \circ \varphi^{(1)}) = (-1)^k \varphi\,\omega(\theta \circ 1_{\mathcal{A}}) = (-1)^k \varphi\,\omega(\theta).$$

And this is exactly the graded commutativity that we had to prove. ∎

The next result is a continuation of Theorem 12.8, so we will use the notation from it without further comment. That theorem already tells us that μ_ω is a left $\mathcal{B}$-module morphism.

Theorem 12.10. *If a quantum connection ω is regular, then the associated map $\mu_\omega : \mathfrak{ver}^1(P) \to \Omega^1(P)$ is a right $\mathcal{B}$-module morphism.*

Proof. To start off, we have to specify precisely what the right $\mathcal{B}$-module structure is on each of these spaces. For $\Omega^1(P)$, we use the identity $\mathcal{B} = \Omega^0(P)$. Then the right action of $\mathcal{B}$ on $\Omega^1(P)$ is defined to be the multiplication m_Ω in the graded algebra $\Omega(P)$:

$$\Omega^1(P) \otimes \mathcal{B} = \Omega^1(P) \otimes \Omega^0(P) \xrightarrow{\,m_\Omega\,} \Omega^1(P).$$

We will come back to the right $\mathcal{B}$-module structure on $\mathfrak{ver}^1(P)$ momentarily.

Since $\mathfrak{hor}^0(P) = \mathcal{B}$, we can apply the defining property of a regular connection for the particular case $\varphi = c \in \mathcal{B}$. This gives

$$\omega(\theta)c = (-1)^0 c^{(0)}\omega(\theta \circ c^{(1)}) = c^{(0)}\omega(\theta \circ c^{(1)}).$$

Recalling that the multiplication m_Ω is written as juxtaposition, we have

$$\mu_\omega(b \otimes \theta)c = b\omega(\theta)c$$
$$= bc^{(0)}\omega(\theta \circ c^{(1)})$$
$$= \mu_\omega(bc^{(0)} \otimes \theta \circ c^{(1)})$$
$$= \mu_\omega\big((b \otimes \theta)(c \otimes 1_A)\big).$$

In the last equality we used the definition of the multiplication in the graded algebra $\mathfrak{vet}(P)$. This immediately motivates the following definition for the right $\mathcal{B}$-module structure on $\mathfrak{vet}^1(P)$:

$$(b \otimes \theta)c := (b \otimes \theta)(c \otimes 1_A).$$

With this definition, μ_ω obviously is a right $\mathcal{B}$-module morphism, which is what we wished to prove. ∎

Remark. The above proof uses just a fraction of the information encoded in the hypothesis, so it seems unlikely that the converse could be true.

Definition 12.13. *A connection displacement* $\lambda : {}_{\text{inv}}\Gamma \to \Omega^1(P)$ *is said to be* regular *if it satisfies the linear equation*

$$\lambda(\theta)\,\varphi = (-1)^k \varphi^{(0)}\,\lambda(\theta \circ \varphi^{(1)})$$

for all $\varphi \in \mathfrak{hor}^k(P)$ *and all* $\theta \in {}_{\text{inv}}\Gamma$.

Clearly, the set of all regular connection displacements forms a vector space. The set of regular connections, if it is not empty, forms an affine space whose associated vector space is the space of all regular connection displacements.

Proposition 12.9. *A quantum connection* ω *is regular if and only if*

$$\varphi\,\omega(\theta) = (-1)^k \omega(\theta \circ \kappa^{-1}(\varphi^{(1)}))\,\varphi^{(0)}$$

for all $\varphi \in \mathfrak{hor}^k(P)$ *and all* $\theta \in {}_{\text{inv}}\Gamma$.

Proof. The condition for the regularity of ω is

$$\omega(\theta)\varphi = (-1)^k \varphi^{(0)}\omega(\theta \circ \varphi^{(1)})$$

for all $\varphi \in \mathfrak{hor}^k(P)$ and $\theta \in {}_{\text{inv}}\Gamma$, where $_{\Omega(P)}\Psi(\varphi) = \varphi^{(0)} \otimes \varphi^{(1)} \in \mathfrak{hor}^k(P) \otimes \mathcal{A}$. We apply the *-operation in $\Omega^{k+1}(P)$ to get

$$\varphi^*\omega(\theta^*) = (-1)^k \omega((\theta \circ \varphi^{(1)})^*)\varphi^{(0)*}$$
$$= (-1)^k \omega((\theta^* \circ \kappa(\varphi^{(1)})^*))\varphi^{(0)*}.$$

Put $\eta = \theta^* \in {}_{inv}\Gamma$ and $\psi = \varphi^* \in \mathfrak{hor}^k(P)$. Notice that η and ψ are arbitrary elements in their respective spaces. Since $\Omega(P)\Psi$ is a $*$-morphism, we have

$$\psi^{(0)} \otimes \psi^{(1)} = \varphi^{(0)*} \otimes \varphi^{(1)*}.$$

In terms of η and ψ, the previous equation then reads as

$$\psi\,\omega(\eta) = (-1)^k \omega(\eta \circ \kappa(\psi^{(1)*})^*)\psi^{(0)}$$
$$= (-1)^k \omega(\eta \circ \kappa^{-1}(\psi^{(1)}))\psi^{(0)},$$

where we used (11.2), only now written as $\kappa(a^*)^* = \kappa^{-1}(a)$. This is the identity we wished to prove, except in other notation.

The reverse implication is proved similarly and is left to the reader. ∎

Notice that this result used the condition that a quantum connection is a $*$-morphism, but not the condition that contains the real geometric content of a quantum connection.

Definition 12.14. *The* regularity deviation measure *is the map*

$$l_\omega : {}_{inv}\Gamma \times \mathfrak{hor}(P) \to \Omega(P)$$

defined for $\theta \in {}_{inv}\Gamma$ and $\varphi \in \mathfrak{hor}^k(P)$ by

$$l_\omega(\theta,\varphi) := \omega(\theta)\,\varphi - (-1)^k \varphi^{(0)}\,\omega(\theta \circ \varphi^{(1)}).$$

Clearly, ω is regular if and only if l_ω is identically zero. Thus, l_ω measures how much ω deviates from being regular. Or we can say l_ω is the obstruction for ω being regular. What may not be obvious is that this deviation is completely horizontal. Again, we recall that the definition of horizontal does not depend on the quantum connection ω under consideration.

Proposition 12.10. *For all $\varphi \in \mathfrak{hor}^k(P)$ and all $\theta \in {}_{inv}\Gamma$, we have that*

$$\Omega(P)\Psi\big(l_\omega(\theta,\varphi)\big) = l_\omega(\theta^{(0)},\varphi^{(0)}) \otimes \theta^{(1)}\varphi^{(1)} \in \Omega^{k+1}(P) \otimes \mathcal{A}, \qquad (12.44)$$

where $\mathrm{ad}(\theta) = \theta^{(0)} \otimes \theta^{(1)}$ and $\mathrm{ad} : {}_{inv}\Gamma \to {}_{inv}\Gamma \otimes \mathcal{A}$ is the right adjoint action on ${}_{inv}\Gamma$. In particular, the image of l_ω lies in $\mathfrak{hor}(P)$.

Proof. The statement about the image of l_ω follows immediately from the formula in (12.44).

Note that $k = |\varphi| = |\varphi^{(0)}|$. We then evaluate as follows:

$$l_\omega(\theta^{(0)},\varphi^{(0)}) \otimes \theta^{(1)}\varphi^{(1)} = \big(\omega(\theta^{(0)})\varphi^{(0)} - (-1)^k \varphi^{(00)}\omega(\theta^{(0)} \circ \varphi^{(01)})\big) \otimes \theta^{(1)}\varphi^{(1)}$$
$$= \big(\omega(\theta^{(0)})\varphi^{(0)} \otimes \theta^{(1)}\varphi^{(1)}\big) - (-1)^k \big(\varphi^{(00)}\omega(\theta^{(0)} \circ \varphi^{(01)}) \otimes \theta^{(1)}\varphi^{(1)}\big)$$
$$= \omega(\theta^{(0)})\varphi^{(0)} \otimes \theta^{(1)}\varphi^{(1)} - (-1)^k \varphi^{(0)}\omega(\theta^{(0)} \circ \varphi^{(1)}) \otimes \theta^{(1)}\varphi^{(2)}.$$

In the last equation we switched to the single-digit Sweedler's notation for ease of comparison with the following calculations.

On the other hand, using the single-digit Sweedler's notation, we have

$$\Omega_{(P)}\Psi\big(l_\omega(\theta,\varphi)\big) = \Omega_{(P)}\Psi\big(\omega(\theta)\,\varphi - (-1)^k \varphi^{(0)}\,\omega(\theta \circ \varphi^{(1)})\big)$$

$$= \Omega_{(P)}\Psi(\omega(\theta))\,\Omega_{(P)}\Psi(\varphi) - (-1)^k \Omega_{(P)}\Psi(\varphi^{(0)})\,\Omega_{(P)}\Psi(\omega(\theta \circ \varphi^{(1)}))$$

$$= \Omega_{(P)}\Psi(\omega(\theta))(\varphi^{(0)} \otimes \varphi^{(1)}) - (-1)^k (\varphi^{(0)} \otimes \varphi^{(1)})\,\Omega_{(P)}\Psi(\omega(\theta \circ \varphi^{(2)}))$$

$$= \big((\omega \otimes id_A)\mathrm{ad}(\theta) + 1_B \otimes \theta\big)(\varphi^{(0)} \otimes \varphi^{(1)})$$

$$\quad - (-1)^k (\varphi^{(0)} \otimes \varphi^{(1)})\big((\omega \otimes id_A)\mathrm{ad}(\theta \circ \varphi^{(2)}) + 1_B \otimes (\theta \circ \varphi^{(2)})\big)$$

$$= \big((\omega(\theta^{(0)}) \otimes \theta^{(1)}) + 1_B \otimes \theta\big)(\varphi^{(0)} \otimes \varphi^{(1)})$$

$$\quad - (-1)^k (\varphi^{(0)} \otimes \varphi^{(1)})\big((\omega \otimes id_A)\mathrm{ad}(\pi(a) \circ \varphi^{(2)}) + 1_B \otimes (\theta \circ \varphi^{(2)})\big)$$

$$= \omega(\theta^{(0)})\varphi^{(0)} \otimes \theta^{(1)}\varphi^{(1)} + (-1)^k \varphi^{(0)} \otimes \theta\varphi^{(1)}$$

$$\quad - (-1)^k (\varphi^{(0)} \otimes \varphi^{(1)})(\omega \otimes id_A)\mathrm{ad}(\pi(a) \circ \varphi^{(2)}) - (-1)^k \varphi^{(0)} \otimes \varphi^{(1)}(\theta \circ \varphi^{(2)})$$

$$= \omega(\theta^{(0)})\varphi^{(0)} \otimes \theta^{(1)}\varphi^{(1)} - (-1)^k (\varphi^{(0)} \otimes \varphi^{(1)})(\omega \otimes id_A)\mathrm{ad}(\pi(a\varphi^{(2)})),$$

$$(12.45)$$

due to the cancellation of the second and fourth terms in the next-to-last expression. These terms cancel because

$$\varphi^{(0)} \otimes \varphi^{(1)}(\theta \circ \varphi^{(2)}) = \varphi^{(0)} \otimes \varphi^{(1)}(\kappa(\varphi^{(2)})\theta\varphi^{(3)})$$

$$= \varphi^{(0)} \otimes \varepsilon(\varphi^{(1)})\theta\varphi^{(2)}$$

$$= \varepsilon(\varphi^{(1)})\varphi^{(0)} \otimes \theta\varphi^{(2)}$$

$$= \varphi^{(0)} \otimes \theta\varphi^{(1)},$$

where we used Proposition 6.8 in the first equality. Continuing with just the last term (12.45) (but without the sign), we calculate as follows:

$$(\varphi^{(0)} \otimes \varphi^{(1)})(\omega \otimes id_A)\mathrm{ad}(\pi(a\varphi^{(2)}))$$

$$= (\varphi^{(0)} \otimes \varphi^{(1)})(\omega \otimes id_A)(\pi \otimes id_A)\mathrm{ad}(a\varphi^{(2)})$$

$$= (\varphi^{(0)} \otimes \varphi^{(1)})(\omega\pi \otimes id_A)\big((a\varphi^{(2)})^{(2)} \otimes \kappa((a\varphi^{(2)})^{(1)})(a\varphi^{(2)})^{(3)}\big)$$

$$= (\varphi^{(0)} \otimes \varphi^{(1)})(\omega\pi \otimes id_A)\big((a^{(2)}\varphi^{(3)}) \otimes \kappa((a^{(1)}\varphi^{(2)}))(a^{(3)}\varphi^{(4)})\big)$$

$$= (\varphi^{(0)} \otimes \varphi^{(1)})\big(\omega\pi(a^{(2)}\varphi^{(3)}) \otimes \kappa((a^{(1)}\varphi^{(2)}))(a^{(3)}\varphi^{(4)})\big)$$

$$= \varphi^{(0)}\omega\pi(a^{(2)}\varphi^{(3)}) \otimes \varphi^{(1)}\kappa(\varphi^{(2)})\kappa(a^{(1)})a^{(3)}\varphi^{(4)}$$

$$= \varphi^{(0)} \omega(\varepsilon(a^{(2)})\varphi^{(3)}) \otimes \varphi^{(1)} \kappa(\varphi^{(2)}) \kappa(a^{(1)}) a^{(3)} \varphi^{(4)}$$

$$= \varphi^{(0)} \omega(\pi(a^{(2)}) \circ \varphi^{(2)}) \otimes \varepsilon(\varphi^{(1)}) \kappa(a^{(1)}) a^{(3)} \varphi^{(3)}$$

$$= \varphi^{(0)} \omega(\pi(a^{(2)}) \circ \varphi^{(1)}) \otimes \kappa(a^{(1)}) a^{(3)} \varphi^{(2)}$$

$$= \varphi^{(0)} \omega(\theta^{(0)} \circ \varphi^{(1)}) \otimes \theta^{(1)} \varphi^{(2)}.$$

The last equality follows from

$$\theta^{(0)} \otimes \theta^{(1)} = \mathrm{ad}(\theta) = \mathrm{ad}(\pi(a)) = (\pi \otimes id)\mathrm{ad}(a) = \pi(a^{(2)}) \otimes \kappa(a^{(1)}) a^{(3)}.$$

Putting all these pieces together gives us (12.44).

Notice that the expression in (12.45) already shows that the image of l_ω lies in $\mathfrak{hor}(P)$. The two terms that canceled to give this expression lie in $\Omega^k(P) \otimes_{\mathrm{inv}} \Gamma$, while the two surviving terms lie in $\Omega^{k+1}(P) \otimes \mathcal{A}$. ∎

The above motivation behind the definition of quantum connection can be made rigorous in the sense that it is possible to interpret any classical Ehresmann connection in a principal bundle (with respect to the standard, complexified de Rham exterior calculi on all three spaces) as a special case of a quantum connection. We note that when this is done, it turns out that every classical Ehresmann connection is both multiplicative and regular. (The reader may wish to prove this. **Hint:** Try writing everything in local coordinates.) However, it is perfectly possible to construct QPBs based on classical spaces (and their commutative algebras of C^∞-functions) but with differential calculi that are quantum, not classical, objects (see [48]). And it can turn out that there are quantum connections on such *quantum* principal bundles that are neither multiplicative nor regular.

12.8   Curvature

The material is this section on curvature is optional; it is not needed in order to read the rest of this book.

In classical differential geometry every connection has a curvature, which is an important geometric structure. Whether or not the curvature is zero turns out to be a critical consideration in the study of the connection. For example, the distribution of the horizontal subspaces of the connection is integrable if and only if its curvature is zero. Since nonzero curvature arises in important cases, it follows that nonintegrable distributions are important as well. There are many different types of curvature in classical differential geometry, but we will only be concerned here with the curvature that is associated with a connection. It turns out that a quantum connection also has its curvature, and this is then an important structure in the theory of QPBs. So in this section we define and study the basic properties of the curvature of a quantum connection on a QPB.

We have some preliminary results to take care of before proceeding to the main topic of this section. First, we introduce a twisting map for $_{\text{inv}}\Gamma$.

Definition 12.15. *For $\eta, \theta \in {}_{\text{inv}}\Gamma$, define $\sigma_{li} : {}_{\text{inv}}\Gamma^{\otimes 2} \to {}_{\text{inv}}\Gamma^{\otimes 2}$ by*

$$\sigma_{li}(\eta \otimes \theta) := \theta^{(0)} \otimes (\eta \circ \theta^{(1)}), \tag{12.46}$$

where ad $: {}_{\text{inv}}\Gamma \to {}_{\text{inv}}\Gamma \otimes \mathcal{A}$ *is the right adjoint co-action of $\mathcal{A}$ on $_{\text{inv}}\Gamma$ and* $\text{ad}(\theta) = \theta^{(0)} \otimes \theta^{(1)}$.

Exercise 12.26. *Show that σ_{li} is the left invariant part of the braid operator $\sigma : \Gamma^{\otimes 2} \to \Gamma^{\otimes 2}$, which was defined in Definition 7.1. This result explains the subscript in the notation σ_{li}.*

Definition 12.16. *Let P be a quantum principal bundle with the differential calculus $\Omega(P)$ (and other structures with standard notation).*

We say that a linear map $\alpha : {}_{\text{inv}}\Gamma \to \Omega^k(P)$ that commutes with the co-actions of $\mathcal{A}$ is a pseudotensorial *k-form on P. We denote the vector space of all such maps α by $\psi^k(P)$.*

We say that a linear map $\beta : {}_{\text{inv}}\Gamma \to \mathfrak{hor}^k(P)$ that commutes with the co-actions of $\mathcal{A}$ is a tensorial *k-form on P. We denote the vector space of all such maps β by $\tau^k(P)$.*

These definitions are the noncommutative (that is, dual) analogs to the definitions in the case of classical differential geometry, as can be found in [49] Chapter II, Section 5.

Clearly, $\tau^k(P)$ is a vector subspace of $\psi^k(P)$.

Proposition 12.11. *Let ω be a regular quantum connection on P, a QPB with a differential calculus. (We continue to use the standard notations for all this.) Suppose that $\varphi \in \psi^k(P)$. Let m_Ω denote the graded multiplication map on $\Omega(P)$. Then we have*

$$m_\Omega(\omega \otimes \varphi) = (-1)^k m_\Omega(\varphi \otimes \omega)\sigma_{li} \tag{12.47}$$

as maps $_{\text{inv}}\Gamma^{\otimes 2} \to \Omega^{k+1}(P)$.

Proof. Take $\eta, \theta \in {}_{\text{inv}}\Gamma$ and write $\text{ad}(\theta) = \theta^{(0)} \otimes \theta^{(1)}$. We start by expanding the right side of (12.47) as follows:

$$m_\Omega(\varphi \otimes \omega)\sigma_{li}(\eta \otimes \theta) = m_\Omega(\varphi \otimes \omega)\big(\theta^{(0)} \otimes (\eta \circ \theta^{(1)})\big)$$

$$= m_\Omega\big(\varphi(\theta^{(0)}) \otimes \omega(\eta \circ \theta^{(1)})\big)$$

$$= \varphi(\theta^{(0)})\omega(\eta \circ \theta^{(1)}).$$

Next, we use the hypothesis on φ, which yields

$$\Omega(P)\Phi(\varphi(\theta)) = (\varphi \otimes id_A)\mathrm{ad}(\theta) = (\varphi \otimes id_A)(\theta^{(0)} \otimes \theta^{(1)}) = \varphi(\theta^{(0)}) \otimes \theta^{(1)}.$$

This can be rewritten as $(\varphi(\theta))^{(0)} = \varphi(\theta^{(0)})$ and $(\varphi(\theta))^{(1)} = \theta^{(1)}$. So, using these identities, we continue with the expansion of the right side of (12.47):

$$
\begin{aligned}
m_\Omega(\varphi \otimes \omega)\sigma_{li}(\eta \otimes \theta) &= \varphi(\theta^{(0)})\omega(\eta \circ \theta^{(1)}) \\
&= (\varphi(\theta))^{(0)}\omega(\eta \circ (\varphi(\theta))^{(1)}) \\
&= (-1)^k \, \omega(\eta)\varphi(\theta) \\
&= (-1)^k \, m_\Omega(\omega(\eta) \otimes \varphi(\theta)) \\
&= (-1)^k \, m_\Omega(\omega \otimes \varphi)(\eta \otimes \theta),
\end{aligned}
$$

where we used the defining property of a regular quantum connection in the third equality. Since this holds for all $\eta \otimes \theta$, we have proved (12.47). ∎

Notice that the next definition is not about the complete structure of a QPB with a differential calculus, but only concerns an FODC Γ of its structure "group" A, which represents the fiber space, and its universal calculus, which is functorially determined by the FODC Γ.

Definition 12.17. *Suppose that* (Γ, d) *is a bicovariant and* $*$-*covariant FODC over* A. *A linear map* $\delta : {}_{inv}\Gamma \to {}_{inv}\Gamma^{\otimes 2}$ *is called an* embedded differential *if*

1. $\mathrm{ad}^{\otimes 2}\,\delta = (\delta \otimes id_A)\,\mathrm{ad}$, *where* $\mathrm{ad} : {}_{inv}\Gamma \to {}_{inv}\Gamma \otimes A$ *is the right adjoint co-action of* A *on* ${}_{inv}\Gamma$ *and* $\mathrm{ad}^{\otimes 2} := M(\mathrm{ad} \otimes \mathrm{ad})$ *with*

$$M : {}_{inv}\Gamma \times A \otimes {}_{inv}\Gamma \otimes A \to {}_{inv}\Gamma \otimes {}_{inv}\Gamma \otimes A$$

given by $M : \theta \otimes a \otimes \eta \otimes b \mapsto \theta \otimes \eta \otimes ab$.

2. $d(\theta) = \theta^{(1)}\theta^{(2)}$ *and* $\delta(\theta^*) = -\theta^{(2)*} \otimes \theta^{(1)*}$ *for all* $\theta \in {}_{inv}\Gamma$, *where we have written* $\delta(\theta) = \theta^{(1)} \otimes \theta^{(2)}$. *[This is a Sweedler-type notation, but not that for the right adjoint co-action that was used above:* $\mathrm{ad}(\theta) = \theta^{(0)} \otimes \theta^{(1)}$.]

In the equation $d(\theta) = \theta^{(1)}\theta^{(2)}$, the differential $d = d^\wedge$ is that of the universal differential calculus $\Gamma^\wedge$. And on the right side of the same equation the product is that of $\Gamma^\wedge$.

Let's note that the set of embedded differentials is not a vector space, because the condition $d(\theta) = \theta^{(1)}\theta^{(2)}$ is not linear in δ in general. All of the other conditions in this definition are linear in δ. In particular, $\delta \equiv 0$ satisfies all the other conditions but does not satisfy $d(\theta) = \theta^{(1)}\theta^{(2)}$ in general.

The next proposition is a consequence of Definition 12.17 and so only refers to a bicovariant and $*$-covariant FODC (Γ, d) over A and its uniquely determined universal differential calculus.

Proposition 12.12. *Let* (Γ, d) *be a bicovariant and *-covariant FODC over* $\mathcal{A}$. *Suppose that* δ *is an embedded differential and* $\theta \in {}_{\mathrm{inv}}\Gamma$. *Then there exists an element* $a \in \ker \varepsilon$ *(not necessarily unique) such that*

1. $\pi(a) = \theta$ *and*
2. $\delta(\theta) = -\pi(a^{(1)}) \otimes \pi(a^{(2)})$.

Here $\phi(a) = a^{(1)} \otimes a^{(2)}$ *is the co-multiplication in Sweedler's notation, and* $\pi :$ $\mathcal{A} \to {}_{\mathrm{inv}}\Gamma$ *is the quantum germs map.*

Proof. We have the notation $\delta(\theta) = \theta^{(1)} \otimes \theta^{(2)}$ and, by using it, the identity $d(\theta) = \theta^{(1)} \theta^{(2)}$. But the surjectivity of π gives us an element $c \in \ker \varepsilon$ such that $\pi(c) = \theta$. So we have

$$\theta^{(1)} \theta^{(2)} = d(\theta) = d\pi(c) = -\pi(c^{(1)}) \pi(c^{(2)}) \in {}_{\mathrm{inv}}\Gamma^{\wedge 2}, \qquad (12.48)$$

where the last equality is the Maurer–Cartan formula. Here $\phi(c) = c^{(1)} \otimes c^{(2)}$. The element $c \in \ker \varepsilon$ we have used is not unique, but it is unique modulo elements in $\mathcal{R}$, the right ideal of $\mathcal{A}$ that is contained in $\ker \varepsilon$ and that defines the FODC Γ. Equation (12.48) implies that

$$\theta^{(1)} \otimes \theta^{(2)} + \pi(c^{(1)}) \otimes \pi(c^{(2)}) \in {}_{\mathrm{inv}}S^{\wedge},$$

where ${}_{\mathrm{inv}}S^{\wedge}$ is the two-sided ideal in ${}_{\mathrm{inv}}\Gamma^{\otimes}$ generated by the elements

$$q = \pi(b^{(1)}) \otimes \pi(b^{(2)}) \in {}_{\mathrm{inv}}\Gamma^{\otimes 2} \qquad (12.49)$$

with $b \in \mathcal{R}$ and $\phi(b) = b^{(1)} \otimes b^{(2)}$. This is so since ${}_{\mathrm{inv}}\Gamma^{\wedge} = {}_{\mathrm{inv}}\Gamma^{\otimes} / {}_{\mathrm{inv}}S^{\wedge}$. But we also obviously know that

$$\theta^{(1)} \otimes \theta^{(2)} + \pi(c^{(1)}) \otimes \pi(c^{(2)}) \in {}_{\mathrm{inv}}\Gamma^{\otimes 2}.$$

Now we observe that ${}_{\mathrm{inv}}S^{\wedge} \cap {}_{\mathrm{inv}}\Gamma^{\otimes 2}$ consists exactly of linear combinations of elements of the form q in (12.49) with complex coefficients. Furthermore, for $\lambda \in \mathbb{C}$ and q as in (12.49), we have that

$$\lambda q = \lambda \pi(b^{(1)}) \otimes \pi(b^{(2)}) = \pi(\lambda b^{(1)}) \otimes \pi(b^{(2)}) = \pi(e^{(1)}) \otimes \pi(e^{(2)}),$$

where $e = \lambda b \in \mathcal{R}$. This says that λq is again in the form of (12.49), and so every element in the intersection ${}_{\mathrm{inv}}S^{\wedge} \cap {}_{\mathrm{inv}}\Gamma^{\otimes 2}$ is a *sum* of elements of the form (12.49). In particular, this gives us

$$\theta^{(1)} \otimes \theta^{(2)} = -\pi(c^{(1)}) \otimes \pi(c^{(2)}) + \sum_{k} \pi(b_k^{(1)}) \otimes \pi(b_k^{(2)}), \qquad (12.50)$$

where each $b_k \in \mathcal{R}$ and $k \in K$, a finite set. Next, we define $a := c - \sum_k b_k$. Since $\sum_k b_k \in \mathcal{R} \subset \ker \varepsilon$ and $c \in \ker \varepsilon$, this implies $a \in \ker \varepsilon$ as well as $\pi(a) = \pi(c) = \theta$, which is property 1. It also follows that

$$a^{(1)} \otimes a^{(2)} = \phi(a) = \phi(c) - \sum_k \phi(b_k) = c^{(1)} \otimes c^{(2)} - \sum_k b_k^{(1)} \otimes b_k^{(2)}.$$

Applying $\pi \otimes \pi$ to both sides of this and using (12.50) yields

$$\pi(a^{(1)}) \otimes \pi(a^{(2)}) = \pi(c^{(1)}) \otimes \pi(c^{(2)}) - \sum_k \pi(b_k^{(1)}) \otimes \pi(b_k^{(2)}) = -\theta^{(1)} \otimes \theta^{(2)}.$$

Then we immediately have property 2:

$$\delta(\theta) = \theta^{(1)} \otimes \theta^{(2)} = -\pi(a^{(1)}) \otimes \pi(a^{(2)}).$$

And this shows everything we set out to prove.   ∎

Remark. I hope the reader has already noticed the marked resemblance of property 2 to the Maurer–Cartan formula.

Definition 12.18. *The* transposed commutator $c^{\mathsf{T}} : {}_{\mathrm{inv}}\Gamma \to {}_{\mathrm{inv}}\Gamma^{\otimes 2}$ *is defined as the composition*

$$ {}_{\mathrm{inv}}\Gamma \xrightarrow{\mathrm{ad}} {}_{\mathrm{inv}}\Gamma \otimes \mathcal{A} \xrightarrow{id \otimes \pi} {}_{\mathrm{inv}}\Gamma \otimes {}_{\mathrm{inv}}\Gamma, $$

that is, $c^{\mathsf{T}} := (id \otimes \pi)\mathrm{ad}$.

Let $\mathcal{G}$ be an arbitrary algebra, possibly graded, with multiplication $m_{\mathcal{G}}$. Let δ be an embedded differential. Also, let $\varphi, \eta : {}_{\mathrm{inv}}\Gamma \to \mathcal{G}$ be linear maps. Then define two linear maps ${}_{\mathrm{inv}}\Gamma \to \mathcal{G}$ by

$$\langle \varphi, \eta \rangle := m_{\mathcal{G}}(\varphi \otimes \eta)\delta,$$

$$[\varphi, \eta] := m_{\mathcal{G}}(\varphi \otimes \eta)c^{\mathsf{T}}.$$

This definition becomes relevant to the theory of QPBs when we take $\mathcal{G}$ to be either $\Omega(P)$ or $\mathfrak{hor}(P)$.

Notice that c^{T} has the same domain and co-domain as an embedded differential. Also, like an embedded differential, it only refers to the part of the structure of a QPB that is associated with the fiber space. And so it might not be too surprising that the identity in the next lemma parallels an identity for an embedded differential.

Lemma 12.3. *The transposed commutator c^{T} satisfies the identity*

$$\mathrm{ad}^{\otimes 2}\, c^{\mathsf{T}} = (c^{\mathsf{T}} \otimes id_{\mathcal{A}})\, \mathrm{ad}.$$

Proof. Consider this diagram:

$$
\begin{array}{ccccc}
{}_{\mathrm{inv}}\Gamma & \xrightarrow{\ ad\ } & {}_{\mathrm{inv}}\Gamma \otimes \mathcal{A} & \xrightarrow{\ id\otimes\pi\ } & {}_{\mathrm{inv}}\Gamma \otimes {}_{\mathrm{inv}}\Gamma \\
{\scriptstyle ad}\downarrow & & {\scriptstyle id\,\otimes\,\phi}\downarrow & & \downarrow{\scriptstyle ad^{\otimes 2}} \\
{}_{\mathrm{inv}}\Gamma \otimes \mathcal{A} & \xrightarrow{\ ad\otimes id_A\ } & {}_{\mathrm{inv}}\Gamma \otimes \mathcal{A} \otimes \mathcal{A} & \xrightarrow{\ id\otimes\pi\otimes id_A\ } & {}_{\mathrm{inv}}\Gamma \otimes {}_{\mathrm{inv}}\Gamma \otimes \mathcal{A}.
\end{array}
$$

The composition of the upper horizontal arrows is c^{T}, while the composition of the lower horizontal arrows is $c^{\mathsf{T}} \otimes id_A$. So the commutativity of the entire diagram is equivalent to the identity that we wish to prove. Therefore, it suffices to show the commutativity of each square. Now the square on the left commutes since ad is a right co-action. But it seems that the square on the right does not necessarily commute.

So we will take an element $\theta = \pi(a) \in {}_{\mathrm{inv}}\Gamma$ on the top left corner and chase it to the bottom right corner. First, this element maps down to $\pi(a^2) \otimes \kappa(a^1)a^3$ as we have seen many times. Next, that element goes to

$$
\begin{aligned}
(ad \otimes id_A)(\pi(a^2) \otimes \kappa(a^1)a^3) &= ad(\pi(a^2)) \otimes \kappa(a^1)a^3 \\
&= \pi(a^{22}) \otimes \kappa(a^{21})a^{23} \otimes \kappa(a^1)a^3 \\
&= \pi(a^3) \otimes \kappa(a^2)a^4 \otimes \kappa(a^1)a^5
\end{aligned}
$$

in the middle bottom, where the last line is in the single-digit Sweedler's notation. Finally, this gets mapped to the lower right corner:

$$
(id \otimes \pi \otimes id_A)(\pi(a^3) \otimes \kappa(a^2)a^4 \otimes \kappa(a^1)a^5) = \pi(a^3) \otimes \pi(\kappa(a^2)a^4) \otimes \kappa(a^1)a^5.
$$

Unfortunately, the across–down chase is much more laborious. First, we start out the same with $\pi(a^2) \otimes \kappa(a^1)a^3$ in the middle top. Then this maps to $\pi(a^2) \otimes \pi(\kappa(a^1)a^3)$ on the top right. We next apply ad $\otimes$ ad to this, getting

$$
\begin{aligned}
&(ad \otimes ad)(\pi(a^2) \otimes \pi(\kappa(a^1)a^3)) \\
&= \pi(a^{22}) \otimes \kappa(a^{21})a^{23} \otimes \pi((\kappa(a^1)a^3)^2) \otimes \kappa\big((\kappa(a^1)a^3)^1\big)(\kappa(a^1)a^3)^3 \\
&= \pi(a^{22}) \otimes \kappa(a^{21})a^{23} \otimes \pi(\kappa(a^1)^2 a^{32}) \otimes \kappa\big((\kappa(a^1)^1 a^{31})\big)\kappa(a^1)^3 a^{33} \\
&= \pi(a^{22}) \otimes \kappa(a^{21})a^{23} \otimes \pi(\kappa(a^{12})a^{32}) \otimes \kappa\big((\kappa(a^{13})a^{31})\big)\kappa(a^{11})a^{33} \\
&= \pi(a^{22}) \otimes \kappa(a^{21})a^{23} \otimes \pi(\kappa(a^{12})a^{32}) \otimes \kappa(a^{31})\kappa(\kappa(a^{13}))\kappa(a^{11})a^{33}.
\end{aligned}
$$

At this point we are going to switch to the single-digit Sweedler's notation. Since there are nine occurrences of a^k, we will have k taking exactly one each of the values in $\{1, 2, \ldots, 9\}$. So we continue

$$
\begin{aligned}
&ad^{\otimes 2}(\pi(a^2) \otimes \pi(\kappa(a^1)a^3)) \\
&= M(ad \otimes ad)(\pi(a^2) \otimes \pi(\kappa(a^1)a^3))
\end{aligned}
$$

$$= M\left(\pi(a^{22}) \otimes \kappa(a^{21})a^{23} \otimes \pi(\kappa(a^{12})a^{32}) \otimes \kappa(a^{31})\kappa(\kappa(a^{13}))\kappa(a^{11})a^{33}\right)$$

$$= M\left(\pi(a^{5}) \otimes \kappa(a^{4})a^{6} \otimes \pi(\kappa(a^{2})a^{8}) \otimes \kappa(a^{7})\kappa(\kappa(a^{3}))\kappa(a^{1})a^{9}\right)$$

$$= \pi(a^{5}) \otimes \pi(\kappa(a^{2})a^{8}) \otimes \kappa(a^{4})a^{6}\kappa(a^{7})\kappa(\kappa(a^{3}))\kappa(a^{1})a^{9}$$

$$= \pi(a^{5}) \otimes \pi(\kappa(a^{2})a^{7}) \otimes \kappa(a^{4})\varepsilon(a^{6})\kappa(\kappa(a^{3}))\kappa(a^{1})a^{8}$$

$$= \pi(a^{5}\varepsilon(a^{6})) \otimes \pi(\kappa(a^{2})a^{7}) \otimes \kappa(a^{4})\kappa(\kappa(a^{3}))\kappa(a^{1})a^{8}$$

$$= \pi(a^{5}) \otimes \pi(\kappa(a^{2})a^{6}) \otimes \kappa(a^{4})\kappa(\kappa(a^{3}))\kappa(a^{1})a^{7}$$

$$= \pi(a^{4}) \otimes \pi(\kappa(a^{2})a^{5}) \otimes \varepsilon(\kappa(a^{3}))\kappa(a^{1})a^{6}$$

$$= \pi(a^{4}) \otimes \pi(\kappa(a^{2})a^{5}) \otimes \varepsilon(a^{3})\kappa(a^{1})a^{6}$$

$$= \pi(a^{3}) \otimes \pi(\kappa(a^{2})a^{4}) \otimes \kappa(a^{1})a^{5},$$

which was the result from chasing the diagram the other way. So the diagram commutes, as claimed. ∎

Exercise 12.27. *The formula* $\phi(\kappa(a)) = \kappa(a)^{1} \otimes \kappa(a)^{2} = \kappa(a^{2}) \otimes \kappa(a^{1})$ *for* $a \in \mathcal{A}$ *has a generalization to higher-order co-products. For example, prove that*

$$\kappa(a)^{1} \otimes \kappa(a)^{2} \otimes \kappa(a)^{3} = \kappa(a^{3}) \otimes \kappa(a^{2}) \otimes \kappa(a^{1}).$$

Then find and prove the full generalization. You should realize that this has been used in the previous proof.

Proposition 12.13. *Suppose that* $\varphi \in \psi^{j}(P)$ *and* $\eta \in \psi^{k}(P)$. *Then*

$$\langle \varphi, \eta \rangle :_{\text{inv}}\Gamma \to \Omega^{j+k}(P)$$

$$[\varphi, \eta] :_{\text{inv}}\Gamma \to \Omega^{j+k}(P)$$

and both of these maps are in $\psi^{j+k}(P)$.

Furthermore, $\langle \varphi, \eta \rangle^{*} = -(-1)^{jk}\langle \eta^{*}, \varphi^{*} \rangle$, *where we defined* T^{*} *above for any linear map* $T : V \to W$ *as* $T^{*}(x) := (T(x^{*}))^{*}$ *for* $x \in V$. *Here* V *and* W *are vector spaces, each with a* ∗*-operation.*

Proof. As for the first paragraph, the only detail that may not be totally obvious is the intertwining property. But this follows because each of these maps is the composition of three maps that intertwine appropriately. The details of that are left to the reader.

As for the identity in the second paragraph, we note that for any $\theta \in {}_{\text{inv}}\Gamma$, we get

$$\langle \varphi, \eta \rangle^{*}(\theta) = \left(m_{\Omega}(\varphi \otimes \eta)\delta\right)^{*}(\theta)$$

$$= \left(m_{\Omega}(\varphi \otimes \eta)\delta(\theta^{*})\right)^{*}$$

$$= \left(-m_\Omega(\varphi \otimes \eta)(\theta^{(2)*} \otimes \theta^{(1)*})\right)^*$$

$$= -\left(m_\Omega\left(\varphi(\theta^{(2)*}) \otimes \eta(\theta^{(1)*})\right)\right)^*$$

$$= -\left(m_\Omega\left(\varphi^*(\theta^{(2)})^* \otimes \eta^*(\theta^{(1)})^*\right)\right)^*$$

$$= -\left(\left(\varphi^*(\theta^{(2)})^* \eta^*(\theta^{(1)})^*\right)\right)^*$$

$$= -(-1)^{jk}\left(\left(\eta^*(\theta^{(1)})\varphi^*(\theta^{(2)})\right)^*\right)^*$$

$$= -(-1)^{jk}\eta^*(\theta^{(1)})\varphi^*(\theta^{(2)})$$

$$= -(-1)^{jk}m_\Omega\left(\eta^*(\theta^{(1)}) \otimes \varphi^*(\theta^{(2)})\right)$$

$$= -(-1)^{jk}m_\Omega(\eta^* \otimes \varphi^*)(\theta^{(1)} \otimes \theta^{(2)})$$

$$= -(-1)^{jk}m_\Omega(\eta^* \otimes \varphi^*)\delta(\theta)$$

$$= -(-1)^{jk}\langle \eta^*, \varphi^* \rangle(\theta).$$

And that proves the desired identity. ∎

Remark. The previous proposition is true if we replace pseudotensorial forms by tensorial forms.

We now have arrived at the central definition of this section.

Definition 12.19. *Let ω be a quantum connection in a QPB P with a differential calculus. Let δ be an embedded differential with corresponding braces $\langle \cdot, \cdot \rangle$. Then the map $R_\omega : {}_{\text{inv}}\Gamma \to \Omega^2(P)$ defined by*

$$R_\omega := d_P \omega - \langle \omega, \omega \rangle \tag{12.51}$$

is called the curvature *of ω.*

Notice that both $d_P \omega$ and $\langle \omega, \omega \rangle$ are linear maps ${}_{\text{inv}}\Gamma \to \Omega^2(P)$. So R_ω is a function taking values in 2-forms. One often shortens this by saying that R_ω is a 2-form, or simply a form. Also, notice that the differential

$$d_P : \Omega^1(P) \to \Omega^2(P)$$

of the HODC $\Omega(P)$ enters in this definition.

Proposition 12.14. *If the quantum connection ω is multiplicative, then the curvature R_ω does not depend on the choice of the embedded differential δ.*

Exercise 12.28. *Prove this proposition.*

Theorem 12.11. *The following diagram commutes:*

$$
\begin{array}{ccc}
{}_{\mathrm{inv}}\Gamma & \xrightarrow{R_\omega} & \Omega^2(P) \\
\text{ad}\downarrow & & \downarrow {}_{\Omega(P)}\Psi \\
{}_{\mathrm{inv}}\Gamma \otimes \mathcal{A} & \xrightarrow{R_\omega \otimes id_A} & \Omega^2(P) \otimes \Gamma^\wedge.
\end{array}
\tag{12.52}
$$

Equivalently, ${}_{\Omega(P)}\Psi\, R_\omega = (R_\omega \otimes id_A)\,\text{ad}$. *In particular, the image of* ${}_{\Omega(P)}\Psi\, R_\omega$ *lies in* $\Omega^2(P) \otimes \mathcal{A}$. *This diagram says that* R_ω *is a pseudotensorial* 2*-form, namely,* $R_\omega \in \psi^2(P)$.

Proof. Take $\theta \in {}_{\mathrm{inv}}\Gamma$. Let $\tilde{d}$ denote the differential in $\Omega(P) \otimes \Gamma^\wedge$. Then on the one hand, we calculate

$$
\begin{aligned}
{}_{\Omega(P)}\Psi\, R_\omega(\theta) &= {}_{\Omega(P)}\Psi\,(d_P\,\omega - \langle \omega, \omega \rangle)(\theta) \\
&= {}_{\Omega(P)}\Psi\,(d_P\,\omega(\theta)) - {}_{\Omega(P)}\Psi\,(\langle \omega, \omega \rangle(\theta)) \\
&= \tilde{d}\big({}_{\Omega(P)}\Psi\,(\omega(\theta))\big) - {}_{\Omega(P)}\Psi\,(m_\Omega(\omega \otimes \omega)\delta(\theta)) \\
&= \tilde{d}\big({}_{\Omega(P)}\Psi\,(\omega(\theta))\big) + {}_{\Omega(P)}\Psi\,(m_\Omega(\omega \otimes \omega)(\pi(a^{(1)}) \otimes \pi(a^{(2)})) \\
&= \tilde{d}\big({}_{\Omega(P)}\Psi\,(\omega(\theta))\big) + {}_{\Omega(P)}\Psi\,(\omega(\pi(a^{(1)}))\omega(\pi(a^{(2)}))) \\
&= \tilde{d}\big({}_{\Omega(P)}\Psi\,(\omega(\theta))\big) + {}_{\Omega(P)}\Psi\,(\omega(\pi(a^{(1)})))\,{}_{\Omega(P)}\Psi\,(\omega(\pi(a^{(2)}))),
\end{aligned}
$$

where we used Proposition 12.12. Working with the second term only gives

$$
\begin{aligned}
&{}_{\Omega(P)}\Psi\,(\omega(\pi(a^{(1)})))\,{}_{\Omega(P)}\Psi\,(\omega(\pi(a^{(2)}))) \\
&= \big((\omega \otimes id)\text{ad}(\pi(a^{(1)})) + 1 \otimes \pi(a^{(1)})\big)\big((\omega \otimes id)\text{ad}(\pi(a^{(2)})) + 1 \otimes \pi(a^{(2)})\big) \\
&= \big((\omega \otimes id)(\pi \otimes id)\text{ad}(a^{(1)}) + 1 \otimes \pi(a^{(1)})\big)\cdot \\
&\quad \big((\omega \otimes id)(\pi \otimes id)\text{ad}(a^{(2)}) + 1 \otimes \pi(a^{(2)})\big) \\
&= \big((\omega \otimes id)(\pi \otimes id)\big(a^{(2)} \otimes \kappa(a^{(1)})a^{(3)}\big) + 1 \otimes \pi(a^{(1)})\big)\cdot \\
&\quad \big((\omega \otimes id)(\pi \otimes id)\big(a^{(5)} \otimes \kappa(a^{(4)})a^{(6)}\big) + 1 \otimes \pi(a^{(2)})\big) \\
&= \big((\omega\pi(a^{(2)}) \otimes \kappa(a^{(1)})a^{(3)}) + 1 \otimes \pi(a^{(1)})\big)\cdot \\
&\quad \big((\omega\pi(a^{(5)}) \otimes \kappa(a^{(4)})a^{(6)}) + 1 \otimes \pi(a^{(2)})\big) \\
&= 1 \otimes \pi(a^{(1)})\pi(a^{(2)}) + \omega\pi(a^{(2)})\omega\pi(a^{(5)}) \otimes \kappa(a^{(1)})a^{(3)}\kappa(a^{(4)})a^{(6)} \\
&\quad + \omega\pi(a^{(2)}) \otimes \kappa(a^{(1)})a^{(3)}\pi(a^{(4)}) - \omega\pi(a^{(3)}) \otimes \pi(a^{(1)})\kappa(a^{(2)})a^{(4)} \\
&= 1 \otimes \pi(a^{(1)})\pi(a^{(2)}) + \omega\pi(a^{(2)})\omega\pi(a^{(5)}) \otimes \kappa(a^{(1)})a^{(3)}\kappa(a^{(4)})a^{(6)} \\
&\quad + \omega\pi(a^{(2)}) \otimes \kappa(a^{(1)})a^{(3)}\kappa(a^{(4)})da^{(5)} - \omega\pi(a^{(4)}) \otimes \kappa(a^{(1)})da^{(2)}\kappa(a^{(3)})a^{(5)}.
\end{aligned}
$$

Now peeking ahead to what this is supposed to be, so that when all is said and done, we calculate the following expression:

$$- (\langle \omega, \omega \rangle \otimes id) \mathrm{ad}(\theta) = -\big(m_\Omega (\omega \otimes \omega) \delta \otimes id\big) \mathrm{ad}(\theta)$$

$$= -\big(m_\Omega (\omega \otimes \omega) \otimes id\big)(\delta \otimes id) \mathrm{ad}(\theta)$$

$$= -\big(m_\Omega (\omega \otimes \omega) \otimes id\big) \mathrm{ad}^{\otimes 2} \delta(\theta)$$

$$= \big(m_\Omega (\omega \otimes \omega) \otimes id\big) \mathrm{ad}^{\otimes 2} \left(\pi(a^{(1)}) \otimes \pi(a^{(2)})\right)$$

$$= \big(m_\Omega (\omega \otimes \omega) \otimes id\big) M \left(\mathrm{ad}(\pi(a^{(1)})) \otimes \mathrm{ad}(\pi(a^{(2)}))\right)$$

$$= \big(m_\Omega (\omega \otimes \omega) \otimes id\big) M \left((\pi(a^{(2)}) \otimes \kappa(a^{(1)}) a^{(3)}) \otimes (\pi(a^{(5)}) \otimes \kappa(a^{(4)}) a^{(6)})\right)$$

$$= \big(m_\Omega (\omega \otimes \omega) \otimes id\big) \big(\pi(a^{(2)}) \otimes \pi(a^{(5)}) \otimes \kappa(a^{(1)}) a^{(3)} \kappa(a^{(4)}) a^{(6)}\big)$$

$$= m_\Omega (\omega \otimes \omega) \big(\pi(a^{(2)}) \otimes \pi(a^{(5)})\big) \otimes \kappa(a^{(1)}) a^{(3)} \kappa(a^{(4)}) a^{(6)}$$

$$= \omega(\pi(a^{(2)})) \omega(\pi(a^{(5)})) \otimes \kappa(a^{(1)}) a^{(3)} \kappa(a^{(4)}) a^{(6)}.$$

But this is the second term in the previous calculation. So we put together all the bits and pieces that we have so far and get that

$$_{\Omega(P)}\Psi\, R_\omega(\theta) = \tilde{d}\big(_{\Omega(P)}\Psi\,(\omega(\theta))\big) + 1 \otimes \pi(a^{(1)}) \pi(a^{(2)})$$

$$- (\langle \omega, \omega \rangle \otimes id)\mathrm{ad}(\theta) + \omega\pi(a^{(2)}) \otimes \kappa(a^{(1)}) a^{(3)} \kappa(a^{(4)}) da^{(5)}$$

$$- \omega\pi(a^{(4)}) \otimes \kappa(a^{(1)}) da^{(2)} \kappa(a^{(3)}) a^{(5)}.$$

But we also have a bit of a simplification of one of these terms:

$$\omega\pi(a^{(2)}) \otimes \kappa(a^{(1)}) a^{(3)} \kappa(a^{(4)}) da^{(5)} = \omega\pi(a^{(2)}) \otimes \kappa(a^{(1)}) \varepsilon(a^{(3)}) da^{(4)}$$

$$= \omega\pi(\varepsilon(a^{(3)}) a^{(2)}) \otimes \kappa(a^{(1)}) da^{(4)}$$

$$= \omega\pi(a^{(2)}) \otimes \kappa(a^{(1)}) da^{(3)}.$$

So now we have arrived at

$$_{\Omega(P)}\Psi\, R_\omega(\theta) = \tilde{d}\big(_{\Omega(P)}\Psi\,(\omega(\theta))\big) + 1 \otimes \pi(a^{(1)}) \pi(a^{(2)})$$

$$- (\langle \omega, \omega \rangle \otimes id)\mathrm{ad}(\theta) + \omega\pi(a^{(2)}) \otimes \kappa(a^{(1)}) da^{(3)}$$

$$- \omega\pi(a^{(4)}) \otimes \kappa(a^{(1)}) da^{(2)} \kappa(a^{(3)}) a^{(5)}.$$

Continuing just with the first term here, we see that

$$\tilde{d}\big(_{\Omega(P)}\Psi\,(\omega(\theta))\big) = \tilde{d}\big((\omega \otimes id_A)\mathrm{ad}(\theta) + 1 \otimes \theta\big)$$

$$= \tilde{d}\big((\omega \otimes id_A)\mathrm{ad}(\pi(a))\big) + \tilde{d}\big(1 \otimes \pi(a)\big)$$

$$\begin{aligned}
&= \tilde{d}\big((\omega \otimes id_A)(\pi \otimes id_A)\mathrm{ad}(a)\big) + 1 \otimes d^{\wedge}\pi(a)\\
&= \tilde{d}\big(\omega\pi(a^{(2)}) \otimes \kappa(a^{(1)})a^{(3)}\big) - 1 \otimes \pi(a^{(1)})\pi(a^{(2)})\\
&= d_P(\omega\pi(a^{(2)})) \otimes \kappa(a^{(1)})a^{(3)} - \omega\pi(a^{(2)}) \otimes d(\kappa(a^{(1)})a^{(3)})\\
&\quad - 1 \otimes \pi(a^{(1)})\pi(a^{(2)})\\
&= (d_P\,\omega \otimes id_A)\mathrm{ad}(\theta) - \omega\pi(a^{(2)}) \otimes d(\kappa(a^{(1)})a^{(3)})\\
&\quad - 1 \otimes \pi(a^{(1)})\pi(a^{(2)}),
\end{aligned}$$

where we used the Maurer–Cartan formula along with many other identities. Substituting this back, we now get

$$\begin{aligned}
{}_{\Omega(P)}&\Psi\,R_\omega(\theta)\\
&= (d_P\,\omega \otimes id)\mathrm{ad}(\theta) - \omega\pi(a^{(2)}) \otimes d(\kappa(a^{(1)})a^{(3)}) - (\langle\omega,\omega\rangle \otimes id_A)\mathrm{ad}(\theta)\\
&\quad + \omega\pi(a^{(2)}) \otimes \kappa(a^{(1)})da^{(3)} - \omega\pi(a^{(4)}) \otimes \kappa(a^{(1)})da^{(2)}\kappa(a^{(3)})a^{(5)}\\
&= ((d_P\,\omega - \langle\omega,\omega\rangle) \otimes id_A)\mathrm{ad}(\theta) - \omega\pi(a^{(2)}) \otimes d(\kappa(a^{(1)}))a^{(3)}\\
&\quad - \omega\pi(a^{(4)}) \otimes \kappa(a^{(1)})da^{(2)}\kappa(a^{(3)})a^{(5)}\\
&= (R_\omega \otimes id_A)\mathrm{ad}(\theta) - \omega\pi(a^{(2)}) \otimes d(\kappa(a^{(1)}))a^{(3)}\\
&\quad - \omega\pi(a^{(4)}) \otimes \kappa(a^{(1)})da^{(2)}\kappa(a^{(3)})a^{(5)},
\end{aligned}$$

which is so close to the result that you can taste it! So now it is a question of showing that the last two terms cancel. And they do, as this shows:

$$\begin{aligned}
-\omega\pi(a^{(4)}) \otimes \kappa(a^{(1)})da^{(2)}\kappa(a^{(3)})a^{(5)} &= \omega\pi(a^{(4)}) \otimes \kappa(a^{(1)})a^{(2)}d\kappa(a^{(3)})a^{(5)}\\
&= \omega\pi(a^{(3)}) \otimes \varepsilon(a^{(1)})d\kappa(a^{(2)})a^{(4)}\\
&= \omega\pi(a^{(2)}) \otimes d\kappa(a^{(1)})a^{(3)},
\end{aligned}$$

where we used the "partial differentiation" identity

$$(db^{(1)})\kappa(b^{(2)}) = -b^{(1)}d\kappa(b^{(2)})$$

for all $b \in \mathcal{A}$. The last identity follows from applying d to $b^{(1)}\kappa(b^{(2)}) = \varepsilon(b)1_A$. So we have indeed that

$$_{\Omega(P)}\Psi\,R_\omega(\theta) = (R_\omega \otimes id_A)\mathrm{ad}(\theta),$$

as claimed.   ∎

12.9   Covariant Derivatives

In this section we will define covariant derivatives in this noncommutative setting. These will be operators that correspond to covariant derivatives in classical
differential geometry. The latter are local differential operators and as such can
also be studied using classical analysis. But in this more general setting covariant
derivatives are not necessarily local operators. How could they be? After all,
the basic idea behind noncommutative geometry is to study geometric concepts
without any reference to points and therefore without local structures such as
neighborhoods of points. What we do have in the noncommutative setting are
infinitesimal structures, namely, differential calculi. So what we are dealing with
here can well be called *noncommutative analysis*. This point of view is essential in
the new way of understanding the nonlocal Dunkl operators as covariant derivatives.
We will take up that topic in Chapter 14.

   We now define the covariant derivative as an operator acting on horizontal vectors
in $\Omega(P)$. With this as a starting point, one can then extend the domain of this
operator to $\Omega(P)$ itself, though we will not present that. For the details on this
extension, see Eq. (4.59) and Theorem 4.15 in [22].

Definition 12.20. *Let ω be a quantum connection on a QPB with differential
calculus. Then for each $\varphi \in \mathfrak{hor}^k(P)$, we define*

$$D_\omega(\varphi) := d_P\,\varphi - (-1)^k \varphi^{(0)} \omega(\pi(\varphi^{(1)})) \in \Omega^{k+1}(P), \qquad (12.53)$$

*where $_\mathfrak{h}\Phi(\varphi) = \varphi^{(0)} \otimes \varphi^{(1)} \in \mathfrak{hor}^k(P) \otimes \mathcal{A}$. We call D_ω the covariant derivative of
the quantum connection ω.*

Remarks. Clearly, D_ω raises the degree by 1 since $\varphi^{(0)} \in \mathfrak{hor}^k(P) \subset \Omega^k(P)$ and
$\omega(\pi(\varphi^{(1)})) \in \Omega^1(P)$ as well as $d_P\,\varphi \in \Omega^{k+1}(P)$. In the next theorem we will show
that D_ω maps horizontal forms to horizontal forms. But neither of the two terms
on the right side of (12.53) need be horizontal in general. So we can think of this
definition as a way of subtracting off a term from $d_P\,\varphi$ in order to get a horizontal
result; that is, $(-1)^k \varphi^{(0)} \omega(\pi(\varphi^{(1)}))$ is the vertical "piece" of $d_P\,\varphi$. Note that this
vertical "piece" depends, of course, on the quantum connection ω being considered,
and therefore the covariant derivative D_ω also depends on ω. We hope that these
comments provide some motivation for the definition of covariant derivative. Also,
it is worth noting that this definition of covariant derivative dovetails well with the
definitions of horizontal and vertical forms.

   The definition of covariant derivative depends on the choices of the HODC for
the total space in the expression d_P in the first term and the product in $\Omega(P)$ in
the second term as well as on the FODC for the fiber space in the expression $\pi =
\pi_{\mathrm{inv}}\,d$ in the second term. These differentials are the appropriate generalizations of
local differential operators in the case of classical differential geometry, but in this
noncommutative setting the very concept of locality is not present. So covariant
derivatives here have nothing to do in general with local operators.

See [22] for much more concerning covariant derivatives, including the extension of the definition of covariant derivative from $\mathfrak{hor}(P)$ to all of $\Omega(P)$.

Theorem 12.12. *For all* $\varphi \in \mathfrak{hor}^k(P)$, *we have that*

$$_{\Omega(P)}\Psi \, D_\omega(\varphi) = D_\omega(\varphi^{(0)}) \otimes \varphi^{(1)}, \tag{12.54}$$

where $_{\mathfrak{h}}\Phi(\varphi) = \varphi^{(0)} \otimes \varphi^{(1)} \in \mathfrak{hor}^k(P) \otimes A$. *In particular,* $D_\omega(\varphi)$ *is horizontal; that is,* D_ω *leaves* $\mathfrak{hor}(P)$ *invariant, that is,*

$$D_\omega : \mathfrak{hor}(P) \to \mathfrak{hor}(P),$$

and so $_{\Omega(P)}\Psi \, D_\omega(\varphi) = {}_{\mathfrak{h}}\Phi \, D_\omega(\varphi)$.

Proof. This reduces to the following lengthy, but standard calculation:

$$_{\Omega(P)}\Psi \, D_\omega(\varphi) = {}_{\Omega(P)}\Psi \, (d\varphi) - (-1)^k \, {}_{\Omega(P)}\Psi \, (\varphi^{(0)}) \, {}_{\Omega(P)}\Psi \, (\omega(\pi(\varphi^{(1)})))$$

$$= d_P({}_{\Omega(P)}\Psi(\varphi))$$
$$\quad - (-1)^k (\varphi^{(00)} \otimes \varphi^{(01)}) \left((\omega \otimes id_A)\mathrm{ad}(\pi(\varphi^{(1)})) + 1_B \otimes \pi(\varphi^{(1)}) \right)$$

$$= d_P({}_{\mathfrak{h}}\Phi(\varphi))$$
$$\quad - (-1)^k (\varphi^{(00)} \otimes \varphi^{(01)}) \left((\omega \otimes id_A)(\pi \otimes id_A)\mathrm{ad}(\varphi^{(1)}) + 1_B \otimes \pi(\varphi^{(1)}) \right)$$

$$= d_P(\varphi^{(0)} \otimes \varphi^{(1)})$$
$$\quad - (-1)^k (\varphi^{(00)} \otimes \varphi^{(01)}) \left(\omega\pi(\varphi^{(12)}) \otimes \kappa(\varphi^{(11)})\varphi^{(13)} + 1_B \otimes \pi(\varphi^{(1)}) \right)$$

$$= d_P(\varphi^{(0)} \otimes \varphi^{(1)}) - (-1)^k \varphi^{(00)} \omega\pi(\varphi^{(12)}) \otimes \varphi^{(01)}\kappa(\varphi^{(11)})\varphi^{(13)}$$
$$\quad - (-1)^k \varphi^{(00)} \otimes \varphi^{(01)}\pi(\varphi^{(1)})$$

$$= d_P(\varphi^{(0)}) \otimes \varphi^{(1)} - (-1)^k \varphi^{(0)} \omega\pi(\varphi^{(3)}) \otimes \varphi^{(1)}\kappa(\varphi^{(2)})\varphi^{(4)}$$
$$\quad - (-1)^k \varphi^{(0)} \otimes \varphi^{(11)}\pi(\varphi^{(12)}) + (-1)^k (\varphi^{(0)} \otimes d\varphi^{(1)})$$

$$= d_P(\varphi^{(0)}) \otimes \varphi^{(1)} - (-1)^k \varphi^{(0)} \omega\pi(\varphi^{(2)}) \otimes \varepsilon(\varphi^{(1)})\varphi^{(3)}$$
$$\quad - (-1)^k \varphi^{(0)} \otimes d\varphi^{(1)} + (-1)^k (\varphi^{(0)} \otimes d\varphi^{(1)})$$

$$= d_P(\varphi^{(0)}) \otimes \varphi^{(1)} - (-1)^k \varphi^{(0)} \omega\pi(\varepsilon(\varphi^{(1)})\varphi^{(2)}) \otimes \varphi^{(3)}$$

$$= d_P(\varphi^{(0)}) \otimes \varphi^{(1)} - (-1)^k \varphi^{(0)} \omega\pi(\varphi^{(1)}) \otimes \varphi^{(2)}$$

$$= d_P(\varphi^{(0)}) \otimes \varphi^{(1)} - (-1)^k \varphi^{(00)} \omega\pi(\varphi^{(01)}) \otimes \varphi^{(1)}$$

$$= \left(d_P(\varphi^{(0)}) - (-1)^k \varphi^{(00)} \omega\pi(\varphi^{(01)}) \right) \otimes \varphi^{(1)}$$

$$= D_\omega(\varphi^{(0)}) \otimes \varphi^{(1)}.$$

Notice that both the iterated Sweedler's notation as well as the simpler one-digit Sweedler's notation have been used together here in one argument. Also, various definitions and standard identities have been used.

Finally, the last statement then follows immediately from this calculation by the definitions of horizontal form and $_\mathfrak{h}\Phi$. ∎

Theorem 12.13. *The covariant derivative D_ω of any quantum connection intertwines the co-action $_\mathfrak{h}\Phi$ in the appropriate way, namely,*

$$_\mathfrak{h}\Phi \, D_\omega = (D_\omega \otimes id_{\mathcal{A}}) \,_\mathfrak{h}\Phi$$

or, equivalently,

$$
\begin{array}{ccc}
\mathfrak{hor}(P) & \overset{_\mathfrak{h}\Phi}{\longrightarrow} & \mathfrak{hor}(P) \otimes \mathcal{A} \\
D_\omega \downarrow & & \downarrow D_\omega \otimes id_{\mathcal{A}} \\
\mathfrak{hor}(P) & \overset{_\mathfrak{h}\Phi}{\longrightarrow} & \mathfrak{hor}(P) \otimes \mathcal{A}
\end{array}
$$

is a commutative diagram.

Proof. Take $\varphi \in \mathfrak{hor}^k P$ and write $_\mathfrak{h}\Phi(\varphi) = \varphi^{(0)} \otimes \varphi^{(1)}$. Then, using (12.54), we immediately have

$$
\begin{aligned}
_\mathfrak{h}\Phi \, D_\omega(\varphi) &= _{\Omega(P)}\Psi \, D_\omega(\varphi) \\
&= D_\omega(\varphi^{(0)}) \otimes \varphi^{(1)} \\
&= (D_\omega \otimes id)(\varphi^{(0)} \otimes \varphi^{(1)}) \\
&= (D_\omega \otimes id) \,_\mathfrak{h}\Phi(\varphi),
\end{aligned}
$$

which proves the result. ∎

Theorem 12.14. *Suppose that the quantum connection ω is regular. Then the covariant derivative D_ω satisfies the graded Leibniz rule,*

$$D_\omega(\varphi\psi) = D_\omega(\varphi)\psi + (-1)^k \varphi D_\omega(\psi),$$

where $\varphi, \psi \in \mathfrak{hor}(P)$ and $\deg \varphi = k$.

Proof. Let $\deg \psi = l$ and $\theta \in {}_{inv}\Gamma$. Since ω is regular, we have the identity

$$\omega(\theta)\psi = (-1)^l \psi^{(0)} \omega(\theta \circ \psi^{(1)}). \tag{12.55}$$

Then we have

$$
\begin{aligned}
D_\omega(\varphi\psi) &= d_P(\varphi\psi) - (-1)^{k+l}(\varphi\psi)^{(0)}\omega\big(\pi((\varphi\psi)^{(1)})\big) \\
&= d_P(\varphi\psi) - (-1)^{k+l}(\varphi^{(0)}\psi^{(0)})\omega\big(\pi(\varphi^{(1)}\psi^{(1)})\big)
\end{aligned}
$$

$$= d_P(\varphi\psi) - (-1)^{k+l}(\varphi^{(0)}\psi^{(0)})\omega\big(\pi(\varphi^{(1)})\circ\psi^{(1)} + \varepsilon(\varphi^{(1)})\pi(\psi^{(1)})\big)$$

$$= d_P(\varphi\psi) - (-1)^{k+l}(\varphi^{(0)}\psi^{(0)})\omega\big(\pi(\varphi^{(1)})\circ\psi^{(1)}\big)$$

$$- (-1)^{k+l}(\varphi^{(0)}\psi^{(0)})\omega\big(\varepsilon(\varphi^{(1)})\pi(\psi^{(1)})\big)$$

$$= d_P(\varphi\psi) - (-1)^{k+l}\varphi^{(0)}(-1)^l\omega(\pi(\varphi^{(1)}))\psi$$

$$- (-1)^{k+l}(\varepsilon(\varphi^{(1)})\varphi^{(0)}\psi^{(0)})\omega\big(\pi(\psi^{(1)})\big)$$

$$= d_P(\varphi)\psi + (-1)^k\varphi d_P(\psi) - (-1)^k\varphi^{(0)}\omega(\pi(\varphi^{(1)}))\psi$$

$$- (-1)^k\varphi(-1)^l\psi^{(0)}\omega\big(\pi(\psi^{(1)})\big)$$

$$= [d_P(\varphi) - (-1)^k\varphi^{(0)}\omega(\pi(\varphi^{(1)}))]\psi$$

$$+ (-1)^k\varphi[d_P(\psi) - (-1)^l\psi^{(0)}\omega\big(\pi(\psi^{(1)})\big)]$$

$$= D_\omega(\varphi)\psi + (-1)^k\varphi D_\omega(\psi),$$

where we used the identity (12.55) in the fifth equality. And this is the desired result. ∎

This result has a converse, which we will not prove. It says that if D_ω satisfies the graded Leibniz rule, then ω is regular. See [26] for a proof. This result provides a pretty way to prove that any Ehresmann connection in classical differential geometry is regular; one just shows that the covariant derivative of the connection satisfies the graded Leibniz rule.

Exercise 12.29 (Optional). *Suppose that a quantum connection ω is regular. Then the covariant derivative D_ω is a $*$-morphism.*

12.10   The Multiplicative Obstruction

The purpose of this short section is to introduce the natural definition of the obstruction for a quantum connection to be multiplicative and to indicate that it possesses a counterintuitive property. This material will not be used later in this book.

Definition 12.21. *Let ω be a quantum connection in a QPB with differential calculus. Let $\mathcal{R}$ be the right ideal in $\mathcal{A}$ that defines the FODC in the structure "group" $\mathcal{A}$ of the QPB. Define the* multiplicative obstruction *of ω to be the function $r_\omega : \mathcal{R} \to \Omega^2(P)$ defined by*

$$r_\omega(a) := \omega\pi(a^{(1)})\omega\pi(a^{(12)})$$

for all $a \in \mathcal{R}$.

Clearly, ω is multiplicative if and only if $r_\omega \equiv 0$. So any deviation of r_ω from being identically equal to zero is a measure of an obstruction for ω being multiplicative. In classical differential geometry $r_\omega \equiv 0$, and so this is not an object of interest in that setting. However, in the noncommutative setting we can have quantum connections that are not multiplicative. There are even quantum connections that are regular and also not multiplicative.

One basic property of r_ω is that it is a product of two vertical vectors, making it also a vertical object. In this regard, the next result is surprising.

Theorem 12.15. *The multiplicative obstruction r_ω of a quantum connection ω satisfies*

$$\Omega^\Psi r_\omega = (r_\omega \otimes id_A) \, \mathrm{ad},$$

where ad *is the right adjoint co-action of A on itself.*

In particular, $r_\omega(a) \in \mathfrak{hor}^2(P)$ *for all* $a \in \mathcal{R}$; *that is,* $r_\omega(a)$ *is horizontal. Therefore,* r_ω *is a tensorial 2-form, namely,* $r_\omega \in \tau^2(P)$.

Of course, this result is *not* surprising in the sense that it accords with the general pattern of quantum structures as seen throughout this chapter. Nonetheless, it runs counter to the intuition that one might acquire from a study of connections in classical differential geometry. Basically, it seems that there are two attitudes that one could take about this situation. One would be that the concepts of "horizontal" and "vertical" in this setting are not all that analogous to the corresponding classical concepts. Another would be to say that the concept of "quantum connection" is not all that analogous to the corresponding classical concept. My personal preference is the first alternative, but the essential consideration is how useful all of these concepts are in applications, especially those in physics.

Exercise 12.30. *Prove Theorem 12.15.*

12.11 Examples

In this section we present various examples of this theory of QPBs. We do not discuss the general frame bundles found in [28] even though that is also an interesting and nontrivial example. This is because that theory involves compact quantum groups (see [84]) in an essential way, and we do not wish to develop all the relevant background material here.

Example 1: The Trivial Bundle

We continue with the study of the explicit product bundles introduced in Example 12.3. So we are considering the explicit product QPB $P = (V \otimes \mathcal{A}, \mathcal{A}, id_V \otimes \phi)$.

We also only consider for now the product differential calculus on this QPB. It is important to realize that this is an additional choice and that within the theory of noncommutative geometry there exist other choices.

In particular, the algebra $\Omega(P)$ of the total space in this case is

$$\Omega(P) = \Omega(M) \otimes \Gamma^\wedge,$$

with $_{\Omega(P)}\Psi = id_{\Omega(M)} \otimes \hat{\phi}$ and $V = \Omega^0(M)$. Then the horizontal forms are given by

$$\mathfrak{hor}(P) = \Omega(M) \otimes \mathcal{A} \tag{12.56}$$

and the vertical algebra is identified (up to isomorphism) as

$$\mathfrak{ver}(P) \cong V \otimes \Gamma^\wedge.$$

Moreover, the differential d_v of $\mathfrak{ver}(P)$ corresponds under this isomorphism to the map $id \otimes d^\wedge$. While all of these are easily established facts, they should help the reader to have more intuition about the various definitions of these structures. Speaking of which, here is another one:

Definition 12.22. *For each integer $j \geq 0$, define*

$$\mathcal{L}^j := \{L : {}_{inv}\Gamma \to \Omega^j(M) \mid L \text{ is a linear map over } \mathbb{C}\}.$$

Then put $\mathcal{L} := \oplus_{j \geq 0} \mathcal{L}^j$.

It follows that $\mathcal{L}$ is a graded $*$-$\Omega(M)$-bimodule, where for $\alpha \in \Omega^i(M), L \in \mathcal{L}^j$, $\beta \in \Omega^k(M)$, and $\theta \in {}_{inv}\Gamma$, we define

$$(\alpha L)(\theta) := \alpha(L(\theta)),$$
$$(L\beta)(\theta) := (L(\theta))\beta,$$
$$L^*(\theta) := (L(\theta^*))^*.$$

The last definition is actually something we have seen before. We check, for example, that for all $\theta \in {}_{inv}\Gamma$, we have

$$(\alpha L)^*(\theta) = (\alpha L(\theta^*))^* = (L(\theta^*))^*\alpha^* = L^*(\theta)\alpha^* = L^*\alpha^*(\theta),$$

and so $(\alpha L)^* = L^*\alpha^*$ follows.

Proposition 12.15. *For each $L \in \mathcal{L}^j$, we define $\varphi_L \in \tau^j(P)$ by*

$$\varphi_L := (L \otimes id_{\mathcal{A}})\mathrm{ad},$$

where ad $: {}_{\mathrm{inv}}\Gamma \to {}_{\mathrm{inv}}\Gamma \otimes \mathcal{A}$ *is the right adjoint co-action. Then the mapping defined by $\varphi(L) := \varphi_L$ gives a grade-preserving isomorphism $\varphi : \mathcal{L} \to \tau(P)$ of graded $*$-$\Omega(M)$-bimodules.*

Remark. This result tells us that the study of all the tensorial j-forms is equivalent to the study of all the linear maps $L : {}_{\mathrm{inv}}\Gamma \to \Omega^j(M)$, a simpler space of maps.

Proof. First, by using (12.56), we note for $\theta \in {}_{\mathrm{inv}}\Gamma$ that we have

$$\varphi_L(\theta) = (L \otimes id_{\mathcal{A}})\mathrm{ad}(\theta) \in \Omega^j(M) \otimes \mathcal{A} = \mathfrak{hor}^j(P).$$

Thus, $\varphi_L : {}_{\mathrm{inv}}\Gamma \to \mathfrak{hor}^j(P)$. A standard calculation shows $(id_{\Omega(M)} \otimes \varepsilon)\varphi_L = L$. It follows that $\varphi_L = \varphi_M$ implies that $L = M$. So the map φ is one-to-one.

Next, to show that $\varphi_L \in \tau^j(P)$, we have to prove that it intertwines the co-actions of $\mathcal{A}$ in the two spaces. And here it is:

$$\begin{aligned}
{}_{\Omega(P)}\Phi(\varphi_L) &= (id \otimes p_0) \, {}_{\Omega(P)}\Psi \, (L \otimes id_{\mathcal{A}})\mathrm{ad} \\
&= (id \otimes p_0)(id_{\Omega(M)} \otimes \hat{\phi})(L \otimes id_{\mathcal{A}})\mathrm{ad} \\
&= (id_{\Omega(M)} \otimes \phi)(L \otimes id_{\mathcal{A}})\mathrm{ad} \\
&= (L \otimes id_{\mathcal{A}} \otimes id_{\mathcal{A}})(id \otimes \phi)\mathrm{ad} \\
&= (L \otimes id_{\mathcal{A}} \otimes id_{\mathcal{A}})(\mathrm{ad} \otimes id_{\mathcal{A}})\mathrm{ad} \\
&= ((L \otimes id_{\mathcal{A}})\mathrm{ad} \otimes id_{\mathcal{A}})\mathrm{ad} \\
&= (\varphi_L \otimes id_{\mathcal{A}})\mathrm{ad}.
\end{aligned}$$

To show that φ is onto, take an arbitrary element $\psi \in \tau^j(P)$. So this means by definition that

$$(id \otimes \phi)(\psi) = (\psi \otimes id_{\mathcal{A}})\mathrm{ad}.$$

Applying $id \otimes \varepsilon \otimes id$ to this equality yields

$$\psi = ((id \otimes \varepsilon)\psi \otimes id_{\mathcal{A}})\mathrm{ad}.$$

Now put $L := (id \otimes \varepsilon)\psi : {}_{\mathrm{inv}}\Gamma \to \Omega^j(M)$, a linear map. So $L \in \mathcal{L}^j$ and

$$\psi = (L \otimes id_{\mathcal{A}})\mathrm{ad} = \varphi_L = \varphi(L).$$

And this proves that φ is onto, thereby establishing that $\varphi : \mathcal{L} \to \tau(P)$ is a grade-0 isomorphism of graded vector spaces.

It is left to the reader to finish this proof by showing that the map φ is a $*$-$\Omega(M)$-bimodule isomorphism. ∎

The study of more properties of QPBs can be carried out in this example. See [22] for details. Here is one such result that the reader may wish to prove as an exercise.

Proposition 12.16. *The map* $\omega_0 : {}_{inv}\Gamma \to \Omega^1(P)$ *defined by*

$$\omega_0(\theta) := 1_B \otimes \theta$$

is a quantum connection on P. *Moreover, it is* flat, *which means that its curvature is zero.*

Example 2: Quantum Homogeneous Bundles

Classically, if a Lie group H has a closed subgroup G, then the homogeneous space of cosets H/G is canonically a differential manifold and the quotient map $H \to H/G$ is a principal fiber bundle with structure group G (for example, see [39]). Dually in the noncommutative setting, the inclusion map $\iota : G \subset H$ corresponds to a linear map

$$j : \mathcal{B} \to \mathcal{A}$$

from the "quantum total space" $\mathcal{B}$ to the "quantum structure group" $\mathcal{A}$. The map j is required to preserve all the possible structures. In particular, it should be a unital $*$-morphism. Reflecting dually the condition that the inclusion map $\iota : G \to H$ is injective, we also require that j is surjective. Both of these $*$-algebras represent groups, so we require that both of them are $*$-Hopf algebras. The "group" structure maps of $\mathcal{A}$ are denoted as usual by $\varepsilon, \kappa, \phi$. The corresponding "group" structure maps of $\mathcal{B}$ are denoted as $\varepsilon', \kappa', \phi'$. Since j is thought of as corresponding to a group morphism, we require that j preserves these three group structures as well. This says that the following three diagrams are commutative:

$$
\begin{array}{ccc}
\mathcal{B} \xrightarrow{j} \mathcal{A} & \quad \mathcal{B} \xrightarrow{\kappa'} \mathcal{B} & \quad \mathcal{B} \xrightarrow{\phi'} \mathcal{B} \otimes \mathcal{B} \\
{\scriptstyle \varepsilon'}\downarrow \quad \downarrow{\scriptstyle \varepsilon} & \quad {\scriptstyle j}\downarrow \quad \downarrow{\scriptstyle j} & \quad {\scriptstyle j}\downarrow \quad \downarrow{\scriptstyle j \otimes j} \\
\mathbb{C} \xrightarrow{id} \mathbb{C} & \quad \mathcal{A} \xrightarrow{\kappa} \mathcal{A} & \quad \mathcal{A} \xrightarrow{\phi} \mathcal{A} \otimes \mathcal{A}.
\end{array}
$$

The corresponding equations are $\varepsilon' = \varepsilon j$, $\kappa j = j\kappa'$, and $\phi j = (j \otimes j)\phi'$.

In short, we start with a pair of $*$-Hopf algebras and a surjective morphism of $*$-Hopf algebras $j : \mathcal{B} \to \mathcal{A}$ from one of them to the other.

The "spaces" $\mathcal{A}$ and $\mathcal{B}$ can be made into part of the structure of a QPB. First of all, they satisfy the condition (qpb1). However, the map j is a structure particular for the case under consideration. And now we use j and the "group" structure of $\mathcal{B}$ in order to define the right co-action for this QPB. Explicitly, we define $_\mathcal{B}\Phi$ to be the composition $_\mathcal{B}\Phi := (id \otimes j)\phi'$, that is,

$$_\mathcal{B}\Phi : \mathcal{B} \xrightarrow{\phi'} \mathcal{B} \otimes \mathcal{B} \xrightarrow{id \otimes j} \mathcal{B} \otimes \mathcal{A}.$$

This is the dual of the right action of G on H defined as the composition

$$H \times G \xrightarrow{id \times \iota} H \times H \xrightarrow{m_H} H,$$

where m_H is the multiplication map of H. Since $_\mathcal{B}\Phi$ is the composition of two unital $*$-morphisms, it follows that it too is a unital $*$-morphism. This is part of the condition (qpb2). But it remains to prove that $_\mathcal{B}\Phi$ is a right co-action of $\mathcal{A}$ on $\mathcal{B}$. For starters, we have

$$
\begin{aligned}
(id_\mathcal{B} \otimes \varepsilon) \, _\mathcal{B}\Phi &= (id_\mathcal{B} \otimes \varepsilon)(id_\mathcal{B} \otimes j)\phi' \\
&= (id_\mathcal{B} \otimes \varepsilon j)\phi' \\
&= (id_\mathcal{B} \otimes \varepsilon')\phi' \\
&\cong id_\mathcal{B},
\end{aligned}
$$

where the last step uses that $\mathcal{B}$ is a Hopf algebra. This shows one of the defining properties of a co-action. The other defining property of a right co-action is the commutativity of this diagram:

$$
\begin{array}{ccc}
\mathcal{B} & \xrightarrow{\,_\mathcal{B}\Phi\,} & \mathcal{B} \otimes \mathcal{A} \\
{\scriptstyle _\mathcal{B}\Phi}\downarrow & & \downarrow{\scriptstyle _\mathcal{B}\Phi \otimes id} \\
\mathcal{B} \otimes \mathcal{A} & \xrightarrow{id \otimes \phi} & \mathcal{B} \otimes \mathcal{A} \otimes \mathcal{A}.
\end{array}
$$

This is a standard diagram chase proof that uses $\phi j = (j \otimes j)\phi'$. So (qpb2) is satisfied.

We chose to verify (qpb4) next, which is equivalent to (qpb3). So pick an arbitrary element $a \in \mathcal{A}$. Then we can write it as $a = j(b)$ for some $b \in \mathcal{B}$ since j is surjective. Write $\phi'(b) = b^{(1)} \otimes b^{(2)}$ in Sweedler's notation for the Hopf algebra $\mathcal{B}$. But then we calculate

$$
\begin{aligned}
\kappa'(b^{(1)}) \, _\mathcal{B}\Phi(b^{(2)}) &= \kappa'(b^{(1)})(id \otimes j)\phi'(b^{(2)}) \\
&= \kappa'(b^{(1)})\left(b^{(2)} \otimes j(b^{(3)})\right)
\end{aligned}
$$

$$= \varepsilon'(b^{(1)})1 \otimes j(b^{(2)})$$

$$= 1 \otimes j(\varepsilon'(b^{(1)})b^{(2)})$$

$$= 1 \otimes j(b)$$

$$= 1 \otimes a.$$

And this shows that $1 \otimes a$ can always be written in the form required in the condition (qpb4).

We summarize this discussion in the following result.

Theorem 12.16. *Let A and B be $*$-Hopf algebras. Suppose that $j : B \to A$ is a surjective, unital morphism of $*$-Hopf algebras, and define $_B\Phi$ to be*

$$_B\Phi := (id \otimes j)\phi',$$

where ϕ' is the co-multiplication map for B. Then $P = (B, A, {_B\Phi})$ is a QPB. This particular case of a QPB is called a quantum homogeneous bundle.

This theorem tells us that there are many quantum homogeneous bundles and consequently also many quantum principal bundles. We comment on this further in the notes.

Exercise 12.31 (Optional). *Show that the assignment from surjective, unital $*$-Hopf algebra morphisms j to the QPB $P = (B, A, {_B\Phi})$, constructed above is functorial.* **Warning:** *Be careful in identifying the domain category of this functor.*

According to the definition, the "base space" of this QPB is

$$V = \{b \in B \mid {_B\Phi}(b) = b \otimes 1_A\} = B_{inv},$$

the space of right invariant elements. This $*$-algebra is called the *quantum homogeneous space* of the quantum homogeneous bundle. It corresponds to the homogeneous space H/G in the classical case. The requirement that $_B\Phi(b) = b \otimes 1_A$, or equivalently, $b^{(1)} \otimes j(b^{(2)}) = b \otimes 1_A$, does not seem very transparent. But here is the result.

Proposition 12.17. *The space of right invariant elements is given by*

$$V = \{b \in K \mid \phi'(b) \in B \otimes K\},$$

where $K := \ker(j - \varepsilon'(\cdot)1_A)$.

Proof. First, suppose that $b \in V$. We start with $b \otimes 1_A = b^{(1)} \otimes j(b^{(2)})$ and apply the map $\varepsilon' \otimes id_A$ to both sides, getting

$$\varepsilon'(b) 1_A = \varepsilon'(b^{(1)})j(b^{(2)}) = j(\varepsilon'(b^{(1)})b^{(2)}) = j(b),$$

which shows that $b \in \mathcal{K}$. Next, we observe that

$$\phi'(b) = b^{(1)} \otimes b^{(2)} \xrightarrow{id \otimes \varepsilon'(\cdot)1_A} b^{(1)} \otimes \varepsilon'(b^{(2)})1_A = b \otimes 1_A$$

and

$$\phi'(b) = b^{(1)} \otimes b^{(2)} \xrightarrow{id \otimes j} b^{(1)} \otimes j(b^{(2)}) = b \otimes 1_A.$$

And this shows that $\phi'(b) \in \mathcal{B} \otimes \mathcal{K}$.

Conversely, take $b \in \mathcal{K}$ with $\phi'(b) \in \mathcal{B} \otimes \mathcal{K}$. Then

$$
\begin{aligned}
{}_\mathcal{B}\Phi(b) &= (id_\mathcal{B} \otimes j)\phi'(b) \\
&= (id_\mathcal{B} \otimes j)(b^{(1)} \otimes b^{(2)}) \\
&= b^{(1)} \otimes j(b^{(2)}) \\
&= b^{(1)} \otimes \varepsilon'(b^{(2)})1_A \\
&= b \otimes 1_A
\end{aligned}
$$

since by hypothesis we can take $b^{(2)} \in \mathcal{K}$. But this just says that $b \in \mathcal{V}$. ∎

Next, after dividing by the right action of G, the quotient space H/G inherits a left action of H. The corresponding structure here would be a left co-action of the total space $\mathcal{B}$ on the base space $\mathcal{V}$, that is, a left co-action

$$\mathcal{V} \to \mathcal{B} \otimes \mathcal{V}.$$

The question now is whether such a left co-action always exists in this noncommutative setting. On the classical side, we take H/G to be the space of cosets of the form hG for some $h \in H$. Then the left action of $k \in H$ sends hG to khG. So the multiplication of H passes to a map

$$H \times H/G \to H/G. \tag{12.57}$$

And without a further hypothesis on G, this is as far as it goes. In other words, we do not in general get a group structure on H/G. In the noncommutative setting, we have the "group structure" map

$$\phi' : \mathcal{B} \to \mathcal{B} \otimes \mathcal{B},$$

which we can restrict to $\mathcal{V}$. But this only gives us a map

$$\phi' i : \mathcal{V} \to \mathcal{B} \otimes \mathcal{B},$$

where $i : \mathcal{V} \to \mathcal{B}$ is the inclusion map. The next result is the noncommutative analog to (12.57).

Lemma 12.4. *The range of $\phi' i$ is contained in $\mathcal{B} \otimes \mathcal{V}$. In other words, we have that*

$$\phi' i : \mathcal{V} \to \mathcal{B} \otimes \mathcal{V}.$$

Remark. In general, dual to the classical case, we do not have that the co-multiplication map ϕ' of $\mathcal{B}$ passes to a co-multiplication map $\mathcal{V} \to \mathcal{V} \otimes \mathcal{V}$.

Proof. Let $b \in \mathcal{V}$ be given. So $_{\mathcal{B}}\Phi(b) = b \otimes 1_{\mathcal{A}}$. To show that $\phi' i(b) = \phi'(b)$ lies in $\mathcal{B} \otimes \mathcal{V}$, we calculate using definitions and the co-associativity of ϕ' to get

$$
\begin{aligned}
(id_{\mathcal{B}} \otimes {}_{\mathcal{B}}\Phi)\phi'(b) &= (id_{\mathcal{B}} \otimes (id_{\mathcal{B}} \otimes j)\phi')\phi'(b) \\
&= (id_{\mathcal{B}} \otimes id_{\mathcal{B}} \otimes j)(id_{\mathcal{B}} \otimes \phi')\phi'(b) \\
&= (id_{\mathcal{B}} \otimes id_{\mathcal{B}} \otimes j)(\phi' \otimes id_{\mathcal{B}})\phi'(b) \\
&= (\phi' \otimes id_{\mathcal{A}})(id_{\mathcal{B}} \otimes j)\phi'(b) \\
&= (\phi' \otimes id_{\mathcal{A}}) \,{}_{\mathcal{B}}\Phi(b) \\
&= (\phi' \otimes id_{\mathcal{A}})(b \otimes 1_{\mathcal{A}}) \\
&= \phi'(b) \otimes 1_{\mathcal{A}},
\end{aligned}
$$

which means that $\phi'(b) \in \mathcal{B} \otimes \mathcal{V}$, as desired. ∎

The proof of the next anticlimactic result is left to the reader.

Proposition 12.18. *The map*

$$\phi' i : \mathcal{V} \to \mathcal{B} \otimes \mathcal{V}.$$

is a left co-action of $\mathcal{B}$ on $\mathcal{V}$.

Now we wish to identify the left invariant elements of $\mathcal{V}$ under the left co-action $\phi' i$.

Proposition 12.19. *An element $v \in \mathcal{V} = \mathcal{B}_{\mathrm{inv}}$ is left invariant under the left co-action $\phi' i$ of $\mathcal{B}$ on $\mathcal{V}$ if and only if $v = \lambda 1_{\mathcal{B}}$ for some unique $\lambda \in \mathbb{C}$.*

Remark. One says that such an element v is a *constant*. It turns out that λ is given by $\lambda = \varepsilon'(v)$.

Recall that by definition $\mathcal{V}$ is the space of *right* invariant elements in $\mathcal{B}$ under the right co-action $_{\mathcal{B}}\Phi$ of $\mathcal{A}$ on $\mathcal{B}$.

Proof. To say that v is left invariant means that

$$\phi' i(v) = v^{(1)} \otimes v^{(2)} = 1_{\mathcal{B}} \otimes v.$$

To say that, $v \in \mathcal{V}$ means, using the previous formula, that

$$v \otimes 1_{\mathcal{A}} = {}_{\mathcal{B}}\Phi(v) = (id_{\mathcal{B}} \otimes j)\phi'(v) = 1_{\mathcal{B}} \otimes j(v).$$

Applying $id_{\mathcal{B}} \otimes \varepsilon$ to both sides of this equation gives

$$v = \varepsilon j(v)1_{\mathcal{B}} = \varepsilon'(v)1_{\mathcal{B}}.$$

And this proves that $v = \lambda 1_{\mathcal{B}}$ with $\lambda = \varepsilon'(v)$, as claimed above.

Conversely, we now suppose that $v = \lambda 1_{\mathcal{B}}$ with $\lambda \in \mathbb{C}$. Then, on the one hand, we see that

$$\phi'i(v) = \phi'i(\lambda 1_{\mathcal{B}}) = \lambda(1_{\mathcal{B}} \otimes 1_{\mathcal{B}}),$$

while, on the other hand, we get

$$1_{\mathcal{B}} \otimes v = 1_{\mathcal{B}} \otimes \lambda 1_{\mathcal{B}} = \lambda(1_{\mathcal{B}} \otimes 1_{\mathcal{B}}).$$

It follows that $\phi'i(v) = 1_{\mathcal{B}} \otimes v$ and so v is left invariant, as we wished to show. The uniqueness of λ is clear. ∎

For a classical left action (say of a Lie group G on a manifold M), the left invariant functions are constant on each orbit of the action. So the action is transitive if and only if there is exactly one orbit, which in turn implies that all the left invariant functions are constants globally. (Recall that a classical left action of G on M is said to be *transitive* if, for every pair $x_1, x_2 \in M$, there exists $g \in G$ such that $g \cdot x_1 = x_2$. Here $\cdot$ denotes the left action.) This remark justifies our saying that ϕ' is a *transitive left co-action*. This is further justification for calling $\mathcal{V}$ a quantum homogeneous space.

As we have noted many times before, the mere construction of a QPB is not enough. We also have to introduce the differential calculi to get a complete theory. In this example the differential calculi on $\mathcal{A}$ and on $\mathcal{B}$ should be related to each other via the structure map $j : \mathcal{B} \to \mathcal{A}$. We start off with the theory of FODCs in this setting. Suppose that we have a left invariant FODC Γ over $\mathcal{A}$ and a left invariant FODC Ψ over $\mathcal{B}$. By the general theory, applied separately to $\mathcal{A}$ and to $\mathcal{B}$, we have that Γ corresponds to a right ideal $\mathcal{R} \subset \ker \varepsilon \subset \mathcal{A}$ while Ψ corresponds to a right ideal $\mathcal{R}' \subset \ker \varepsilon' \subset \mathcal{B}$. Using the relation $\varepsilon' = \varepsilon j$, we see that

$$b \in \ker \varepsilon' \implies \varepsilon'(b) = 0 \implies \varepsilon j(b) = 0 \implies j(b) \in \ker \varepsilon,$$

which says that $j(\ker \varepsilon') \subset \ker \varepsilon$. So this motivates the requirement on the two FODCs that

$$j(\mathcal{R}') \subset \mathcal{R}.$$

Given this, we have the following commutative diagram:

$$
\begin{array}{ccccc}
\mathcal{R}' & \xrightarrow{i'} & \ker \varepsilon' & \xrightarrow{\pi'} & {}_{\mathrm{inv}}\Psi \\
{\scriptstyle j}\downarrow & & {\scriptstyle j}\downarrow & & \downarrow{\scriptstyle \rho} \\
\mathcal{R} & \xrightarrow{i} & \ker \varepsilon & \xrightarrow{\pi} & {}_{\mathrm{inv}}\Gamma,
\end{array}
$$

where i and i' are the inclusion maps, π and π' are the quantum germs maps, and ρ is the induced map in the quotient.

Exercise 12.32. *Show that* Φ_j *defined as the composition*

$$
\mathcal{B} \xrightarrow{\mathrm{ad}'} \mathcal{B} \otimes \mathcal{B} \xrightarrow{\mathrm{id} \otimes j} \mathcal{B} \otimes \mathcal{A}
$$

is a right co-action of $\mathcal{A}$ *on* $\mathcal{B}$*. (Here* ad' *denotes the right co-action of* $\mathcal{B}$ *on itself. Note that the subscript in* Φ_j *is not an integer, but rather a map.)*

Show that $\ker \varepsilon'$ *is an invariant subspace under this right co-action.*

To have this co-action on $\ker \varepsilon'$ pass to the quotient ${}_{\mathrm{inv}}\Psi$, we also require this compatibility condition:

$$
\Phi_j(\mathcal{R}') \subset \mathcal{R}' \otimes \mathcal{A}.
$$

Given this, we denote the quotient co-action by

$$
\chi : {}_{\mathrm{inv}}\Psi \to {}_{\mathrm{inv}}\Psi \otimes \mathcal{A}.
$$

(As an aside, let's note that this *is* a right co-action.) Then we have this commutative diagram:

$$
\begin{array}{ccc}
\mathcal{B} & \xrightarrow{\pi'} & {}_{\mathrm{inv}}\Gamma \\
{\scriptstyle \mathrm{ad}'}\downarrow & & \downarrow{\scriptstyle \chi} \\
\mathcal{B} \otimes \mathcal{B} & \xrightarrow{\pi' \otimes j} & {}_{\mathrm{inv}}\Gamma \otimes \mathcal{A}.
\end{array}
$$

This is a consequence of the above discussion if we replace the upper left corner with the space $\ker \varepsilon' \subset \mathcal{B}$. However, on the complementary subspace $\mathbb{C}1_{\mathcal{B}} \subset \mathcal{B}$, this diagram also commutes since both compositions are zero.

We package up the previous motivational discussion as follows:

Definition 12.23. *Suppose that* $j : \mathcal{B} \to \mathcal{A}$ *is a surjective, $*$-Hopf morphism and that* $\mathcal{B}$ *(resp., $\mathcal{A}$) has a left invariant FODC* Ψ *(resp., Γ). Then we say that this pair of FODCs is a* homogeneous FODC *on the associated quantum homogeneous bundle if*

$$j(\mathcal{R}') \subset \mathcal{R},$$

$$\Phi_j(\mathcal{R}') \subset \mathcal{R}' \otimes \mathcal{A}.$$

Exercise 12.33. *Prove that ρ and χ are $*$-morphisms.*

Proposition 12.20. *For all $\eta \in {}_{\mathrm{inv}}\Psi$ and $b \in \mathcal{B}$,*

$$\rho(\eta \circ b) = \rho(\eta) \circ j(b).$$

This says that ρ is a right $\mathcal{B}$-module morphism, where the right action of $\mathcal{B}$ on ${}_{\mathrm{inv}}\Gamma$ is the pullback of the right action of $\mathcal{A}$ on ${}_{\mathrm{inv}}\Gamma$ via j, namely,

$$\theta \circ b := \theta \circ j(b)$$

for $\theta \in {}_{\mathrm{inv}}\Gamma$ and $b \in \mathcal{B}$.
   Also, the following diagram commutes:

$$
\begin{array}{ccc}
{}_{\mathrm{inv}}\Psi & \xrightarrow{\ \chi\ } & {}_{\mathrm{inv}}\Psi \otimes \mathcal{A} \\
{\scriptstyle\rho}\downarrow & & \downarrow{\scriptstyle\rho \otimes id} \\
{}_{\mathrm{inv}}\Gamma & \xrightarrow{\ \mathrm{ad}\ } & {}_{\mathrm{inv}}\Gamma \otimes \mathcal{A}.
\end{array}
$$

This says ρ is covariant with respect to the two right co-actions χ and ad.

Proof. For the first statement, we write

$$\eta = \pi'(c)$$

for some $c \in \ker \varepsilon'$. Then using the definition of the action $\circ$, we compute

$$\eta \circ b = \pi'(c) \circ b = \pi'(cb - \varepsilon'(c)b) = \pi'(cb),$$

which immediately implies

$$\rho(\eta \circ b) = \rho(\pi'(cb)). \tag{12.58}$$

On the other hand, we can write $\rho(\eta)$ on general principles as

$$\rho(\eta) = \pi(a)$$

for some $a \in \ker \varepsilon$. But in this context we can actually take $a = j(c) \in \ker \varepsilon$, where $c \in \ker \varepsilon'$ is the element chosen in the first part of this argument. This is so because with this choice of $a \in \ker \varepsilon$, we have

$$\pi(a) = \pi(j(c)) = \rho(\pi'(c)) = \rho(\eta).$$

But then we calculate

$$\rho(\eta) \circ j(b) = \pi(a) \circ j(b)$$
$$= \pi(aj(b) - \varepsilon(a)j(b))$$
$$= \pi(aj(b))$$
$$= \pi(j(c)j(b))$$
$$= \pi(j(cb))$$
$$= \rho(\pi'(cb))$$
$$= \rho(\eta \circ b),$$

where we used standard identities, Eq. (12.58), and the fact that j is a multiplicative morphism. The proves the first statement of the proposition.

The proof of the next part of the proposition is a diagram chase. So we pick an element $\eta \in {}_{\text{inv}}\Psi$ in the upper left corner. As in the first part of this proof, we write $\eta = \pi'(c)$ for some $c \in \ker \varepsilon'$ and $\rho(\eta) = \pi(a)$ with $a = j(c) \in \ker \varepsilon$. Then chasing down and then across, we see that

$$\text{ad}(\rho(\eta)) = \text{ad}(\pi(a))$$
$$= (\pi \otimes id)\text{ad}(a)$$
$$= (\pi \otimes id)\text{ad}(j(c))$$
$$= (\pi \otimes id)(j \otimes j)\text{ad}'(c)$$
$$= (\pi j \otimes j)\text{ad}'(c)$$
$$= (\rho\pi' \otimes j)\text{ad}'(c).$$

The across-and-down chase yields

$$(\rho \otimes id)\chi(\eta) = (\rho \otimes id)\chi(\pi'(c))$$
$$= (\rho \otimes id)(\pi' \otimes j)\text{ad}'(c)$$
$$= (\rho\pi' \otimes j)\text{ad}'(c).$$

And this shows the commutativity of the diagram.   ∎

In the preceding diagram chase argument, we used (without mentioning it explicitly) a property of the right adjoint co-action of a Hopf algebra. This is something that we could have discussed as early as in Chapter 4, but somehow it was not that important until now. So here it is.

Exercise 12.34. *Suppose that* $j : B \to A$ *is a morphism of Hopf algebras. Let* ad_A
(resp., ad_B*) denote the right adjoint co-action of* A *(resp.,* B*) on itself. Then prove
that this diagram is commutative:*

$$
\begin{array}{ccc}
B & \xrightarrow{\ j\ } & A \\
{\scriptstyle \mathrm{ad}_B} \downarrow & & \downarrow {\scriptstyle \mathrm{ad}_A} \\
B \otimes B & \xrightarrow{\ j \otimes j\ } & A \otimes A.
\end{array}
\tag{12.59}
$$

The result of this exercise can be interpreted as saying that the right adjoint co-action
is a natural transformation. The details of this interpretation are left for the reader's
consideration.

For the interested reader, we note that HODCs appropriate for this example are
considered in [22].

Example 3: Quantum Hopf Bundle over Podles Sphere

This example is a special case of a quantum homogeneous bundle. So we have to
define the $*$-Hopf algebras A and B as well as the surjective morphism of $*$-Hopf
algebras

$$
j : B \to A.
$$

First, we let $B = SU_q(2)$, the $*$-Hopf algebra defined in Section 11.6. Recall that
this is simply $SL_q(2)$ with a specific choice of $*$-structure. In particular, $SU_q(2)$ is
generated as a $*$-algebra by two elements a and c subject to the relations given in
Section 11.6. The "group" structures are also given in that section though now they
will be denoted as ϕ', ε', and κ' in order to be consistent with the notation used
in the previous subsection. For the reader's convenience, we repeat these formulas
with the current notation:

$$
\varepsilon'(a) = 1,
$$
$$
\varepsilon'(c) = 0,
$$
$$
\phi'(a) = a \otimes a - qc^* \otimes c,
$$
$$
\phi'(c) = c \otimes a + a^* \otimes c,
$$
$$
\kappa'(a) = a^*,
$$
$$
\kappa'(a^*) = a,
$$
$$
\kappa'(c) = -qc,
$$
$$
\kappa'(c^*) = -q^{-1}c^*.
$$

As we have seen, the $*$-Hopf algebra $\mathcal{B} = SU_q(2)$ is not co-commutative and when $q \neq 1$, it is also not commutative. However, the "subgroup" $\mathcal{A}$ will be a commutative $*$-Hopf, which will represent the classical group $U(1) \cong S^1$, the unit circle in the complex plane $\mathbb{C}$, or, equivalently, the matrix algebra of 1×1 unitary matrices with complex entries. Classical coordinates on S^1 are given by the functions z and z^{-1}, which are conjugates of each other. Staying within the context of polynomial algebras and their quotients, we define the $*$-algebra

$$\mathcal{A} = \mathbb{C}[z, z^{-1}] := \mathbb{C}\{z, w\}/\langle zw = wz = 1\rangle$$

whose $*$-structure is defined by $z^* := z^{-1}$. The algebra $\mathcal{A} = \mathbb{C}[z, z^{-1}]$ is called the *Laurent polynomial algebra*. Here $\mathbb{C}\{z, w\}$ denotes the noncommutative free $\mathbb{C}$-algebra on the two generators $\{z, w\}$. Note that the algebra $\mathbb{C}[z, z^{-1}]$ is commutative.

Recall that the matrix

$$u = \begin{pmatrix} a & -qc^* \\ c & a^* \end{pmatrix}$$

is unitary. In the classical case $q = 1$, the group $U(1)$ sits inside $SU(2)$ as the diagonal matrices

$$D = \begin{pmatrix} z & 0 \\ 0 & z^* \end{pmatrix}$$

with $zz^* = 1$. So this motivates the definition of $j : \mathcal{B} \to \mathcal{A}$ dually in the noncommutative case to be given on the $*$-generators by

$$j(c) = 0 \quad \text{and} \quad j(a) = z.$$

Furthermore, j is supposed to be a morphism of $*$-Hopf algebras. If this is to be the case, then we necessarily have certain conditions imposed on the "group" structure maps of $\mathcal{A}$. For example, we must have that

$$\varepsilon(z) = \varepsilon(j(a)) = \varepsilon'(a) = 1$$

as well as

$$\phi(z) = \phi(j(a)) = (j \otimes j)\phi'(a) = (j \otimes j)(a \otimes a - qc^* \otimes c) = z \otimes z.$$

Next, we note that we must have

$$\kappa(z) = \kappa(j(a)) = j(\kappa'(a)) = j(a^*) = z^*.$$

Similarly, we must have $\kappa(z^*) = z$.

Now making a virtue of necessity, we take the above formula as definitions. The maps ε and ϕ must be $*$-morphisms, while κ is defined independently on z and z^*. And all's well that ends well, as the next exercise asserts.

Exercise 12.35. *With these definitions of ε, ϕ, and κ, the $*$-algebra $\mathcal{A}$ becomes a $*$-Hopf algebra and $j : \mathcal{B} \to \mathcal{A}$ is a morphism of $*$-Hopf algebras. Moreover, in this case κ is a $*$-morphism and satisfies $\kappa^2 = id_{\mathcal{A}}$.*

So by the general results of the previous section, $j : \mathcal{B} \to \mathcal{A}$ defines a quantum homogeneous bundle. It is known as the *quantum Hopf bundle*. For the special case $q = 1$, it can be identified with the classical Hopf bundle over the standard sphere S^2. For example, see [6]. But now we are interested in identifying the base "space" for general $q \in \mathbb{C} \setminus \{0\}$. One question is whether this is a classical space. As we know from the general theory of quantum homogeneous bundles, the base "space" is represented by the $*$-algebra

$$\mathcal{V} = \{v \in \mathcal{B} \mid {}_\mathcal{B}\Phi(v) = v \otimes 1_{\mathcal{A}}\},$$

where ${}_\mathcal{B}\Phi = (id \otimes j)\phi'$ and $\mathcal{B} = SU_q(2)$. Equivalently,

$$\mathcal{V} = \{b \in \mathcal{K} \mid \phi'(b) \in \mathcal{B} \otimes \mathcal{K}\},$$

where $\mathcal{K} := \ker(j - \varepsilon'(\cdot)1_{\mathcal{A}})$.

Proposition 12.21. *The space of left invariant elements $\mathcal{V}$ is generated as a $*$-algebra by the elements $\{aa^*, ac^*, ca^*\}$.*

Proof. We start with these two simple calculations using ${}_\mathcal{B}\Phi = (id \otimes j)\phi'$:

$$ {}_\mathcal{B}\Phi(a) = a \otimes z, $$
$$ {}_\mathcal{B}\Phi(c) = c \otimes z. $$

These immediately conjugate to give

$$ {}_\mathcal{B}\Phi(a^*) = a^* \otimes z^*, $$
$$ {}_\mathcal{B}\Phi(c^*) = c^* \otimes z^*. $$

Since ${}_\mathcal{B}\Phi$ is a multiplicative morphism, one then sees that

$$ {}_\mathcal{B}\Phi(aa^*) = aa^* \otimes 1, $$
$$ {}_\mathcal{B}\Phi(ac^*) = ac^* \otimes 1, $$
$$ {}_\mathcal{B}\Phi(ca^*) = ca^* \otimes 1. $$

This shows that $\{aa^*, ac^*, ca^*\} \subset \mathcal{V}$, and so the $*$-algebra generated by $\{aa^*, ac^*, ca^*\}$ is a $*$-subalgebra of $\mathcal{V}$.

Finally, one must show the opposite inclusion, that is, that every element $v \in \mathcal{V}$ lies in the $*$-algebra generated by $\{aa^*, ac^*, ca^*\}$. We leave this part of the proof to the interested reader. ∎

Theorem 12.17. *The base "space" $\mathcal{V}$ of this QPB is a quantum sphere in the sense of Podles as defined in [66].*

Proof. There are many equivalent ways of defining the quantum spheres as $*$-algebras in terms of generators and relation. In this regard see [63] as well as [66]. However, for this proof we will use the generators and relations given in [6]. These generators are

$$b_+ = ca^*,$$
$$b_- = -qac^*,$$
$$b_3 = aa^*.$$

Then one can check that these elements satisfy the relations as given in [6] just before Proposition 5.11. This is a bit of a drudge and will be left for the reader to consider in detail. ∎

One can introduce various different differential calculi on this quantum principal bundle. There are such calculi that extend both the 3D FODC and the 4D FODC on $SU_q(2)$. These calculi lead to extensive calculations, especially in the case of the 4D FODC. Details can be found in [22].

12.12   Notes

This particular topic in noncommutative geometry is rather new and most probably not developed yet fully to a mature form. Certainly, there are several alternative approaches being actively studied at present and while they have elements in common, it is also true that there are incompatibilities among these approaches. None of these approaches is any more "right" than it is "wrong." It is a question of delineating defining properties that lead to interesting and applicable theories. This process is still continuing with regards to the very definition of *quantum group* itself, which is just a part of what we are dealing with in the theory of QPBs. So a more detailed exposition of the approaches not presented here, and their relation to what is presented here, is clearly called for.

To start with, the very definition of QPB varies. Our basic reference for this is [22], though many of the concepts already appear in the special case studied in [21]. However, the definition of QPB presented here is not exactly that given in [22], where the setting is that of compact quantum groups as defined in [84]. In that case, condition (qpb5) is automatically satisfied (see [24]) and so is not presented explicitly as part of the definition in [22]. Here, instead of a compact quantum group, we use any Hopf algebra with invertible antipode, and we do not require the Hopf–Galois condition (qpb5). For some researchers working in a setting more general than compact quantum groups, the Hopf–Galois condition is imposed as part of the definition of a QPB. For example, see [37]. In that approach a lot of the theory is developed from this condition, and so the material in Section 12.2 is quite relevant. This makes for a theory that is complementary to that presented here. Other papers, such as [3, 4, 6] and [7], have other definitions of a QPB, which we will discuss below.

Whether or not the Hopf–Galois condition (qpb5) should be included in the definition of a QPB is an issue with advocates on both sides. However, the dominant opinion at present is that it should be included though we do not need it to do a lot of the basic theory, as our exposition shows. Neither option is any more right than it is wrong, and only time will tell how this will play out. It is mostly a question of convenience as far as I am concerned, but others take the issue more to heart. An analogy—perhaps inappropriate—is whether one should include the concept of connection in the definition of a principal bundle. In the classical case, this issue has long since been decided in favor of leaving a connection as an auxiliary structure that a principal bundle may be equipped with. In the quantum case, as far as I am aware, all authors have chosen to follow this tradition set in the classical case. But some researchers might consider these choices to be misguided since from their point of view the only reason to study a principal bundle is because it is the scenario for introducing a connection. Other researchers wish to study principal bundles on their own without considering their connections. And it is this second group that has won the day as far as the basic definitions go. But both groups proceed unheeded in their research.

An alternative approach to QPBs is to define them as principal co-module algebras over a Hopf algebra with bijective antipode. This definition includes the Hopf–Galois condition as well as a technical splitting condition. See [9, 10, 38, 52] and Definition 4.1 in [4]. This may well be the best definition currently in use as far as some experts are concerned. For a first reading of this material I find it to be way over the top. But my kind readers are invited to consult the above references and see what they think of it for themselves.

One philosophy is to define QPBs in terms of something corresponding to the local trivializations of the base space used in the classical case. For example, this is a main guiding idea behind the approaches in [21] and [65]. To be perfectly honest, these approaches seem to me to nullify the basic idea behind noncommutative geometry itself, namely, to implement geometric concepts in contexts where points in spaces are rendered irrelevant to the theory. So I have avoided this sort of approach altogether. In part, this is also to keep the presentation limited to a reasonable length. However, from a pedagogical point of view this could be a mistake, since the approach using local trivializations can provide a stepping stone to the more abstract theory presented here. Nonetheless, for me locality is a way of hanging on to the concept of nearby points, no matter how carefully this might be disguised in abstract terms, say, of general topology, sheaves, and so on. A completely opposing viewpoint holds that locality is a general principle of physics, and especially of quantum field theory, and so should be reflected somehow in the structure of noncommutative geometry. See [65] in this regard. But infinitesimal structures, strangely enough, do make sense in the noncommutative context without implicitly contradicting its basic idea. This is a miracle, as the saying goes. However, it is a limited miracle, since even though differential calculi do exist, they are not unique! So Leibniz would be pleased, I guess, to see infinitesimals being taken seriously but might not be so happy to find out that they are not given to us by nature.

As for Section 12.4 on the differential calculus of QPBs, our reference is [22]. Again, let me emphasize the central importance of having a theory of differential calculus adapted to QPBs. In my opinion, the mere definition of a QPB—whichever variant one wishes to use—is inadequate for developing an interesting theory with mathematical as well as physical content. The differential calculus links geometry, topology, analysis, and algebra to the study of noncommutative spaces, making for a theory with a lot of potential (as well as actual!) applications. A *leit motif* of this book, as the reader already knows, is the importance of differential calculi.

If we adapt the philosophy that a QPB should satisfy the condition (qpb5), then we should also impose a corresponding condition on any differential calculus associated to it. Hence, we say in that case that the differential calculus on a QPB is *compatible with translations* provided that the following holds:

- *(diff3) There is a given isomorphism* $\Omega(P) \otimes_{\Omega(M)} \Omega(P) \xrightarrow{\cong} \Omega(P) \otimes \Gamma^\wedge$ *extending the isomorphism X associated with the translation map τ.*

Recall that $\Omega(M) = \{\omega \in \Omega(P) \mid {}_{\Omega(P)}\Psi(\omega) = \omega \otimes 1\}$. However, we have left such a fancier, more advanced theory to a side in this decidedly introductory book.

A useful reference for understanding how to construct differential calculi for QPBs is [23]. In that paper Đurđevich starts with a graded $*$-algebra, denoted by $\mathfrak{hor}_P$, and a $*$-homomorphism $F_\wedge : \mathfrak{hor}_P \to \mathfrak{hor}_P \otimes \mathcal{A}$. Then, for certain bicovariant, $*$-covariant HODCs over $\mathcal{A}$ (including the braided exterior algebra and the universal algebra of an FODC), a QPB with differential calculus is explicitly constructed such that its horizontal algebra is $\mathfrak{hor}_P$ and such that the co-action of the HODC on the QPB projects to $F_\wedge$. This construction is carried out in general in the case of compact quantum groups. And a modification of this construction is also presented that works for arbitrary (noncompact) quantum groups. Using this, several explicit examples of QPBs with differential calculus are then constructed in [23].

Also, the definitions of horizontal and vertical forms used here are not universal. Our reference in this matter has been [22]. While these concepts are quite transparent in the classical case, there are different, nonequivalent definitions in use in the quantum case. For example, the authors of [6] have a different definition of horizontal, which turns out to be a subspace of what we denote as $\mathfrak{hor}(P)$. In their approach the sequence corresponding to that in Theorem 12.7 is not necessarily exact, and so they add a condition to their definition of a QPB which requires that sequence to be exact. In the present approach, this exactness is a theorem. Further discussion of this can be found in [22], Section 6.3.

Remarkably enough, it seems that all of the definitions of a quantum connection in the literature essentially agree. However, they are usually called connections without any adjective, a minor point, and often are not required to be $*$-morphisms, since the setting has no $*$-operation. Nonetheless, some researchers have focused on a particular class of connections, called strong connections, for particular study. See [37] in this regard. Since our reference for quantum connection is [22], we have presented the classes of connections given there, namely, regular and multiplicative connections. So there are different approaches used in the study of connections. A curious side note here is that a co-module algebra is principal if and only if it has at least one strong connection on it. See [9] and references given there for the proof. Recall that a principal co-module algebra is regarded, according to some experts, as being the correct formulation of the concept of a QPB.

The survey paper [3] gives an extensive discussion of how ideas from noncommutative geometry are manifest is classical principal bundles and their associated bundles. Their definition of a principal bundle as a certain type of monoidal functor might not sit well with the beginner. However, abstract approaches often have their merits, though usually on a second pass over the material. In any event, I have chosen what I feel to be a more accessible approach to QPBs for someone with only some familiarity with the classical case and who is viewing this noncommutative material for the first time. The ideas of QPBs that are presented in [3] come from the works of the various authors of that paper and of their collaborators without making any reference at all to the works of Đurđevich, some of which appeared 10 years prior to [3]. As such, it is a valuable, though limited, discussion of the alternative approaches to these topics.

I have included the topics of curvature and covariant derivative to give more substance to the theory of quantum connections on QPBs. Again, our reference is [22]. However, the definitions of curvature and covariant derivative used here are not the same as in other references, such as [6]. So the theory presented here follows very much Đurđevich's approach. The application of these structures to the study of Dunkl operators in Chapter 14 showcases the utility of this approach in an area of mathematics that had nothing to do with noncommutative geometry before the appearance of [31].

Another rather nice topic concerning QPBs is the definition and study of characteristic classes. This is presented in [22] in terms analogous to the classical Weil theory and (nonequivalently in general) in [25] and [27] by a construction corresponding to the classifying space approach in the classical case. Also, see [3] for their take on this interesting topic.

The example of the trivial bundle with the trivial differential calculi comes from [22], where even more details have been worked out. The example of quantum homogeneous spaces is also based directly on [22]. A similar approach to quantum homogeneous spaces is presented in [6].

Moreover, quantum homogeneous spaces are also covered in [48], though with a weaker definition of quantum homogeneous space and no mention at all of quantum bundles. So in [48] the authors take the point of view that quantum homogeneous spaces are objects of their own intrinsic interest without regard to their possible relation to quantum bundle theory. Not only is the definition in [48] logically weaker than that used here, it is actually *strictly* weaker in the sense that the class of quantum homogeneous spaces according to [48] is strictly larger than the class of quantum homogeneous spaces as defined here. In particular, certain q-deformations of symmetric spaces are quantum homogeneous spaces in the definition used in [48], while they are not in the definition used here. See [48] for more details on this. What one makes of this situation could depend on various criteria that are not within the realm of mathematics itself. Of course, if I weaken someone's definition of certain objects, I will get a (in general, strictly) larger class of objects satisfying the new definition. Similarly, if I strengthen someone's definition of certain objects, I will get a (in general, strictly) smaller class of objects. The tricky bit is when I change someone's definition of certain objects in a way that is neither weaker nor stronger. Whether a mathematical definition is "good" or "right" is just not a question that can be answered always within the scope of mathematical analysis itself. The use of q-deformations as a "litmus test" in this situation has value to the extent that one credits q-deformations as being important objects, either in mathematics or in related areas. And it should be noted that [48] is heavily oriented toward q-mathematics, and so the reader will find many very specific examples of homogeneous spaces given there in terms of q-deformations of classical spaces. More on q-deformations in a moment.

But first, here is another way to think about this situation. There is a very profound theorem is classical differential geometry that roughly says that every homogeneous space arises as the base space of a principal fiber bundle. When I first learned this theorem, I thought it to be quite amazing, quite unexpected. After all, a homogeneous space is basically an algebraic object, a set of cosets, while a locally trivial principal bundle is a topological and geometric object. Why should there be such a close relation between these two types of objects? My attitude today is pretty much the same. But it is well known that noncommutative geometry differs in dramatic ways from its classical origins. Maybe this is another case in point. More specifically, in noncommutative geometry we may have the possibility that there are quantum homogeneous spaces that do not arise from QPBs. In other words, the amazing classical theorem might just not go over to the noncommutative setting. In this point of view, we would accept the definition of quantum homogeneous space as given in [48] and rephrase the results given here as being about those particular quantum homogeneous spaces (those associated with "subgroups") that do arise from QPBs. And then we would be left to wonder how the category of smooth manifolds turns out to be the home for such a profound and wonderful theorem, as noted above.

Another large class of examples of homogeneous spaces can be found in [21], where the base space is a classical differential manifold and the QPB has a structure "consistent" (in a sense precisely described in [21]) with the local trivializations of the base space. As indicated earlier, I have decided not to discuss formalisms that invoke locality in such a direct manner.

Let's note that one moral of the story told in Theorem 12.16 is that there are lots and lots of examples of quantum homogeneous bundles—and thus of quantum principal bundles. This actually is an embarrassment of riches. There are far and away more examples than one might know what to do with! (However, it is well known that there are fewer subgroups of a given quantum group than there are subgroups of a given group. The reader is invited to investigate what this vague statement actually means.) One needs to understand various more specific examples within the rubric of this quite general example. This is often done by studying q-deformations, which more than anything is a general, almost philosophic approach that involves taking some well-known structure in standard mathematics and then introducing a "parameter" q to get a new "deformed" structure. There are two conditions usually imposed on a q-deformation. The first is that for some particular value of q (typically $q = 1$), the q-deformed theory reduces to the original, motivating theory from standard mathematics. While this condition is quite precise in its formulation, it says nothing at all about the properties of the q-deformation for other values of q (typically $q \neq 1$). The second condition speaks to this issue; it requires that the q-deformed theory maintain the "basic" properties of the original, motivating theory from standard mathematics. However, this condition is vague since it is not clear what the general criteria are for a property to be "basic." For example, commutativity in the original theory is typically lost in its q-deformed version. This vagueness might be inherent. For example, as far as I know, there seems to be no *q-deformation* algorithm nor functor.

The example of the quantum Hopf bundle is, of course, both a particular case of a quantum homogeneous bundle and a q-deformation. It is presented in even greater detail in [6], where it is shown how the standard Hopf bundle over the standard sphere S^2 arises as the special case $q = 1$. This initially might make one feel rather confident that we are seeing a rather important generalization of the classical, standard Hopf bundle. In other words, the good news is that the quantum Hopf bundle is related in a rather pretty way to the classical case. But another way to view this situation is that the quantum Hopf bundle is too closely related to the classical case and so does not reflect the full power of the noncommutative theory. My point is that while q-deformations do have their charm in pure mathematics, it may well be the case that the more important applications of noncommutative structures exist on new "continents" that are not so directly linked to the "continent" of classical mathematics via a parameter. I am thinking of applications in physics in particular. This purely speculative comment must face the test of difficult calculations in specific examples before being credited as anything other than a call for more research.

The Holy Grail behind all of these theories of QPBs is the construction of physically relevant quantum gauge theories. A mathematically rigorous formulation of quantum electrodynamics (QED), a massless quantum gauge theory based on the abelian Lie group S^1 (the unit circle in the complex plane), remains a challenge yet to be realized. However, see the book [35] by Folland, where the rigorous free quantum field theory is presented and then used to motivate the nonrigorous physics theory of QED. And more ambitious researchers wish to formalize in rigorous mathematics a massive $SU(2)$ quantum gauge theory in dimension $3 + 1$. Of course, the group $SU(2)$ (of 2×2 unitary matrices with complex entries and determinant 1) is a nonabelian Lie group. Minimally, "quantum" in this context means a Hilbert space–based theory with Planck's constant $\hbar > 0$ appearing as a parameter. Moreover, that theory should have a reasonable semiclassical behavior as $\hbar \to 0^+$. Some references for this difficult, important problem are [6, 13, 18], and [29]. Due to the very preliminary state of the science of these topics, we do not discuss them further.

It is also safe to say that this lengthy chapter is a lot shorter than it could have been, but it is only meant as an introduction to the literature.

Chapter 13
Finite Classical Groups

This chapter is devoted to the simplest class of examples of this theory. Nonetheless, many concepts are clearly illustrated and so should become more intuitive for the reader. Despite the simplicity of this material, it also serves as motivation for the general theory.

13.1   The Associated Hopf Algebra

Let G be a finite group. The only Hausdorff topology on G is the discrete topology; that is, every subset of G is open. Since any cover of G is finite, G is compact. So the Gel'fand–Naimark theory for compact, Hausdorff spaces applies to the topological space G with the discrete topology.

According to this theory, the study of the space G is equivalent to the study of its dual C^*-algebra of continuous functions $G \to \mathbb{C}$. Since *any* function $G \to \mathbb{C}$ is automatically continuous when G has the discrete topology, we have that this C^*-algebra is

$$\mathcal{A} = \{\alpha \,|\, \alpha : G \to \mathbb{C} \text{ is any function}\}.$$

The product of functions according to the Gel'fand–Naimark theory is the usual pointwise product defined by

$$\alpha\beta(g) := \alpha(g)\beta(g),$$

where $\alpha, \beta \in \mathcal{A}$ and $g \in G$. This is a commutative multiplication.

For every $g \in G$, the characteristic function of the subset $\{g\}$ of G is in $\mathcal{A}$. This function is denoted as δ_g since it is also the Kronecker delta function associated with the point $g \in G$. In short, by definition, $\delta_g(g) := 1$ and $\delta_g(h) := 0$ for $h \in G$ with $h \neq g$. It is easy to show that $\{\delta_g \,|\, g \in G\}$ is a linearly independent set over $\mathbb{C}$ that

© Springer International Publishing Switzerland 2015

S.B. Sontz, *Principal Bundles*, Universitext, DOI 10.1007/978-3-319-15829-7_13

spans $\mathcal{A}$. So it is a basis of $\mathcal{A}$, which is therefore a finite-dimensional vector space over $\mathbb{C}$ with dimension equal to the number of elements in G. Some properties of these elements of $\mathcal{A}$ are

$$\sum_{g \in G} \delta_g = 1, \quad (\delta_g)^* = \delta_g, \quad (\delta_g)^2 = \delta_g, \quad \delta_g \delta_h = 0 \text{ for } g \neq h.$$

This is a simple, commutative case of a resolution of the identity by self-adjoint, pairwise disjoint projections.

Now the fact that G carries a group structure allows us to define more structures on the algebra $\mathcal{A}$. These are dual operations to the group multiplication $m : G \times G \to G$ (a binary operation) given by $m(g, h) = gh$, the group inversion $\iota : G \to G$ (a unary operation) given by $\iota(g) = g^{-1}$, and the group identity element $e \in G$ or equivalently as the map from a one-point space $i : \{*\} \to G$ (a 0-ary operation) given by $i(*) = e$. By pullback these three operations give operations on $\mathcal{A}$. There are the co-multiplication $\phi := m^*$, the co-inverse (or antipode) $\kappa := \iota^*$, and the co-unit $\varepsilon := i^*$. In the following, the symbol $\circ$ denotes composition of functions. So the co-multiplication

$$\phi = m^* : \mathcal{A} = C(G) \to C(G \times G) \cong C(G) \otimes C(G) = \mathcal{A} \otimes \mathcal{A}$$

is simply $\phi(\alpha) = m^*(\alpha) = \alpha \circ m$. Also, the co-inverse

$$\kappa := \iota^* : \mathcal{A} \to \mathcal{A}$$

is given by $\kappa(\alpha) = \iota^*(\alpha) = \alpha \circ \iota$. For example, on the standard basis elements, this gives

$$\kappa(\delta_g) = \delta_g \circ \iota = \delta_{g^{-1}}.$$

The last equality here is a simple exercise. Finally, the co-unit

$$\varepsilon = i^* : \mathcal{A} \to C(\{*\}) \cong \mathbb{C}$$

is given by

$$\varepsilon(\alpha) = i^*(\alpha) = \alpha \circ i = \alpha(e).$$

The last equality is a consequence of the identification of elements in $C(\{*\})$ with elements in $\mathbb{C}$ by using the correspondence that sends any function $\beta \in C(\{*\})$ to $\beta(*) \in \mathbb{C}$.

It only remains to find a formula for ϕ on a typical standard basis element δ_g. Now, $\phi(\delta_g) = m^*(\delta_g) = \delta_g \circ m$. So we evaluate for $(h, k) \in G \times G$

$$\delta_g \circ m(h, k) = \delta_g(hk),$$

which is a function that takes only two values, namely, 0 and 1. So it is the characteristic function of some subset of $G \times G$. But that subset is clearly $\{(h, k) \in G \times G | hk = g\}$. Thus, letting $\delta_{h,k}$ denote the characteristic function of the point (h, k), we have

$$\phi(\delta_g) = \sum_{h,k \,|\, hk=g} \delta_{h,k} = \sum_{h,k \,|\, hk=g} \delta_h \times \delta_k = \sum_{h \in G} \delta_h \times \delta_{h^{-1}g} = \sum_{h \in G} \delta_h \otimes \delta_{h^{-1}g},$$

where the last equality comes from the identification

$$C(G \times G) \cong C(G) \otimes C(G) = \mathcal{A} \otimes \mathcal{A}.$$

There is also the adjoint map adj : $G \times G \to G$ given for $(x, h) \in G \times G$ by

$$\mathrm{adj}(x, h) := h^{-1}xh.$$

We also use the notation $x \cdot h = \mathrm{adj}(x, h)$. This is a right action of G on itself, since $x \cdot e = x$ and

$$(x \cdot h_1) \cdot h_2 = (h_1^{-1}xh_1) \cdot h_2 = h_2^{-1}h_1^{-1}xh_1h_2 = x \cdot (h_1h_2).$$

It induces by pullback $\mathrm{ad} := \mathrm{adj}^* : \mathcal{A} \to \mathcal{A} \otimes \mathcal{A}$, which is also called the *adjoint map*. By duality this is a right co-action of $\mathcal{A}$ on itself. This is expressed by the commutativity of the following two diagrams:

$$
\begin{array}{ccc}
\mathcal{A} & \xrightarrow{\ \mathrm{ad}\ } & \mathcal{A} \otimes \mathcal{A} \\
\mathrm{ad} \downarrow & & \downarrow \mathrm{ad} \otimes id \\
\mathcal{A} \otimes \mathcal{A} & \xrightarrow{\ id \otimes \phi\ } & \mathcal{A} \otimes \mathcal{A} \otimes \mathcal{A}
\end{array}
\qquad (13.1)
$$

and

$$
\begin{array}{ccc}
\mathcal{A} & \xrightarrow{\ \mathrm{ad}\ } & \mathcal{A} \otimes \mathcal{A} \\
id_{\mathcal{A}} \downarrow & & \downarrow id \otimes \varepsilon \\
\mathcal{A} & \xrightarrow{\ \cong\ } & \mathcal{A} \otimes \mathbb{C}.
\end{array}
\qquad (13.2)
$$

These diagrams correspond to the diagrams (3.7) and (3.8) in the definition of a right co-action. This will also be shown ahead with diagram chases, using the explicit formula for ad derived next.

We evaluate this on a standard basis element, getting

$$\mathrm{ad}(\delta_g) = \mathrm{adj}^*(\delta_g) = \delta_g \circ \mathrm{adj}$$

so that for $x, h \in G$, we have

$$\text{ad}(\delta_g)(x, h) = \delta_g \circ \text{adj}(x, h) = \delta_g(h^{-1}xh).$$

Again, this is a function that only takes the values 0 and 1 and so is the characteristic function of some subset of $G \times G$. It assumes the value 1 if and only if $g = h^{-1}xh$ or, equivalently, $x = hgh^{-1}$. But

$$\left(\sum_k \delta_{kg^{-1}k} \times \delta_k \right)(x, h) = 1$$

if and only if $hg^{-1}h = x$. Identifying $C(G \times G)$ with $C(G) \otimes C(G)$, we end up with

$$\text{ad}(\delta_g) = \sum_{k \in G} \delta_{kgk^{-1}} \otimes \delta_k. \tag{13.3}$$

Exercise 13.1. *Use the fact that* adj *is a composite of functions whose pullbacks are already known in order to prove the previous formula (13.3).*

Even though the diagrams (13.1) and (13.2) clearly commute (being the duals of commuting diagrams), let's explicitly show their commutativity by using (13.3) in diagram chases. For (13.1) we see that going across and down gives

$$(ad \otimes id)ad(\delta_g) = (ad \otimes id)\left(\sum_h \delta_{hgh^{-1}} \otimes \delta_h \right)$$

$$= \sum_h ad(\delta_{hgh^{-1}}) \otimes \delta_h$$

$$= \sum_h \left(\sum_k \delta_{k(hgh^{-1})k^{-1}} \otimes \delta_k \right) \otimes \delta_h$$

$$= \sum_{h,k} \delta_{khgh^{-1}k^{-1}} \otimes \delta_k \otimes \delta_h,$$

while going down and across gives

$$(id \otimes \phi)ad(\delta_g) = (id \otimes \phi)\left(\sum_l \delta_{lgl^{-1}} \otimes \delta_l \right)$$

$$= \sum_l \delta_{lgl^{-1}} \otimes \phi(\delta_l)$$

$$= \sum_l \delta_{lgl^{-1}} \otimes \sum_h \delta_{lh^{-1}} \otimes \delta_h$$

$$= \sum_{h,k} \delta_{khg(kh)^{-1}} \otimes \delta_k \otimes \delta_h$$

using the change of variables $k = lh^{-1}$ (and so, $l = kh$) in the last equality. This establishes the commutativity of (13.1). For (13.2), we note that

$$(id \otimes \varepsilon)ad(\delta_g) = (id \otimes \varepsilon)\Big(\sum_h \delta_{hgh^{-1}} \otimes \delta_h\Big)$$

$$= \sum_h \delta_{hgh^{-1}} \otimes \varepsilon(\delta_h)$$

$$= \sum_h \delta_{hgh^{-1}} \otimes \delta_h(e)$$

$$= \delta_g \otimes 1$$

since $\delta_h(e) = 0$ if $h \neq e$, while $\delta_e(e) = 1$. And this establishes the desired commutativity of (13.2).

The formula (13.3) for ad agrees with (4.1) (see also [86]), which says

$$ad(a) = a^{(2)} \otimes \kappa(a^{(1)})a^{(3)}$$

using Sweedler's notation. To see that this is the case, we calculate (in ordinary notation)

$$(id \otimes \phi)\phi(\delta_g) = (id \otimes \phi) \sum_h \delta_{gh^{-1}} \otimes \delta_h$$

$$= \sum_h \delta_{gh^{-1}} \otimes \phi(\delta_h)$$

$$= \sum_h \delta_{gh^{-1}} \otimes \sum_k (\delta_{hk^{-1}} \otimes \delta_k)$$

$$= \sum_{h,k} \delta_{gh^{-1}} \otimes \delta_{hk^{-1}} \otimes \delta_k.$$

Matching this up with Sweedler's notation, we can write this as

$$\delta_g^{(1)} = \delta_{gh^{-1}} \qquad \delta_g^{(2)} = \delta_{hk^{-1}} \qquad \delta_g^{(3)} = \delta_k. \qquad (13.4)$$

Hence, for $a = \delta_g$, we have

$$a^{(2)} \otimes \kappa(a^{(1)})a^{(3)} = \sum_{h,k} \delta_{hk^{-1}} \otimes \kappa(\delta_{gh^{-1}})\delta_k = \sum_{h,k} \delta_{hk^{-1}} \otimes \delta_{hg^{-1}}\delta_k.$$

The last term is nonzero exactly when $hg^{-1} = k$ or, equivalently, $h = kg$. So summing over h, we obtain

$$a^{(2)} \otimes \kappa(a^{(1)})a^{(3)} = \sum_k \delta_{kgk^{-1}} \otimes \delta_k = ad(\delta_g),$$

which proves that (4.1) agrees with the definition of ad we are using here.

Exercise 13.2. *This exercise is meant to be a test of your level of confidence with Sweedler's notation. Expand $(\phi \otimes id)\phi(\delta_g)$ in Sweedler's notation to get expressions that are not identical to the three expressions in (13.4). Explain carefully why this is not a contradiction.*

Let's pause for a moment and notice that we have shown most of the details for the next exercise.

Exercise 13.3. *Let G be a finite group. Then $C(G)$ is a Hopf algebra with a commutative multiplication and the co-inverse $\kappa : C(G) \to C(G)$ is an invertible map with $\kappa^{-1} = \kappa$.*

And another exercise:

Exercise 13.4. *Show the co-multiplication in $\mathcal{A} = C(G)$ is co-commutative if and only if G is an abelian group.*

13.2 Compact Quantum Group

Next, we would like to consider whether this construction produces a compact quantum group (pseudogroup) in the sense of Woronowicz (see [84]). That definition has several virtues to recommend it. For example, such a quantum group always has a unique Haar state and a well-developed co-representation theory (see [84]). Here is that definition.

Definition 13.1 (See [84].). *We define a* compact matrix quantum group (pseudogroup) *to be a pair (A, u), where A is a C^*-algebra with unit and $u = (u_{kl})$ is an $N \times N$ matrix, such that the following holds:*

1. *$u_{kl} \in A$ for all k, l.*
2. *The $*$-subalgebra $\mathcal{A}$ of A generated by the elements u_{kl} is dense in A.*
3. *There exists a C^*-morphism $\phi : A \to A \otimes A$ such that*

$$\phi(u_{kl}) = \sum_{j=1}^{N} u_{kj} \otimes u_{jl}$$

for all k, l.
4. *There exists a linear antimultiplicative map $\kappa : \mathcal{A} \to \mathcal{A}$ satisfying $\kappa(\kappa(a^*)^*) = a$ for $a \in \mathcal{A}$ and*

$$\sum_{j=1}^{N} \kappa(u_{kj})u_{jl} = \delta_{k,l}1 \tag{13.5}$$

$$\sum_{j=1}^{N} u_{kj}\kappa(u_{jl}) = \delta_{k,l}1 \qquad (13.6)$$

for all k,l, where $\delta_{k,l}$ is the Kronecker delta.

For a brief introduction to C^*-algebras, see Appendix 12.1 in [80]. Or consult the rather complete exposition in [5].

We now consider the Hopf algebra $\mathcal{A} = C(G)$ as introduced above and ask whether this is a compact matrix quantum group (pseudogroup) as just defined. To do this, we need to find an integer N and an $N \times N$ matrix u with entries in $\mathcal{A}$ with the properties in this definition. We first note that since $\mathcal{A}$ is finite dimensional, it can only be a dense subalgebra in $A = \mathcal{A}$ itself. It is well known that $A = \mathcal{A} = C(G)$ is a commutative C^*-algebra with multiplication defined pointwise and with norm defined by $||f|| := \max_{g \in G} |f(g)|$. Its conjugation is defined by $f^*(g) := (f(g))^*$ for all $g \in G$. Moreover, A has a unit element, namely, the constant function 1.

We will take N to be the number of elements in G and will use pairs of elements in G to index the elements of the matrix u. We define this matrix $u = (u_{g,h})_{g,h \in G}$ by

$$u_{g,h} := \delta_{gh^{-1}} \in \mathcal{A}.$$

So property 1 of Definition 13.1 is satisfied. Taking $h = e$, we see that these matrix elements give all of the elements in the standard basis of $\mathcal{A}$, and so the subalgebra generated by all of these matrix elements is $\mathcal{A}$. This proves property 2 of Definition 13.1.

We already have a map named ϕ that sends $\mathcal{A}$ to $\mathcal{A} \otimes \mathcal{A}$. We claim this is the very same ϕ as appears in property 3. So we evaluate

$$\phi(u_{g,h}) = \phi(\delta_{gh^{-1}}) = \sum_{k \in G} \delta_k \otimes \delta_{k^{-1}gh^{-1}} = \sum_{r \in G} \delta_{gr^{-1}} \otimes \delta_{rh^{-1}} = \sum_{r \in G} u_{g,r} \otimes u_{r,h},$$

where we used the substitution $r = k^{-1}g$ or, equivalently, $k = gr^{-1}$. This proves property 3 of Definition 13.1. Next, we use the map κ already defined to evaluate

$$\kappa(\kappa(\delta_g^*)^*) = \kappa(\kappa(\delta_g)^*) = \kappa((\delta_{g^{-1}})^*) = \kappa(\delta_{g^{-1}}) = \delta_g = \delta_g 1.$$

This is part of property 4 of Definition 13.1. Next, for $k,l \in G$, we consider

$$\sum_{j \in G} \kappa(u_{k,j})u_{j,l} = \sum_{j \in G} \kappa(\delta_{kj^{-1}})\delta_{jl^{-1}} = \sum_{j \in G} \delta_{jk^{-1}}\delta_{jl^{-1}} = \delta_{k,l} = \delta_{k,l}1$$

since the term $\delta_{jk^{-1}}\delta_{jl^{-1}}$ is $\delta_{jk^{-1}}$ if $k = l$ and is 0 otherwise. Here we are using $\sum_{j \in G} \delta_{jk^{-1}} = 1$, the resolution of the identity. Again, $\delta_{k,l}$ is the Kronecker delta symbol for the pair of elements $k,l \in G$.

Similarly, we have

$$\sum_{j \in G} u_{k,j} \kappa(u_{j,l}) = \sum_{j \in G} \delta_{kj^{-1}} \kappa(\delta_{jl^{-1}}) = \sum_{j \in G} \delta_{kj^{-1}} \delta_{lj^{-1}} = \delta_{k,l} 1.$$

So the second part of property 4 of Definition 13.1 also holds. We leave one final detail to the reader.

Exercise 13.5. *Show that κ is a linear antimultiplicative map.*

Thus, $(\mathcal{A}, u)$ is a compact quantum group in the sense of Woronowicz.

13.3 Gel'fand–Naimark Theory

We now go into a few details of the Gel'fand–Naimark theory in this case. Strictly speaking, it is not necessary to introduce this theory here since everything can be done without using the language of Gel'fand–Naimark. However, for those readers who have heard a bit of that language, this is an opportunity to see it in use in an admittedly simple case. And for the rest of my readers, this is an opportunity to learn their first words of a new language.

We first note that we use the terminology (resp. *left, right, two-sided) ideal* of any algebra $\mathcal{A}$ to mean a subset $\mathcal{I}$ of $\mathcal{A}$ that is a (resp. left, right, two-sided) ideal in the usual sense of ring theory (that is, $a \in \mathcal{A}$ and $f, g \in \mathcal{I}$ imply $f + g \in \mathcal{I}$ and $af \in \mathcal{I}$ for left ideals, say). If $\mathcal{A}$ also has an identity element (as is the case throughout this book), then it automatically follows that $\mathcal{I}$ is also a vector subspace of $\mathcal{A}$ (by taking $a = \lambda 1$ in the previous parenthetical remark, where $\lambda \in \mathbb{C}$).

The key remark is that there is a bijective correspondence between the two-sided ideals in $\mathcal{A} = C(G)$ and the subsets of G. Note that every one-sided ideal (left or right) in $\mathcal{A}$ is automatically two-sided since $\mathcal{A}$ is commutative. Also, every subset of G is closed since G has the discrete topology. Hence, every subset of G is also compact since G is compact, being finite. For any $K \subset G$, we define à la Gel'fand–Naimark

$$\mathcal{I}_K := \{\alpha \in \mathcal{A} \,|\, \alpha(g) = 0, \forall g \in K\}.$$

Then $\mathcal{I}_K$ is an ideal in $\mathcal{A}$ called the *ideal dual to* K. Similarly, given any ideal $\mathcal{I} \subset \mathcal{A}$, we define

$$K_{\mathcal{I}} := \{g \in G \,|\, \alpha(g) = 0, \forall \alpha \in \mathcal{I}\},$$

which is a subset of G called the *subset dual to* $\mathcal{I}$.

Exercise 13.6. *These are bijective, order-reversing maps, each the inverse of each other, between the set of all ideals $\mathcal{I}$ in $\mathcal{A} = C(G)$ and the set of all subsets K of G. This is known as the* Gel'fand–Naimark duality.

Show that under this bijection the maximal ideals of $\mathcal{A}$ correspond to the singleton (one-point) subsets of G, and so ultimately with the points of G.

Let's note that for any $K \subset G$, the ideal $\mathcal{I}_K$ is a $*$-*ideal*; that is, $\alpha \in \mathcal{I}_K$ implies that $\alpha^* \in \mathcal{I}_K$, where α^* is the complex conjugate of the function α. So we have that every ideal $\mathcal{I}$ in $\mathcal{A}$ is automatically a $*$-ideal as well.

Proposition 13.1. *Let $\mathcal{I} \subset \mathcal{A}$ be an ideal in $\mathcal{A} = C(G)$ and K be a subset of G. Take any element $g \in G$. Then*

1. $\delta_g \in \mathcal{I}$ *if and only if* $g \notin K_\mathcal{I}$.
2. $\delta_g \in \mathcal{I}_K$ *if and only if* $g \notin K$.

Proof. We first consider statement 1.

Suppose $\delta_g \in \mathcal{I}$. Since $\delta_g(g) = 1$, we conclude that there exists some $\alpha \in \mathcal{I}$ (namely δ_g) such that $\alpha(g) \neq 0$. Therefore, $g \notin K_\mathcal{I}$ by the definition of $K_\mathcal{I}$.

Conversely, suppose that $g \notin K_\mathcal{I}$. By the definition of $K_\mathcal{I}$, this means that there exists some $\alpha \in \mathcal{I}$ such that $\lambda := \alpha(g) \neq 0$. We define $\beta := \lambda^{-1}\delta_g \in \mathcal{A}$. Since $\mathcal{I}$ is an ideal, we have that $\beta\alpha \in \mathcal{I}$. But $\beta\alpha = \lambda^{-1}\delta_g\alpha = \delta_g$. So $\delta_g \in \mathcal{I}$.

When we use Exercise 13.6, we see that statement 2 is equivalent to statement 1. ∎

Proposition 13.2. *Suppose that $\mathcal{I} \subset \mathcal{A} = C(G)$ is an ideal. Then we have that $B_\mathcal{I} := \{\delta_h \mid h \notin K_\mathcal{I}\}$ is a basis for $\mathcal{I}$.*

Proof. By the previous proposition, we see that $B_\mathcal{I} \subset \mathcal{I}$. Also, it is a subset of the standard basis $\{\delta_h \mid h \in G\}$ of $\mathcal{A}$ and so is linearly independent. It remains only to show that $B_\mathcal{I}$ spans $\mathcal{I}$. So take any $\alpha \in \mathcal{I}$ and expand it in the standard basis of $\mathcal{A}$ to get

$$\alpha = \sum_{g \in G} \lambda_g \delta_g,$$

where $\lambda_g \in \mathbb{C}$ for all $g \in G$. We would like to identify those coefficients in its expansion that possibly could be distinct from 0. So suppose $\lambda_h \neq 0$. Much as in the previous proposition, we define $\beta := (\lambda_h)^{-1}\delta_h$. Then $\beta\alpha \in \mathcal{I}$. But $\beta\alpha = (\lambda_h)^{-1}\delta_h\alpha = \delta_h$. Thus, $\delta_h \in \mathcal{I}$. Or, equivalently, by the previous proposition, $h \notin K_\mathcal{I}$. So we can write the above expression as

$$\alpha = \sum_{h \notin K_\mathcal{I}} \lambda_h \delta_h,$$

which gives the desired expansion of α in terms of the elements of $B_\mathcal{I}$. ∎

Corollary 13.1. *Suppose that $\mathcal{I} \subset \mathcal{A} = C(G)$ is an ideal. Then*

$$\text{span}\{\delta_h \mid h \notin K_\mathcal{I}\} = \mathcal{I}.$$

13.4 FODCs over $C(G)$

Next, we wish to consider first-order differential calculi (FODCs) over $\mathcal{A} = C(G)$. The existence of any nontrivial FODC whatsoever associated with G is surprising from the point of view of classical differential geometry since G is a differential manifold of dimension 0. Thus, the classical tangent space at every point $g \in G$ is zero-dimensional. This is the same "infinitesimal" structure at $e \in G$ for every finite group G. So this way of viewing G cannot distinguish it from any other finite group. But using the definition of an FODC as a certain type of map $d : \mathcal{A} \to \Gamma$, we will see that such nontrivial objects do exist in this case. However, we do lose uniqueness; namely, in general, we have for a given algebra $\mathcal{A} = C(G)$ for a finite group G that there are many nonisomorphic associated FODCs.

As far as higher-order differential calculi (HODCs) over $\mathcal{A} = C(G)$ are concerned, we can take any FODC and extend it using the braided exterior calculus, the universal differential calculus, or any other enveloping calculus as we wish. The first step here, as always, is to select an FODC. Then the rest is an application of the theory of HODCs.

We wish to construct all irreducible, bicovariant, $*$-covariant FODCs over $\mathcal{A} = C(G)$. We will define irreducible later. We start with bicovariance. According to Theorem 6.10 (also see Theorem 1.8 in [86]), the necessary and sufficient condition for (Γ, d) to be a bicovariant FODC is $\mathrm{ad}(\mathcal{R}) \subset \mathcal{R} \otimes \mathcal{A}$, where $\mathcal{R}$ is the (right) ideal determining the FODC.

Proposition 13.3. $\mathrm{ad}(\mathcal{R}) \subset \mathcal{R} \otimes \mathcal{A}$ *is equivalent to* $g^{-1}(K_{\mathcal{R}})g = K_{\mathcal{R}}$ *for all* $g \in G$.

Proof. In the following, we use the formula $\mathrm{ad}(\delta_g) = \sum_{h \in G} \delta_{hgh^{-1}} \otimes \delta_h$ and the fact that $\{\delta_h \mid h \in G\}$ is a basis for $\mathcal{A}$. Then the following are equivalent:

1. $\mathrm{ad}(\mathcal{R}) \subset \mathcal{R} \otimes \mathcal{A}$.
2. $\mathrm{ad}(\delta_g) \in \mathcal{R} \otimes \mathcal{A}$ for all $g \notin K_{\mathcal{R}}$.
3. $\sum_h \delta_{hgh^{-1}} \otimes \delta_h \in \mathcal{R} \otimes \mathcal{A}$ for all $g \notin K_{\mathcal{R}}$.
4. $\delta_{hgh^{-1}} \otimes \delta_h \in \mathcal{R} \otimes \mathcal{A}$ for all $h \in G$ and all $g \notin K_{\mathcal{R}}$.
5. $\delta_{hgh^{-1}} \in \mathcal{R}$ for all $h \in G$ and all $g \notin K_{\mathcal{R}}$.
6. $hgh^{-1} \notin K_{\mathcal{R}}$ for all $h \in G$ and all $g \notin K_{\mathcal{R}}$.
7. $h(G \setminus K_{\mathcal{R}})h^{-1} \subset G \setminus K_{\mathcal{R}}$ for all $h \in G$.
8. $h(G \setminus K_{\mathcal{R}})h^{-1} = G \setminus K_{\mathcal{R}}$ for all $h \in G$.
9. $hK_{\mathcal{R}}h^{-1} = K_{\mathcal{R}}$ for all $h \in G$. ∎

Going back to the left co-action $\Phi_\Gamma : \Gamma \to \mathcal{A} \otimes \Gamma$, which every left covariant (and in particular, every bicovariant) FODC has, we recall the definition of the left invariant elements of this co-action:

Definition 13.2. $_{\mathrm{inv}}\Gamma := \{\omega \in \Gamma \mid \Phi_\Gamma(\omega) = 1 \otimes \omega\}$.

Then we proved this result:

Proposition 13.4. $_{\text{inv}}\Gamma \cong (\text{Ker}\,\varepsilon)/\mathcal{R}$, *where* $\mathcal{R} \subset \text{ker}\,\varepsilon$ *is the right ideal in* $\mathcal{A}$ *associated with* Γ.

We also recall the definition of the quantum germs map π.

Definition 13.3. *For any* $\alpha \in \mathcal{A}$, *we define* $\pi(\alpha) := \kappa(\alpha^{(1)})d(\alpha^{(2)}) \in _{\text{inv}}\Gamma$.

And we have already shown the following:

Proposition 13.5. $\pi : \mathcal{A} \to _{\text{inv}}\Gamma$ *is an epimorphism with* $\text{ker}\,\pi = \mathcal{R} + \mathbb{C}1$.

Proposition 13.6. *Let* $\mathcal{I}$ *be an ideal in* $C(G)$ *and* $K \subset G$. *Then we have*

- $\mathcal{I} \subset \text{Ker}\,\varepsilon$ *if and only if* $e \in K_{\mathcal{I}}$.
- $\mathcal{I}_K \subset \text{Ker}\,\varepsilon$ *if and only if* $e \in K$.

Proof. Recall that $e \in G$ is the identity element and $\varepsilon : C(G) \to \mathbb{C}$ is defined by $\varepsilon(\alpha) = \alpha(e)$. Now the following statements are clearly equivalent:

- $\mathcal{I} \subset \text{Ker}\,\varepsilon$.
- $\alpha \in \mathcal{I}$ implies that $\alpha \in \text{Ker}\,\varepsilon$.
- $\alpha \in \mathcal{I}$ implies that $\varepsilon(\alpha) = 0$.
- $\varepsilon(\alpha) = 0$ for all $\alpha \in \mathcal{I}$.
- $\alpha(e) = 0$ for all $\alpha \in \mathcal{I}$.
- $e \in K_{\mathcal{I}}$.

This proves the first statement. The second statement is equivalent to the first by the duality between ideals in $C(G)$ and subsets of G. And so the proposition is proved. ∎

Exercise 13.7. *Prove that* $\mathcal{I}_{\{e\}} = \text{ker}\,\varepsilon$.

Theorem 13.1. *A vector space basis for* $_{\text{inv}}\Gamma$ *is given by*

$$\{\pi(\delta_g) \,|\, g \in K_{\mathcal{R}} \setminus \{e\}\}.$$

In particular, $\dim(_{\text{inv}}\Gamma) = \text{card}(K_{\mathcal{R}}) - 1$. *This is also a basis for the free left* $\mathcal{A}$-module $\Gamma \cong \mathcal{A} \otimes _{\text{inv}}\Gamma$.

Proof. The last affirmation is left to the reader; we continue with the vector space result. Since $\{\delta_g \,|\, g \in G\}$ is a vector space basis for $C(G)$, we clearly can decompose $C(G)$ into two vector subspaces:

$$C(G) = \text{span}\{\delta_g \,|\, g \in K_{\mathcal{R}}\} + \text{span}\{\delta_g \,|\, g \notin K_{\mathcal{R}}\}$$
$$= \text{span}\{\delta_g \,|\, g \in K_{\mathcal{R}}\} + \mathcal{R},$$

where we have used Corollary 13.1. Now $\text{ker}\,\pi = \mathbb{C}1 + \mathcal{R}$, so that π maps the last term to zero. Let π' denote π restricted to $\text{span}\{\delta_g \,|\, g \in K_{\mathcal{R}}\}$. It follows that $\text{ker}\,\pi'$ has dimension 1. Clearly, $\text{Ran}\,\imath' = \text{Ran}\,\imath = _{\text{inv}}\Gamma$.

We now identify a nonzero element in $\ker \pi'$, that is, a basis of it. To do this, we note that $1 \in C(G)$, the constant function 1, can be written as

$$1 = \sum_{g \in K_\mathcal{R}} \delta_g + \sum_{g \notin K_\mathcal{R}} \delta_g.$$

Applying π to this equation and using the facts that the last term maps to zero and $\pi(1) = 0$, we see that

$$\pi \left(\sum_{g \in K_\mathcal{R}} \delta_g \right) = 0.$$

This implies $\sum_{g \in K_\mathcal{R}} \delta_g \in \ker \pi'$. This linear combination contains the term δ_e, since $e \in K_\mathcal{R}$. And so

$$\sum_{g \in K_\mathcal{R}} \delta_g \neq 0.$$

Now we consider the restriction π'' of π to the subspace

$$\mathrm{span}\{\delta_g \mid g \in K_\mathcal{R} \setminus \{e\}\}.$$

Then π'' must have kernel 0, since otherwise we would have a linear combination of the $\delta_g s$ (but without a term containing δ_e) in $\ker \pi'' \subset \ker \pi'$ and so the dimension of $\ker \pi'$ would be strictly greater than 1.

Summing up, $\pi'' : \mathrm{span}\{\delta_g \mid g \in K_\mathcal{R} \setminus \{e\}\} \to {}_{\mathrm{inv}}\Gamma$ is an isomorphism of vector spaces. This immediately gives us the vector space result. ∎

We single out some things proved in the course of showing Theorem 13.1.

Corollary 13.2. *With the notation and hypotheses of Theorem 13.1, the following statements hold:*

1. *For all $g \notin K_\mathcal{R}$, we have $\pi(\delta_g) = 0$.*
2. $\pi \left(\sum_{g \in K_\mathcal{R}} \delta_g \right) = 0.$

In the following we will also use this notation: $[\alpha] = \pi(\alpha)$. We have the right action, denoted by $\circ$, of $\mathcal{A}$ on ${}_{\mathrm{inv}}\Gamma$. This has been introduced in (6.28). On the basis elements $[\delta_g]$ of ${}_{\mathrm{inv}}\Gamma$ (so $g \neq e$), we evaluate this right action as follows:

$$[\delta_g] \circ \delta_h = [\delta_g \delta_h - \varepsilon(\delta_g)\delta_h]$$

$$= [\delta_g \delta_h] = \begin{cases} [\delta_g] & \text{if } g = h, \\ 0 & \text{if } g \neq h. \end{cases}$$

Theorem 11.5 (also see Theorem 1.10 in [86]) says that a left covariant FODC (given by a right ideal $\mathcal{R} \subset \text{Ker}\,\varepsilon$ in any $*$-algebra $\mathcal{A}$) is $*$-covariant if and only if $\kappa(\mathcal{R})^* \subset \mathcal{R}$.

Proposition 13.7. *For the $*$-algebra $\mathcal{A} = C(G)$, the $*$-covariance condition $\kappa(\mathcal{I})^* \subset \mathcal{I}$ is equivalent to $K_{\mathcal{I}} = K_{\mathcal{I}}^{-1}$.*

Proof. $\mathcal{I} = \text{span}\{\delta_g \mid g \notin K_{\mathcal{I}}\}$. So $\kappa(\mathcal{I})^* = \text{span}\{\delta_{g^{-1}} \mid g \notin K_{\mathcal{I}}\}$. Then the following statements are immediately seen to be equivalent:

1. $\kappa(\mathcal{I})^* \subset \mathcal{I}$.
2. $\text{span}\{\delta_{g^{-1}} \mid g \notin K_{\mathcal{I}}\} \subset \text{span}\{\delta_g \mid g \notin K_{\mathcal{I}}\}$.
3. $\text{span}\{\delta_g \mid g^{-1} \notin K_{\mathcal{I}}\} \subset \text{span}\{\delta_g \mid g \notin K_{\mathcal{I}}\}$.
4. $\text{span}\{\delta_g \mid g \notin K_{\mathcal{I}}^{-1}\} \subset \text{span}\{\delta_g \mid g \notin K_{\mathcal{I}}\}$.
5. $\text{span}\{\delta_g \mid g \in G \setminus K_{\mathcal{I}}^{-1}\} \subset \text{span}\{\delta_g \mid g \in G \setminus K_{\mathcal{I}}\}$.
6. $G \setminus K_{\mathcal{I}}^{-1} \subset G \setminus K_{\mathcal{I}}$.
7. $K_{\mathcal{I}} \subset K_{\mathcal{I}}^{-1}$.
8. $K_{\mathcal{I}} = K_{\mathcal{I}}^{-1}$.

Perhaps the last equivalence requires more comment. The point is that by applying the operation $(\cdot)^{-1}$ of taking the inverse to statement 7, we obtain $K_{\mathcal{I}}^{-1} \subset K_{\mathcal{I}}$. And this together with statement 7 itself is statement 8. This completes the proof. ∎

We now summarize our results so far.

Theorem 13.2. *The bicovariant and $*$-covariant FODCs over the Hopf algebra $\mathcal{A} = C(G)$ are determined by the subsets K of G that satisfy these three conditions:*

- $e \in K$.
- $g^{-1} K g = K$ *for all* $g \in G$.
- $K = K^{-1}$.

The ideal that determines the FODC associated to K is $\mathcal{R} = \mathcal{I}_K$.

Equivalently, the bicovariant and $*$-covariant FODCs over the Hopf algebra $\mathcal{A} = C(G)$ are determined by the subsets J of $G \setminus \{e\}$ that satisfy these two conditions:

- $g^{-1} J g = J$ *for all* $g \in G$.
- $J = J^{-1}$.

The ideal that determines the FODC associated to J is $\mathcal{R} = \mathcal{I}_{J \cup \{e\}}$.

Remark. The FODC is 0 if and only if $\mathcal{R} = \ker\varepsilon$ if and only if $K = \{e\}$ if and only if $J = \emptyset$, the empty set.

Proof. The first set of three conditions has already been established. For the second set of two conditions, it suffices to use the mutually inverse correspondences $K \mapsto K \setminus \{e\}$ and $J \mapsto J \cup \{e\}$. ∎

Notice that we have not yet addressed a very basic issue, namely, what is the formula for the differential d of a bicovariant, $*$-covariant FODC over $C(G)$? We already know the general formula

$$d(a) = a^{(1)}\pi(a^{(2)})$$

for $a \in \mathcal{A}$. Here $\mathcal{A} = C(G)$, which has the canonical basis $\{\delta_g \,|\, g \in G\}$. So it suffices to give a formula for $d(\delta_g)$ for each $g \in G$. Now we also already know that

$$\phi(\delta_g) = \sum_{h \in G} \delta_{gh^{-1}} \otimes \delta_h.$$

So, taking $a = \delta_g$, we get

$$d(\delta_g) = \sum_{h \in G} \delta_{gh^{-1}} \pi(\delta_h),$$

which qualifies as the general formula for which we are looking. But this formula simplifies by using the two facts proved in Corollary 13.2, which imply that

$$
\begin{aligned}
d(\delta_g) &= \sum_{h \in G} \delta_{gh^{-1}} \pi(\delta_h) \\
&= \sum_{h \in K} \delta_{gh^{-1}} \pi(\delta_h) \\
&= \sum_{h \in K \setminus \{e\}} \delta_{gh^{-1}} \pi(\delta_h) + \delta_g \pi(\delta_e) \\
&= \sum_{h \in K \setminus \{e\}} \delta_{gh^{-1}} \pi(\delta_h) - \sum_{h \in K \setminus \{e\}} \delta_g \pi(\delta_h) \\
&= \sum_{h \in K \setminus \{e\}} (\delta_{gh^{-1}} - \delta_g) \pi(\delta_h).
\end{aligned}
\tag{13.7}
$$

We have just shown that a vector space and left free $\mathcal{A}$-module basis of $_{\mathrm{inv}}\Gamma$ are given by

$$\{\pi(\delta_h) \,|\, h \in K \setminus \{e\}\}.$$

So we now have an expansion of $d(\delta_g)$ in (13.7) over these basis elements with coefficients in $\mathcal{A}$. The formula (13.7) is reminiscent of a finite-difference operator. However, the differential d here satisfies Leibniz's rule, while finite-difference operators typically do not. For example, for any function $\alpha : G \to \mathbb{C}$, we can define the finite-difference operator ∂ for $g \in G$ by

$$\partial \alpha(g) := \sum_h (\alpha(gh) - \alpha(g)),$$

where h ranges over some subset H of G. A typical example would be to take H to be a set of "nearest neighbors" of the identity element $e \in G$. Thus, $\partial\alpha : G \to \mathbb{C}$. One calculates that

$$\partial\delta_g = \sum_h (\delta_{gh^{-1}} - \delta_g).$$

Moreover, as the reader can check, ∂ does not satisfy Leibniz's rule. Also, the differential d that we constructed here has 1-forms, not functions, as its co-domain. So $d : \mathcal{A} \to \Gamma$ while, on the other hand, $\partial : \mathcal{A} \to \mathcal{A}$.

It is important to note that the differential d is defined on functions $G \to \mathbb{C}$ and not on elements of G itself. However, one notation (that we are not using in this book) is to write g for the function δ_g. Then (13.7) becomes

$$d(g) = \sum_{h \in K \setminus \{e\}} (gh^{-1} - g)[h],$$

where we also are using the abbreviated form $[a] = \pi(a)$. This is a very appealing notation, provided one remembers what it really means!

Let's review a bit of group theory (see [41]). Say G is a group and $x, y \in G$. Then we say that x is *conjugate* to y if there exists some $g \in G$ such that $g^{-1}xg = y$, and we denote this relation as $x \sim y$. One proves that $\sim$ is an equivalence relation. Its equivalence classes are called *conjugacy classes*. This gives a partition of the underlying set of G; that is, every element $g \in G$ lies in a unique conjugacy class, which is exactly the set of all elements of G that are conjugate to g. The identity element $e \in G$ is only conjugate to itself, and so its conjugacy class is $\{e\}$. If S is any subset of G, then the condition $g^{-1} S g = S$ for all $g \in G$ is equivalent to saying that for any element $s \in S$, all the conjugates of s also lie in S. Yet another equivalent way to say this is that S is a union of conjugacy classes. Obviously, this is a disjoint union.

The relation $S = S^{-1}$ for a subset S of G is more transparent. It merely means that whenever $s \in S$, it follows that also $s^{-1} \in S$. More importantly, we have this:

Exercise 13.8. *Let R be any subset of G. Then define $S := R \cup R^{-1}$. Show that $R \subset S$ and $S = S^{-1}$.*

Also, show that S is the smallest subset with these two properties; that is, if T is a subset of G satisfying $R \subset T$ and $T = T^{-1}$, then $S \subset T$.

We now ask what happens if we have $J_1 \subset J_2 \subset G \setminus \{e\}$ and both J_1 and J_2 satisfy the two conditions for determining an FODC over $C(G)$?

By the duality of the Gel'fand–Naimark bijection, we see that $\mathcal{I}_{J_2} \subset \mathcal{I}_{J_1}$ and therefore $\mathcal{A} \otimes \mathcal{I}_{J_2} \subset \mathcal{A} \otimes \mathcal{I}_{J_1}$. Letting Γ_1 and Γ_2 denote the FODCs corresponding to $\mathcal{I}_{J_1}$ and $\mathcal{I}_{J_2}$, respectively, we have a quotient map

$$\Gamma_2 \cong \frac{U_L}{\mathcal{A} \otimes \mathcal{I}_{J_2}} \to \frac{U_L}{\mathcal{A} \otimes \mathcal{I}_{J_1}} \cong \Gamma_1$$

induced by the inclusion $\mathcal{A} \otimes \mathcal{I}_{J_2} \subset \mathcal{A} \otimes \mathcal{I}_{J_1}$. This leads us to the concept of irreducibility. Any FODC Γ has two *trivial quotients*, which are obtained by dividing either by the zero subspace or by Γ itself. These quotients are Γ and 0, respectively. Of course, Γ could have many other quotients, which are called *nontrivial quotients*. This should motivate this:

Definition 13.4. *An FODC is* irreducible *if it has no nontrivial quotient FODC. A bicovariant, ∗-covariant FODC is said to be* irreducible *if it has no nontrivial quotient FODC that is bicovariant and ∗-covariant.*

While it is bad manners to introduce terminology with two distinct meanings, we will only use the second definition. So a subset J of $G \setminus \{e\}$ corresponds to an irreducible, bicovariant, ∗-covariant FODC provided that no proper subset of J also corresponds to a bicovariant, ∗-covariant FODC. (Recall that a *proper subset* of a set S is a subset that is neither S nor empty.) So we are looking for minimal subsets J of $G \setminus \{e\}$ that satisfy $g^{-1} J g = J$ for all $g \in G$ and $J = J^{-1}$.

Next, we note that $J_1 \subset J_2$ means that all the conjugacy classes whose union gives J_1 appear in the representation of J_2 as a union of conjugacy classes. So we wish to minimize the number of conjugacy classes that give us an adequate J while maintaining the condition that $J = J^{-1}$. This analysis depends on this little tidbit:

Exercise 13.9. *If $C \subset G$ is a conjugacy class, then C^{-1} is a conjugacy class. Actually, if $x \in C$, then C^{-1} is the conjugacy class of x^{-1}.*

Now J must contain at least one conjugacy class C. And by an exercise, $C \cup C^{-1}$ is closed under the operation $^{-1}$ of taking the inverse. Since we want $J^{-1} = J$, we also know by the same exercise that $C \cup C^{-1} \subset J$. Now, $C \cup C^{-1}$ is a union of conjugacy classes and closed under taking inverses. So if we assume that J is minimal with respect to these properties, we must have $J = C \cup C^{-1}$.

Exercise 13.10. *Prove the converse: Every set of the form $C \cup C^{-1}$, where C is a conjugacy class, is closed under inverses, is a union of conjugacy classes, and is minimal with respect to these properties.*

Thus, $J = C \cup C^{-1}$, where C is a conjugacy class of G. But watch out! There are two cases here: $C = C^{-1}$ and $C \cap C^{-1} = \emptyset$, the empty set. We have proved the next important result:

Theorem 13.3. *Let G be a finite group. Then the irreducible bicovariant, ∗-covariant FODCs over $\mathcal{A} = C(G)$ are determined by individual conjugacy classes C that satisfy $C = C^{-1}$ or by pairs (C, C^{-1}) of conjugacy classes satisfying $C \cap C^{-1} = \emptyset$.*

In general, the task of identifying the conjugacy classes of a given finite group G is by no means trivial. However, much is known in specific cases, such as that of the symmetric groups S_n. But the point of this theorem is that given this information, one can more or less read off the irreducible bicovariant, ∗-covariant FODCs over $C(G)$.

13.5   Coxeter Groups

In this section we introduce a special class of finite groups, known as Coxeter groups. These have been studied extensively and have found applications in many areas of mathematics. Good general references are the texts [36] and [40]. In Chapter 14, we will see Coxeter groups used to define Dunkl operators. This section serves mainly to establish notation, definitions, and basic properties. But we will also introduce an FODC over $C(G)$, where G is a Coxeter group.

We first have to define Coxeter groups. We start by considering a real vector space E, which is a finite-dimensional Euclidean space with inner product $\langle \cdot, \cdot \rangle$ and associated norm $\|\beta\| := \langle \beta, \beta \rangle^{1/2}$ for $\beta \in E$ such that $\dim_{\mathbb{R}} E \geq 1$. If the reader likes to think of this as $\mathbb{R}^n$ for some integer $n \geq 1$, that is okay. Now, for any $0 \neq \alpha \in E$, we define a linear map $\sigma_\alpha : E \to E$, which is nothing other than the orthogonal reflection in the hyperplane H_α orthogonal to α, namely,

$$H_\alpha := \{\beta \in E \mid \langle \alpha, \beta \rangle = 0\}.$$

Then for any $x \in E$, we write $x = y + r\alpha$ for unique elements $y \in H_\alpha$ and $r \in \mathbb{R}$. So we then have

$$\sigma_\alpha : x \mapsto y - r\alpha.$$

In analysis one typically writes a linear operator to the left of the vector on which it acts. And σ_α is a linear operator. But for various historical reasons, we write this action on the right:

$$x\sigma_\alpha = (x)\sigma_\alpha.$$

In part, this notation reflects the convention that in classical geometry the action of the structure group on the total space of a principal bundle is a right action.

The explicit formula for σ_α is

$$x\sigma_\alpha = x - 2\frac{\langle \alpha, x \rangle}{\|\alpha\|^2}\alpha$$

for all $x \in E$.

We also use the usual definition of the *orthogonal group* of E, namely, $O(E) := \{T : E \to E \mid T$ is linear and $\langle Tx, Ty \rangle = \langle x, y \rangle \; \forall x, y \in \mathbb{R}^n\}$. And we say that a T is an *orthogonal* map if $T \in O(E)$.

Exercise 13.11. *Prove the following:*

1. $\sigma_\alpha : E \to E$ *is an orthogonal, linear map and so* $\sigma_\alpha \in O(E)$.
2. $\det \sigma_\alpha = -1$.

3. $\sigma_\alpha^2 = I_E$, the identity map on E, but $\sigma_\alpha \neq I_E$.
4. $\alpha \sigma_\alpha = -\alpha$.
5. $\sigma_{r\alpha} = \sigma_\alpha$ for all $r \in \mathbb{R}$ with $r \neq 0$.

Definition 13.5. *A* root system *in* E *is a finite set* R *of nonzero vectors in* E *satisfying*

1. *$\beta \sigma_\alpha \in R$ for all $\alpha, \beta \in R$. Equivalently, $(R)\sigma_\alpha = R$ for all $\alpha \in R$.*
2. *If $\alpha \in R$ and $t\alpha \in R$ for some $t \in \mathbb{R}$, then either $t = 1$ or $t = -1$.*
3. *For all $\alpha \in R$, one has the normalization condition $\|\alpha\|^2 = 2$.*

Since $\alpha\sigma_\alpha = -\alpha$, the first property in this definition immediately gives us that $\alpha \in R \implies -\alpha \in R$.

Definition 13.6. *The* Coxeter group $G = G(R)$ *associated to a root system R in E is the subgroup of $O(E)$ generated by the set $\{\sigma_\alpha \mid \alpha \in R\}$.*

One of the first results about Coxeter groups is that they are finite groups. See [36] or [40] for an elementary proof. The idea is to embed G as a subgroup of the permutation group of the elements of R. But the latter group is finite since R is finite.

This is a bare-bones presentation of Coxeter groups; almost nothing really has been said. One extra comment is that these groups have been classified (up to isomorphism, of course). Another is that they generalize the Weyl groups used in Lie group theory. Most importantly, there have been many applications in analysis, mathematical physics, probability, and Euclidean geometry, where it all started with Coxeter's work in the 20th century. Yes, amazing and beautiful results were being found in Euclid's geometry that recently. Probably still are. See [14] for a collections of articles on some of this work.

The next result is one of the ingredients used in Chapter 14 in order to study Dunkl operators.

Theorem 13.4. *Let $S := \{\sigma_\alpha \mid \alpha \in R\}$ be the set of the defining generators in the Coxeter group G associated to the root system R. Then S is a union of conjugacy classes in G and satisfies $S^{-1} = S$. Consequently, S determines a bicovariant, $*$-covariant FODC over $\mathcal{A} = C(G)$.*

Proof. Since $\sigma_\alpha^{-1} = \sigma_\alpha$, we see that $S^{-1} = S$. Also, $g^{-1}\sigma_\alpha g = \sigma_{\alpha g}$ for all $g \in G$ and all $\alpha \in R$ (see [36] or [40]), so that S is closed under conjugation. So S is a union of conjugacy classes. Consequently, Theorem 13.2 gives the result. ∎

Exercise 13.12. *Without looking at [36] or [40], prove that $g^{-1}\sigma_\alpha g = \sigma_{\alpha g}$.*

As a final point of this section, we now write down the differential d of the FODC associated to the Coxeter group G by the set S. To do this, we use the general formula (13.7). We simply note that $K \setminus \{e\} = S$ in this case, and so (13.7) reads as

$$d(\delta_g) = \sum_{s \in S}(\delta_{gs^{-1}} - \delta_g)\,\pi(\delta_s).$$

13.6 Notes

The spaces of functions on a finite group serve more as very simple motivating examples for the theory of abstract Hopf algebras and the theory of compact quantum groups than anything else. Naturally, these general theories have other motivating examples. Also, these theories have led to the discovery and study of other, more complicated examples. Similarly, the Gel'fand–Naimark theory presented here is just the tip of the iceberg of a much more profound theory. However, a general theory can provide a handy language even for simple examples. The study of FODCs in the setting of finite groups is discussed further in [26]. While these give us simple examples of nontrivial FODCs, and consequently nontrivial HODCs as well, the special case of the differential calculus based on the set of reflections of a Coxeter group has an interesting application, as we shall see in the next chapter. It turns out that the symmetric group S_n on n objects is a Coxeter group (see [36] or [40] or prove it yourself). In [26] the example of the FODC associated to the set of 2-cycles in S_3 is worked out. But Micho Đurđevich has told me that he did not realize at the time of writing [26] that he was studying a Coxeter group with the FODC associated to its set of reflections.

Chapter 14
Dunkl Operators as Covariant Derivatives in a QPB

14.1 Overview

Dunkl operators are quantum objects. But at first this was not how they were understood. Dunkl operators are a relatively new feature of modern mathematics. They were introduced by C. Dunkl in 1989 (see [19]) in harmonic analysis. They are neither local operators nor differential operators. One immediate use was to define and study new special functions on the Euclidean spaces $\mathbb{R}^n$ for $n \geq 2$. Actually, to the best of this author's knowledge, this was the first occurrence of the use of nonlocal operators to define special functions on $\mathbb{R}^n$ for $n \geq 2$. It is well known that there had been an extensive study of special functions on $\mathbb{R}$ with a lot of work done in the 19th century, but also some results dating back even earlier. A *leit motif* of these previous works is the ubiquitous relationship of special functions to differential equations, that is, with local operators.

Dunkl operators were subsequently used in other areas of mathematics, such as exactly solvable models (see [33] and [77]), mathematical physics (see [69] and [75]), and probability (see [70]. And as the reader will see from the very first glance of their definition, Dunkl operators are superficially similar to covariant derivatives. This is because a Dunkl operator is a partial derivative in Euclidean space plus "something." And that is what a covariant derivative looks like in classical differential geometry. However, the devil lies in the details. And the details are telling. Covariant derivatives in classical differential geometry are local operators. So since that "something" in the definition of Dunkl operators is not local, they cannot be viewed as covariant derivatives in the setting of classical differential geometry. That is simply out of the question.

But the play continues. Entering on the stage is the theory of quantum principal bundles (QPBs). Is this a *deus ex machina*? That is, could one view the Dunkl operators as covariant derivatives in this noncommutative setting? The answer is yes, as we shall see in this chapter. This result gives a relationship between the world of Dunkl operators (including all of their applications) and the world of noncommutative geometry. There had never been any conjecture, let alone a word,

© Springer International Publishing Switzerland 2015
S.B. Sontz, *Principal Bundles*, Universitext, DOI 10.1007/978-3-319-15829-7_14

in the published literature that there might be any kind of a relationship between Dunkl operators and noncommutative geometry. This is the achievement of a paper [31] written in collaboration with Micho Đurđevich. To give my gentle reader a peek behind the scenes of this collaboration, let me simply say that is actually what we were trying to achieve! So we had two theories ready for our use. And between these theories we built a bridge, which we hope will be of value for researchers working on either side of that bridge.

As is done in many research papers, we also proved in [31] some secondary results. And as happens with many research papers, various readers have confused a secondary result for the main purpose of the paper. In the case of [31], we also prove that the Dunkl operators commute among themselves, a result that dates back to Dunkl's original paper [19]. We did this to show how that well-known result fits into the new setting of QPBs, namely, that it is a consequence of the appropriate connection having zero curvature. It surely should not be considered the main point of the paper [31]. Actually, I had long thought that the commutativity of the Dunkl operators had something to do with some curvature being zero, but I also knew that this could not be a classical curvature. This premonition has now been confirmed.

The purpose of this chapter is to present in detail the main result of [31]. To do this, we will construct a specific QPB (including its ever-important quantum differential calculus) as well as a specific (quantum, of course) connection on this bundle. Then we can identify the covariant derivatives of this quantum connection with the Dunkl operators. Of course, the motivation for this construction was to arrive at the Dunkl operators. The goal is known in advance! While we will not use the full generality of the theory of QPBs, that is a detail of minor importance. The point is that even with *all* of the strength of the theory of classical principal bundles, the Dunkl operators cannot be viewed as classical covariant derivatives. Dunkl operators are quantum objects.

14.2 Dunkl Operators: Basics

In this section we present the basic definitions and properties of the Dunkl operators. The seminal paper [19] is the main reference, but another good place to read more about this material is [68].

We start with any fixed root system R in a finite-dimensional Euclidean space E with dim $E = n \geq 1$. Since R is invariant under the action of the generators σ_α of the Coxeter group G associated to R, we have that R is also invariant under the action of G.

Definition 14.1. *A G-invariant function* $\mu : R \to \mathbb{C}$ *is called a* multiplicity function.

Such a multiplicity function is constant on each G orbit in R and takes any arbitrary value in $\mathbb{C}$ on each orbit. So the set of all multiplicity functions on R is a finite-dimensional complex vector space whose dimension is the number of orbits of the action of G on R.

We now come to the fundamental definition. But first, we comment that $C^1(E)$ denotes the vector space of all complex-valued functions on E with continuous first partial derivatives with respect to one (and hence to any) orthonormal basis of E.

Definition 14.2. *Suppose that E is a finite-dimensional Euclidean space with a given root system R. Let $\xi \in E$ be an arbitrary vector and let $\mu : R \to \mathbb{C}$ be a multiplicity function. Then we define the* Dunkl operator $T_{\xi,\mu}$ *by*

$$T_{\xi,\mu} f(x) := \partial_\xi f(x) + \frac{1}{2} \sum_{\alpha \in R} \mu(\alpha) \frac{\langle \alpha, \xi \rangle}{\langle \alpha, x \rangle} (f(x) - f(x \sigma_\alpha)), \tag{14.1}$$

where $f \in C^1(E)$ and $x \in E$ is namely, $\langle \alpha, x \rangle \neq 0$ for all $\alpha \in R$. Also, $\partial_\xi = \langle \xi, \text{grad} \rangle$ is the usual directional derivative associated to the "direction" $\xi \in E$. The gradient operator grad is defined in this context with respect to some orthonormal basis in the usual way and then one shows that the definition does not depend on the choice of orthonormal basis.

Now Definition (14.1) is equivalent to the usual definition of a Dunkl operator, which is

$$T_{\xi,\mu} f(x) := \partial_\xi f(x) + \sum_{\alpha \in R^+} \mu(\alpha) \frac{\langle \alpha, \xi \rangle}{\langle \alpha, x \rangle} (f(x) - f(x \sigma_\alpha)). \tag{14.2}$$

Note that the only differences are that the factor of $1/2$ has been eliminated and the sum is now taken over a subset R^+ of R, which we must define. The point here is that the expression

$$\mu(\alpha) \frac{\langle \alpha, \xi \rangle}{\langle \alpha, x \rangle} (f(x) - f(x \sigma_\alpha))$$

is invariant under the mapping $\alpha \mapsto -\alpha$, which is a bijection of R with itself. This is so since $\sigma_{-\alpha} = \sigma_\alpha$, as already noted, and $\mu(-\alpha) = \mu(\alpha)$ by the assumed G-invariance of μ. So we will take R^+ to be a set with half the cardinality of R (and yes, the cardinality of R is even) in such a way that for each $\alpha \in R$ we will pick exactly one of the two elements $\{\alpha, -\alpha\}$.

While the two expressions (14.1) and (14.2) for a Dunkl operator are trivially equivalent, the way of viewing Dunkl operators in the setting of noncommutative geometry singles out (14.2) as the principal formula. From this point of view, the equivalence with (14.1) turns out to be an accidental feature of a particular example of a general theory. We will see the details of this later on.

So here are the details for defining R^+. We define an *order* on E (or on any real vector space for that matter) to be a binary relation, denoted $u < v$ for $u, v \in E$, which is a partial order that also satisfies $u < v \Rightarrow u + w < v + w$ and $ru < rv$ for all $u, v, w \in E$ and all reals $r > 0$. We say that an order is *total* if the so-called law of trichotomy holds; namely, for all $u, v \in E$, we have that exactly one of these three is true: $u < v$ or $v < u$ or $u = v$. We say that u is *positive* if $0 < u$, and we denote the set of positive elements in R with respect to some total order on E by R^+. The definition of R^+ does depend on the choice of total order on E.

It is now an exercise to show that there do exist total orders on E (but there are many of them!) and that the set R^+ defined from a total order divides R into two disjoint subsets R^+ and $-R^+$. Finally, one shows that Definition (14.2) does not depend on the choice of total order on E. This is then a long and winding road that leads to a definition that is equivalent to that given in (14.1). This discussion has been presented in part because many references use (14.2) as the definition of a Dunkl operator.

Now let's get back to the business of discussing the basic properties of Dunkl operators. First of all, there are two trivial cases. One is obtained by taking $R = \emptyset$, the empty set. This is a root system according to our definition. In this case, the summation on the right side of (14.1) is 0, and so the Dunkl operator in this trivial case is the directional derivative $\partial_\xi f(x)$.

The other trivial case is when the multiplicity function is identically zero, in which case we again have that the Dunkl operator is $\partial_\xi f(x)$, the same as in the first trivial case. In any event, we do have one special, but trivial, example of a Dunkl operator that is a local operator. All the other examples of Dunkl operators are clearly not local operators because of the summation term, which includes the expression $f(x\,\sigma_\alpha)$ that is used to evaluate $T_{\xi,\mu} f$ at the typically distinct point x.

There are two linearity properties here. One is that $T_{\xi,\mu} f$ is linear in f. The other is that it is linear in $\xi \in E$. Because of this second linearity, it suffices to study the finite family of operators $T_{\xi,\mu} f$, where ξ runs over an orthonormal basis of E.

Of course, Definition (14.1) only makes sense for $x \notin \cup_{\alpha \in R} H_\alpha$, where H_α is the hyperplane orthogonal to α. But we can extend the domain of definition of the Dunkl operator to all of E by using the identity

$$\frac{f(x) - f(x\,\sigma_\alpha)}{\langle \alpha, x \rangle} = \int_0^1 dt\, \partial_\alpha f(tx + (1-t)x\,\sigma_\alpha),$$

which holds for all $x \notin H_\alpha$. Given that $f \in C^1(E)$, we know that the directional derivative $\partial_\alpha f$ is continuous on E, so the integral is well defined for *all* $x \in E$. In particular, for $x \in H_\alpha$, we have $x\,\sigma_\alpha = x$, and so the integral yields $\partial_\alpha f(x)$ in that case. We then take this to be the value of the expression on the left side for $x \in H_\alpha$.

Dunkl operators originally were studied in harmonic analysis, where one defines a generalization of the standard Laplacian, usually called the *Dunkl Laplacian*, and an associated generalization of the Fourier transform, known as the *Dunkl transform*, which among other things diagonalizes the Dunkl Laplacian. This is realized in a Hilbert space $\mathcal{H} = L^2(E, w\,d_{Leb})$, where w is a specific weight function on E and d_{Leb} is the Lebesgue measure on E. (Any orthonormal basis of E defines a unit n-cube, which then is taken to have Lebesgue measure 1.) In this setting the Dunkl operators are realized as *unbounded antihermitian operators* acting on appropriate *dense domains* in $\mathcal{H}$. Since the subset $\cup_{\alpha \in R} H_\alpha$ has measure zero, in this setting the lack of a defining formula on that subset is not so important. The purpose of this brief paragraph is merely to indicate that a mature theory of Dunkl operators exists in a useful analytic context. But since this is not a book on analysis, we leave the matter here. The interested reader can pursue and peruse the relevant literature.

Another approach is based on the observation that when f is a polynomial, the expression

$$q(x) = \frac{f(x) - f(x\,\sigma_\alpha)}{\langle \alpha, x\rangle} \quad \text{for } x \notin \cup_{\alpha\in R} H_\alpha$$

is the quotient of polynomials (that is, a rational function), which is actually a polynomial itself since the linear polynomial in the denominator divides the polynomial in the numerator. So the extension of the domain of the definition of $q(x)$ to all of E is straightforward in this case. Of course, this is just a special case of the extension procedure used above for any function in $C^1(E)$. And $T_{\xi,\mu}f$ in this case is again a polynomial. So the complex vector space of all polynomials on E is an *invariant domain* for the Dunkl operators. This provides an algebraic setting in which one can study these operators.

Returning to our algebraic–geometric approach, one natural question is whether the Dunkl operators, being generalizations of directional derivatives, also satisfy Leibniz's rule. The answer is more complicated than one might like, but it still is a relatively manageable identity.

Theorem 14.1. *Let $f, g \in C^1(E)$. Suppose that R is a root system in E with multiplicity function μ. Then for any $\xi, x \in E$, we have*

$$T_{\xi,\mu}(fg)(x) = (T_{\xi,\mu}f)(x)g(x) + f(x)(T_{\xi,\mu}g)(x)$$
$$- \frac{1}{2}\sum_{\alpha\in R}\mu(\alpha)\frac{\langle\alpha,\xi\rangle}{\langle\alpha,x\rangle}\big(f(x) - f(x\,\sigma_\alpha)\big)\big(g(x) - g(x\,\sigma_\alpha)\big),$$

which is a generalization (or deformation) of Leibniz's rule in this setting.

Proof. First, for $x \notin \cup_{\alpha\in R} H_\alpha$, we calculate that

$$T_{\xi,\mu}(fg)(x) - (T_{\xi,\mu}f)(x)g(x) - f(x)(T_{\xi,\mu}g)(x)$$
$$= \frac{1}{2}\sum_{\alpha\in R}\mu(\alpha)\frac{\langle\alpha,\xi\rangle}{\langle\alpha,x\rangle}\Big(f(x)g(x) - f(x\,\sigma_\alpha)g(x\,\sigma_\alpha)$$
$$- (f(x) - f(x\,\sigma_\alpha))g(x) - f(x)(g(x) - g(x\,\sigma_\alpha))\Big)$$
$$= \frac{1}{2}\sum_{\alpha\in R}\mu(\alpha)\frac{\langle\alpha,\xi\rangle}{\langle\alpha,x\rangle}$$
$$\cdot\big(-f(x)g(x) + f(x\,\sigma_\alpha)g(x) + f(x)g(x\,\sigma_\alpha) - f(x\,\sigma_\alpha)g(x\,\sigma_\alpha)\big)$$
$$= -\frac{1}{2}\sum_{\alpha\in R}\mu(\alpha)\frac{\langle\alpha,\xi\rangle}{\langle\alpha,x\rangle}\big(f(x) - f(x\,\sigma_\alpha)\big)\big(g(x) - g(x\,\sigma_\alpha)\big). \tag{14.3}$$

And this is the desired formula for $x \notin \cup_{\alpha \in R} H_\alpha$. If $x \in H_\alpha$ for some $\alpha \in R$, then the corresponding term in the summation on the right side of (14.3) is interpreted as $\partial_\alpha f(x)(g(x) - g(x \sigma_\alpha))$ [or as $(f(x) - f(x \sigma_\alpha)) \partial_\alpha g(x)$] evaluated at $x \in H_\alpha$, which yields 0. But this term should be equal to $\partial_\alpha (fg)(x) - \partial_\alpha f(x)g(x) - f(x)\partial_\alpha g(x)$, which is indeed 0 by Leibniz's rule for ∂_α. So we have the desired formula for all $x \in E$. ∎

Corollary 14.1. *Suppose the same hypotheses as in the previous theorem. Also, suppose either f or g is G-invariant. Then we have Leibniz's rule*

$$T_{\xi,\mu}(fg)(x) = (T_{\xi,\mu}f)(x)g(x) + f(x)(T_{\xi,\mu}g)(x).$$

We cite an important result in the theory of Dunkl operators, which already was proved in Dunkl's paper [19] in 1989. Later we will discuss this result from the viewpoint of noncommutative geometry.

Theorem 14.2. *The family of operators $\{T_{\xi,\mu} \mid \xi \in E\}$ is commutative.*

Of course, the composition of two of these operators is not always defined if we take their domain to be $C^1(E)$. So in this theorem we take the domain to be $C^2(E)$, the space of functions with second-order continuous partial derivatives.

14.3   The QPB and Its Properties

In this section we are going to construct a QPB *with differential calculus* associated with the Coxeter group G of a fixed root system R in a given Euclidean space E. Since in noncommutative geometry we take the dual approach, we replace the finite group G, a classical space with the discrete topology, with the function space

$$\mathcal{A} = \mathcal{F}(G) = \{f : G \to \mathbb{C}\},$$

which we already have seen is a Hopf algebra with invertible antipode κ, since $\kappa^2 = id_G = 1_\mathcal{A}$ in this case. Since the group G is finite, we also already have a complete characterization of all the bicovariant, $*$-covariant FODCs Γ over $\mathcal{A}$. One of these FODCs is the trivial case $\Gamma = 0$, which is the case that corresponds to taking the classical 1-forms on the zero-dimensional differential manifold G. We will not be interested in this particular FODC. Rather, we use the fact that there are many bicovariant, $*$-covariant FODCs over $\mathcal{A}$ in order to pick one of them that is suited to the study of Dunkl operators.

For any finite group G, we define

$$S := \{s \in G \mid s \neq e \quad \text{and} \quad s^2 = e\},$$

the subset of elements of order 2 in G. By the way, this is never a subgroup since $e \notin S$. Thus, the elements of S are the nontrivial "reflections" in G. Since the conjugation of such a reflection is again a reflection, this immediately implies that the set S is closed under conjugation. Also, we know that $s^{-1} = s$, so that the set S satisfies $S^{-1} = S$. Finally, by definition we have that $e \notin S$. All these properties of S, according to our previous result, Theorem 13.2, means that S determines a bicovariant, $*$-covariant FODC over $\mathcal{A} = \mathcal{F}(G)$.

Now we apply the construction of the previous paragraph to a Coxeter group. And at this point in the argument we use a very special property of Coxeter groups, namely, that

$$S = \{\sigma_\alpha \mid \alpha \in R\} = \{\sigma_\alpha \mid \alpha \in R^+\},$$

the set of reflections that generate G (see [36] or [40]. Or prove it yourself). The point here is that an arbitrary reflection in a Coxeter group G must be one of the generators. We denote by Γ the bicovariant, $*$-covariant FODC over $\mathcal{A}$ associated to this set S. We will not be considering any other FODC over $\mathcal{A}$, so we suppress S from the notation. Suppose that $\text{card}(R) \neq 0$, which means that the associated Coxeter group G is nontrivial. Then

$$\text{card}(S) = (1/2)\text{card}(R) = \text{card}(R^+) \neq 0.$$

Applying Theorem 13.1, we have that the set

$$\{ [\delta_s] \mid s \in S \}$$

is a basis of the complex vector space $_{\text{inv}}\Gamma$, where we denote the quantum germs map $\pi : \mathcal{A} \to {}_{\text{inv}}\Gamma$ by $\pi(a) = [a]$. Therefore, $\dim({}_{\text{inv}}\Gamma) = \text{card}(S)$, and so the FODC associated to S satisfies $\Gamma \neq 0$. So for any nontrivial root system R, this means that the FODC Γ is a quantum object because it is not the zero-dimensional space of classical 1-forms on the discrete manifold G. So already we have that the differential calculus of the QPB is going to be a nonclassical, quantum object provided that the Coxeter group is nontrivial. From now on, we assume that $\text{card}(R) \neq 0$.

Exercise 14.1. *For those not willing to accept the theorem that S coincides with the set of generators of G, here is an alternative approach. First, define S to be the subset $\{\sigma_\alpha \mid \alpha \in R\}$ of the Coxeter group G associated with the root system R. Then show directly that S has the three properties required to use it to define a bicovariant, $*$-covariant FODC over $\mathcal{A}$.*

Hint : Use $g^{-1} \sigma_\alpha g = \sigma_{\alpha g}$.

Now, we construct the QPB by first noting that the Coxeter group G associated to R leaves P invariant and acts freely on P, where

$$P := E \setminus \left(\cup_{\alpha \in R} H_\alpha \right),$$

the complement of all the reflecting hyperplanes of G in the Euclidean space E. Then we have the classical principal bundle $P \to M$ with structure group G, where the C^∞-manifold $M = P/G$ is the quotient of P by the right action of G on P. And this is another motivation for writing the action of G on E as a right action, since right actions are what one considers in the classical theory of principal bundles.

More specifically, the algebra for the space P is $\mathcal{B} = C^\infty(P)$, the algebra for the structure group G is $\mathcal{A} = C^\infty(G) = \mathcal{F}(G)$, and the right co-action $\mathcal{B} \to \mathcal{B} \otimes \mathcal{A}$ is the dual of the right action $P \times G \to P$. The subalgebra of right invariant elements in $\mathcal{B}$ is identified with the smooth functions on the base space, that is, $V = C^\infty(M)$. (Recall that for any C^∞-manifold X, the commutative algebra of C^∞-functions $X \to \mathbb{C}$ is denoted as $C^\infty(X)$.) So far this part of the construction is classical. But this is a QPB as well although it seems so far that nothing much is gained by viewing it as such.

However, when we continue to the next step, we will use the power of noncommutative geometry. In this step we construct a (nonunique!) differential calculus for this QPB. First, for the algebra $\mathcal{A}$ associated to the Coxeter group G, we use the universal HODC $\Gamma^\wedge$ that comes from the FODC Γ introduced above. This differential calculus is nonzero and so is *not* the differential calculus of exterior forms on the zero-dimensional manifold G, which is zero in every degree $k \geq 1$. For the total space, we will define a differential calculus $\Omega(P)$ that is *not* the classical differential calculus of exterior forms on P. The definition of $\Omega(P)$ as a graded vector space is

$$\Omega(P) := \mathfrak{D} \otimes_{\mathbb{C} \text{ inv}} \Gamma^\wedge,$$

where $\mathfrak{D}$ is the *complexification* of the standard de Rham calculus of exterior forms on P. All of the algebraic operations in $\Omega(P)$ to be defined presently will "twist" the two nontrivial factors in $\Omega(P)$ so that this can be considered a "twisted" tensor product. The upshot is that $\Omega(P)$ simply will not be the classical differential calculus on P, but rather a quantum object.

Now we go into the details of defining the algebraic operations on $\Omega(P)$. We let D denote the usual de Rham differential (complexified, of course) on $\mathfrak{D}$ and $_\mathfrak{D}\Phi : \mathfrak{D} \to \mathfrak{D} \otimes \mathcal{A}$ denote the right co-action of $\mathcal{A}$ on $\mathfrak{D}$ defined as

$$_\mathfrak{D}\Phi(\eta) := \sum_{g \in G} (g \cdot \eta) \otimes \delta_g \in \mathfrak{D} \otimes \mathcal{A}, \tag{14.4}$$

where $g \cdot \eta$ denotes the left action of g on a classical, complexified differential r-form η that comes from the right action of G on P. Explicitly, letting $R_g : P \to P$ be the right action map $x \mapsto xg$ for $x \in P$ and $g \in G$, we have

$$(g \cdot \eta)(X_1, \ldots, X_r) := \eta(T(R_g)X_1, \ldots, T(R_g)X_r),$$

where $X_1, \ldots, X_r$ are complexified vector fields on the classical manifold P and $T(R_g) : T(P) \to T(P)$ is the map induced by the smooth map R_g (basically its derivative) on the complexified tangent bundle $T(P)$ of P. So $g \cdot \eta \in \mathfrak{D}$ is a classical, complexified r-form.

Proposition 14.1. *The map* $_\mathfrak{D}\Phi$ *is a right co-action of* $\mathcal{A}$ *on* $\mathfrak{D}$.

Proof. On the one hand,

$$(_\mathfrak{D}\Phi \otimes id_\mathcal{A})_\mathfrak{D}\Phi(\eta) = \sum_{g \in G} {}_\mathfrak{D}\Phi(g \cdot \eta) \otimes \delta_g$$

$$= \sum_{g,h \in G} (h \cdot (g \cdot \eta)) \otimes \delta_h \otimes \delta_g$$

$$= \sum_{g,h \in G} ((hg) \cdot \eta) \otimes \delta_h \otimes \delta_g,$$

while on the other hand,

$$(id \otimes \phi)_\mathfrak{D}\Phi(\eta) = \sum_{k \in G} k \cdot \eta \otimes \phi(\delta_k)$$

$$= \sum_{g,k \in G} (k \cdot \eta) \otimes \delta_{kg^{-1}} \otimes \delta_g$$

$$= \sum_{g,h \in G} ((hg) \cdot \eta) \otimes \delta_h \otimes \delta_g,$$

where we used the change of variable $h = kg^{-1}$ as well as the formula for $\phi(\delta_k)$. The equality of these two expressions is one of the conditions for being a right co-action. For the other condition, we calculate

$$(id \otimes \varepsilon)_\mathfrak{D}\Phi(\eta) = \sum_{g \in G} (g \cdot \eta) \otimes \varepsilon(\delta_g) = e \cdot \eta \otimes 1 \cong \eta$$

since $\varepsilon(\delta_g) = \delta_g(e) = 0$ for $g \neq e$ and $\varepsilon(\delta_e) = \delta_e(e) = 1$. And this verifies the second condition for a right co-action. ∎

In the following we use Sweedler's notation

$$_\mathfrak{D}\Phi(\varphi) = \varphi^{(0)} \otimes \varphi^{(1)} \in \mathfrak{D} \otimes \mathcal{A}$$

for $\varphi \in \mathfrak{D}$ and similarly for $\psi \in \mathfrak{D}$. Recalling that $\hat{\phi} : \Gamma^\wedge \to \Gamma^\wedge \otimes \Gamma^\wedge$ restricts to $\hat{\phi} : {}_{\mathrm{inv}}\Gamma^\wedge \to {}_{\mathrm{inv}}\Gamma^\wedge \otimes \Gamma^\wedge$, we also put

$$\hat{\phi}(\theta) = \theta^{(1)} \otimes \theta^{(2)} \in {}_{\mathrm{inv}}\Gamma^\wedge \otimes \Gamma^\wedge$$

for $\theta \in {}_{\mathrm{inv}}\Gamma^\wedge$.

Then for $\varphi \otimes \theta, \psi \otimes \eta \in \Omega(P)$ with $\deg(\theta) = k$ and $\deg(\psi) = j$, we define their product by

$$(\varphi \otimes \theta)(\psi \otimes \eta) := (-1)^{jk} \varphi \psi^{(0)} \otimes (\theta \circ \psi^{(1)})\eta.$$

The $*$-operation is defined for $\deg(\varphi) = l$ by

$$(\varphi \otimes \theta)^* := (-1)^{kl} \varphi^{(0)*} \otimes (\theta^* \circ \varphi^{(1)*}).$$

And the differential in $\Omega(P)$ is defined by

$$d_P(\varphi \otimes \theta) := D(\varphi) \otimes \theta + (-1)^l \varphi^{(0)} \otimes \pi(\varphi^{(1)})\theta + (-1)^l \varphi \otimes d^\wedge(\theta),$$

where $d^\wedge : {}_{\mathrm{inv}}\Gamma^\wedge \to {}_{\mathrm{inv}}\Gamma^\wedge$ is the restriction of the differential d defined on the universal algebra $\Gamma^\wedge$.

Then $\Omega(P)$ is a graded differential, unital $*$-algebra, which is generated as a graded differential calculus by $\Omega^0(P) = \mathcal{B}$. The differential d_P satisfies as well the graded Leibniz rule with respect to the product on $\Omega(P)$ and is a $*$-morphism. Moreover, there is a right co-action of $\Gamma^\wedge$ on $\Omega(P)$

$$_{\Omega(P)}\Psi : \Omega(P) \to \Omega(P) \otimes \Gamma^\wedge$$

that is explicitly defined by

$$_{\Omega(P)}\Psi(\varphi \otimes \theta) := \varphi^{(0)} \otimes \theta^{(1)} \otimes \varphi^{(1)}\theta^{(2)},$$

where $\hat{\phi}(\theta) = \theta^{(1)} \otimes \theta^{(2)} \in {}_{\mathrm{inv}}\Gamma^\wedge \otimes \Gamma^\wedge$. This right co-action extends $_{\mathfrak{D}}\Phi$ and is a differential, unital $*$-morphism of graded algebras. All of this together says that we have defined a differential calculus on the QPB.

The reader may have noted a certain similarity with the construction of the vertical algebra $\mathfrak{ver}(P)$. This is not an accident. If one thinks of the algebra $\mathcal{B}$ [as used in the construction of $\mathfrak{ver}(P)$, not that of this section] as a graded, differential algebra with only grade 0 being nontrivial (and equal to $\mathcal{B}$) and with differential $D = 0$, then the construction given here reduces to that given for $\mathfrak{ver}(P)$. In the present setting we have various signs due to the grading as well as an extra term in the differential that corresponds to the differential of the first factor. So the verification of all the properties of the product, $*$-operation, differential d_P, and the right co-action of $\Omega(P)$ are proved analogously to the proofs given for $\mathfrak{ver}(P)$. A reference for this construction in a general setting is [23].

But here is something that we did not consider in the case of $\mathfrak{ver}(P)$. We first remark that the algebra $\Omega(P)$ has an associated subalgebra of horizontal forms $\mathfrak{hor}(P)$, defined as usual in terms of the right co-action $_{\Omega(P)}\Psi$ on $\Omega(P)$. The next result is quite important.

Theorem 14.3. $\mathfrak{hor}(P) = \mathfrak{D} \otimes 1_{\mathcal{A}}$.

Remark. One typically uses the canonical identification of $\mathfrak{D} \otimes 1_{\mathcal{A}}$ with $\mathfrak{D}$ so that this result then reads as $\mathfrak{hor}(P) = \mathfrak{D}$. So $\mathfrak{D}$, the standard, complexified de Rham calculus of exterior forms on P, turns out to be the algebra of *horizontal forms* of the algebra $\Omega(P)$ of the total space. Recall that the tensor product symbol $\otimes$ without subscript refers to the tensor product over the field $\mathbb{C}$ of complex numbers.

Proof. First, we suppose that $\varphi \in \mathfrak{D}$. Then by using $\hat{\phi}(1_{\mathcal{A}}) = 1_{\mathcal{A}} \otimes 1_{\mathcal{A}}$, we see that

$$_{\Omega(P)}\Psi(\varphi \otimes 1_{\mathcal{A}}) = (\varphi^{(0)} \otimes 1_{\mathcal{A}}) \otimes \varphi^{(1)} \in (\mathfrak{D} \otimes 1_{\mathcal{A}}) \otimes \mathcal{A} \subset \Omega(P) \otimes \mathcal{A}$$

since $_{\mathfrak{D}}\Phi(\varphi) = \varphi^{(0)} \otimes \varphi^{(1)} \in \mathfrak{D} \otimes \mathcal{A}$. And this says that $\varphi \otimes 1_{\mathcal{A}} \in \mathfrak{hor}(P)$. Thus, $\mathfrak{D} \otimes 1_{\mathcal{A}} \subset \mathfrak{hor}(P)$.

Conversely, suppose that $\alpha = \sum_k \varphi_k \otimes \theta_k \in \mathfrak{hor}(P) \subset \Omega(P) = \mathfrak{D} \otimes_{\mathrm{inv}} \Gamma^\wedge$, where $\varphi_k \in \mathfrak{D}$ and $\theta_k \in {}_{\mathrm{inv}}\Gamma^\wedge$. Then the hypothesis says that

$$_{\Omega(P)}\Psi(\alpha) = \sum_k {}_{\Omega(P)}\Psi(\varphi_k \otimes \theta_k) = \sum_k \varphi_k^{(0)} \otimes \theta_k^{(1)} \otimes \varphi_k^{(1)}\theta_k^{(2)} \in \Omega(P) \otimes \mathcal{A}.$$

Here $\varphi_k^{(0)} \in \mathfrak{D}$, $\varphi_k^{(1)} \in \mathcal{A}$, $\theta_k^{(1)} \in {}_{\mathrm{inv}}\Gamma^\wedge$, and $\theta_k^{(2)} \in \mathcal{A}$ hold. But then

$$\theta_k = \varepsilon(\theta_k^{(1)})\theta_k^{(2)} \in \mathcal{A},$$

which implies that $\deg \theta_k = 0$. Combining this with $\theta_k \in {}_{\mathrm{inv}}\Gamma^\wedge$, we see that $\theta_k \in \mathbb{C}$. And this means that

$$\alpha = \sum_k \varphi_k \otimes \theta_k = \sum_k \theta_k\varphi_k \otimes 1_{\mathcal{A}} \in \mathfrak{D} \otimes 1_{\mathcal{A}}.$$

So the inclusion $\mathfrak{hor}(P) \subset \mathfrak{D} \otimes 1_{\mathcal{A}}$ holds too. ∎

It turns out that we do not need any abstract theory to guarantee the existence of a quantum connection on $\Omega(P)$. We can actually write a simple, explicit formula for one quantum connection on $\Omega(P)$.

Theorem 14.4. *The map* $\tilde{\omega} : {}_{\mathrm{inv}}\Gamma \to \Omega^1(P)$ *defined for* $\theta \in {}_{\mathrm{inv}}\Gamma$ *by*

$$\tilde{\omega}(\theta) := 1_{\mathcal{B}} \otimes \theta \in \mathcal{B} \otimes {}_{\mathrm{inv}}\Gamma = \mathfrak{D}^0 \otimes {}_{\mathrm{inv}}\Gamma^{\wedge 1} \subset \Omega^1(P)$$

is a quantum connection on $\Omega(P)$. Its covariant derivative is given by

$$D_{\tilde{\omega}}(\varphi) = d_P(\varphi) - (-1)^k \varphi^{(0)} \otimes \pi(\varphi^{(1)}),$$

where $\varphi \in \mathfrak{hor}^k(P) \cong \mathfrak{D}^k$. We also have

$$D_{\tilde{\omega}} = D,$$

where D as defined earlier is the complexified de Rham differential in $\mathfrak{D}$.

Proof. Take $\theta \in {}_{\mathrm{inv}}\Gamma$. First, we show that $\tilde{\omega}$ is a $*$-morphism by computing

$$\tilde{\omega}(\theta)^* = (1 \otimes \theta)^* = (-1)^0 1^* \otimes \theta^* \circ 1^* = 1 \otimes \theta^* = \tilde{\omega}(\theta^*).$$

This is the first defining property of a connection.

Next, we write $\theta = \pi(a)$ for some $a \in \ker \varepsilon$. Then we have

$$
\begin{aligned}
(\tilde{\omega} \otimes id_A)\mathrm{ad}(\theta) &= (\tilde{\omega} \otimes id_A)\mathrm{ad}(\pi(a)) \\
&= (\tilde{\omega} \otimes id_A)(\pi \otimes id_A)\mathrm{ad}(a) \\
&= (\tilde{\omega} \otimes id_A)(\pi(a^{(2)}) \otimes \kappa(a^{(1)})a^{(3)}) \\
&= 1 \otimes \pi(a^{(2)}) \otimes \kappa(a^{(1)})a^{(3)} \\
&= 1 \otimes \mathrm{ad}(\pi(a)) \\
&= 1 \otimes \mathrm{ad}(\theta).
\end{aligned}
$$

On the other hand, we compute

$$
\begin{aligned}
{}_{\Omega(P)}\Psi(\tilde{\omega}(\theta)) &= {}_{\Omega(P)}\Psi(1 \otimes \theta) \\
&= 1 \otimes \theta^{(1)} \otimes \theta^{(2)} \\
&= 1 \otimes \hat{\phi}(\theta) \\
&= 1 \otimes (1_A \otimes \theta + \mathrm{ad}(\theta)) \\
&= 1 \otimes 1_A \otimes \theta + 1 \otimes \mathrm{ad}(\theta).
\end{aligned}
$$

And this shows

$$_{\Omega(P)}\Psi(\tilde{\omega}(\theta)) = (\tilde{\omega} \otimes id_A)\mathrm{ad}(\theta) + 1 \otimes 1 \otimes \theta,$$

which is the second defining property of a connection. So this proves that $\tilde{\omega}$ is a quantum connection.

Next, we write $_\mathfrak{D}\Phi(\varphi) = \varphi^{(0)} \otimes \varphi^{(1)} \in \mathfrak{D} \otimes \mathcal{A}$ for $\varphi \in \mathfrak{hor}^k(P)$ and use the definition of $\Omega_{(P)}\Psi$ to calculate

$$\Omega_{(P)}\Psi(\varphi) \cong \Omega_{(P)}\Psi(\varphi \otimes 1) = (\varphi^{(0)} \otimes 1) \otimes \varphi^{(1)} \cong \varphi^{(0)} \otimes \varphi^{(1)}$$

since $\hat{\phi}(1) = 1 \otimes 1$. Using this, the definition of covariant derivative, and the definition of $\tilde{\omega}$, we get

$$\begin{aligned} D_{\tilde{\omega}}(\varphi) &= d_P\,\varphi - (-1)^k (\varphi^{(0)} \otimes 1)\,\tilde{\omega}(\pi(\varphi^{(1)})) \\ &= d_P\,\varphi - (-1)^k (\varphi^{(0)} \otimes 1)(1_B \otimes \pi(\varphi^{(1)})) \\ &= d_P\,\varphi - (-1)^k \varphi^{(0)} \otimes \pi(\varphi^{(1)}), \end{aligned}$$

where the definition of the product in $\Omega(P)$ was used in the last equality. We recall that $k = \deg \varphi$. So the second statement of the theorem has been proved.

Using the definition of d_P, we evaluate

$$\begin{aligned} d_P(\varphi \otimes 1) &= D(\varphi) \otimes 1 + (-1)^k \varphi^{(0)} \otimes \pi(\varphi^{(1)})1 + (-1)^k \varphi \otimes d(1) \\ &= D(\varphi) \otimes 1 + (-1)^k \varphi^{(0)} \otimes \pi(\varphi^{(1)}). \end{aligned}$$

Then applying the identification $\mathfrak{D} \cong \mathfrak{D} \otimes 1$, we can rewrite this as

$$D(\varphi) = d_P(\varphi) - (-1)^k \varphi^{(0)} \otimes \pi(\varphi^{(1)}) = D_{\tilde{\omega}}(\varphi),$$

which proves the last statement of the theorem. ∎

We now turn to the question of defining quantum connections in this QPB with differential calculus. These also will not be classical connections in general, so that we will now be entering a truly quantum theory. In particular, the quantum connection that gives us the Dunkl operators will not be classical. In the following we drop the condition of being a *-morphism from the definitions of a quantum connection and of a quantum connection displacement. Even with these modifications of the definitions, we still have that all the quantum connections ω on the QPB P are given by

$$\omega = \tilde{\omega} + \lambda,$$

where λ is a quantum connection displacement (QCD). So we now propose to study QCDs on the QPB. We recall for the reader's convenience the remaining defining property of a QCD $\lambda : {}_{\text{inv}}\Gamma \to \Omega^1(P)$, which is

$$\Omega_{(P)}\Psi(\lambda(\theta)) = (\lambda \otimes id_A)\mathrm{ad}(\theta).$$

For the rest of this section, we let $\varphi \in \mathfrak{hor}^k(P)$ be an arbitrary horizontal k-form on P, which is to say a classical, complexified de Rham k-form on P. Writing an arbitrary quantum connection ω as $\omega = \tilde{\omega} + \lambda$ for some QCD λ and using $\mathfrak{hor}(P) = \mathfrak{D} \otimes 1 \cong \mathfrak{D}$ implicitly, we have

$$
\begin{aligned}
D_{\tilde{\omega}+\lambda}(\varphi) &= D_\omega(\varphi) \\
&= d_P(\varphi) - (-1)^k \varphi^{(0)} \omega(\pi(\varphi^{(1)})) \\
&= d_P(\varphi) - (-1)^k \varphi^{(0)} \tilde{\omega}(\pi(\varphi^{(1)})) - (-1)^k \varphi^{(0)} \lambda(\pi(\varphi^{(1)})) \\
&= D_{\tilde{\omega}}(\varphi) - (-1)^k \varphi^{(0)} \lambda(\pi(\varphi^{(1)})) \\
&= D_{\tilde{\omega}}(\varphi) - (-1)^k T_\lambda(\varphi),
\end{aligned} \tag{14.5}
$$

where we define

$$
T_\lambda(\varphi) := \varphi^{(0)} \lambda(\pi(\varphi^{(1)})) \cong (\varphi^{(0)} \otimes 1)(1 \otimes \lambda(\pi(\varphi^{(1)}))) = \varphi^{(0)} \otimes \lambda(\pi(\varphi^{(1)})).
$$

Note that the tensor product $\otimes$ can be written or omitted between the two factors $\varphi^{(0)}$ and $\lambda(\pi(\varphi^{(1)}))$, depending on how those factors are interpreted. Since we already know that both $D_\omega(\varphi)$ and $D_{\tilde{\omega}}(\varphi)$ are horizontal, it follows that $T_\lambda(\varphi)$ is also horizontal. (Recall that horizontal forms were defined without reference to any connection.)

Then we obtain

$$
\begin{aligned}
T_\lambda(\varphi) &= \varphi^{(0)} \lambda(\pi(\varphi^{(1)})) \\
&= \sum_{g \in G} \varphi_g \lambda(\pi(\delta_g)) \\
&= \sum_{g \in S \cup \{e\}} \varphi_g \lambda(\pi(\delta_g)) \\
&= \sum_{g \in S} \varphi_g \lambda(\pi(\delta_g)) + \varphi_e \lambda(\pi(\delta_e)) \\
&= \sum_{s \in S} \varphi_s \lambda(\pi(\delta_s)) - \sum_{s \in S} \varphi_e \lambda(\pi(\delta_s)) \\
&= \sum_{s \in S} (\varphi_s - \varphi) \lambda(\pi(\delta_s)),
\end{aligned}
$$

since $\pi(\delta_g) = 0$ for $g \notin S \cup \{e\}$ and $\pi(\delta_e) = -\sum_{s \in S} \pi(\delta_s)$. We also used $\varphi_e = e \cdot \varphi = \varphi$. Note that we are abandoning Sweedler's notation here in favor of an explicit summation. This may seem strange to the reader, but we will soon see that this is an essential step in this argument.

Putting this into (14.5), we immediately have

$$D_{\tilde{\omega}+\lambda}(\varphi) = D_{\tilde{\omega}}(\varphi) - (-1)^k \sum_{s \in S} (\varphi_s - \varphi)\lambda(\pi(\delta_s))$$

$$= D_{\tilde{\omega}}(\varphi) + (-1)^k \sum_{s \in S} (\varphi - \varphi_s)\lambda(\pi(\delta_s)). \tag{14.6}$$

This equation gives the first indication of something resembling the definition of a Dunkl operator. It is a question now of finding the appropriate λ since so far this works for any quantum connection $\omega = \tilde{\omega} + \lambda$ on the QPB, that is, for any QCD λ. Now we will put restrictions on the QCD, eventually arriving at a specific QCD that will give us the Dunkl operators. Since $\lambda : {}_{\mathrm{inv}}\Gamma \to \Omega^1(P)$ is linear, it is sufficient to define λ on the basis on ${}_{\mathrm{inv}}\Gamma$ introduced earlier:

$$\{\pi(\delta_s) \,|\, s \in S\}.$$

We wish to emphasize that this basis does not depend on the particular choice of R^+, since $\sigma_{-\alpha} = \sigma_\alpha$ for all $\alpha \in R$. With respect to the $*$-operation on ${}_{\mathrm{inv}}\Gamma$, these basis elements are imaginary since

$$\pi(\delta_s)^* = -\pi(\kappa(\delta_s)^*) = -\pi(\kappa(\delta_s^*)) = -\pi(\kappa(\delta_s)) = -\pi(\delta_{s^{-1}}) = -\pi(\delta_s).$$

Here we used that κ is a $*$-morphism in this case and $s^{-1} = s$. This shows that π is not a $*$-morphism in this case. In general, the elements of the canonical basis ${}_{\mathrm{inv}}\Gamma$ are neither real nor imaginary. The fact that they are imaginary here is a special property of this particular case. In particular, the linear map $\pi : \mathcal{A} \to {}_{\mathrm{inv}}\Gamma$ is not a $*$-morphism in this case since it maps the real elements δ_s to imaginary elements. Actually, $(\pi(a))^* = -\pi(a^*)$ in this case. However, the $\mathbb{C}$-linear map

$$\tilde{\pi} := -i\pi : \mathcal{A} \to {}_{\mathrm{inv}}\Gamma$$

is a $*$-morphism. Since we wish to write just about all maps in terms of $*$-morphisms in order to preserve reality, we rewrite (14.6) as

$$D_{\tilde{\omega}+\lambda}(\varphi) = D_{\tilde{\omega}}(\varphi) + (-1)^k \sum_{s \in S} (\varphi - \varphi_s)\tilde{\lambda}(\tilde{\pi}(\delta_s)),$$

where we require $\tilde{\lambda} := i\lambda$ to be a $*$-morphism. Note that the approach here differs from that in [31], where we took λ to be a $*$-morphism. Nonetheless, the results here are consistent with those in [31].

We let each $\alpha \in R$ also denote the 1-form on P given by

$$x \mapsto (x, \alpha^\sharp) \in P \times E^* = T^*(P), \text{ the cotangent bundle of } P,$$

for all $x \in P$, where $\alpha^\sharp : E \to \mathbb{R}$ is the linear functional $v \mapsto \langle \alpha, v \rangle$ for all $v \in E$. Here we have used the canonical trivialization of the cotangent bundle on P as $T^*(P) = P \times E^*$, where E^* is the (real) dual space to E. In particular, we note that this 1-form α is real.

While the canonical basis of $_{inv}\Gamma$ is uniquely indexed by the elements $s \in S$, or equivalently by the elements $\alpha \in R^+$, it is more convenient for the next definition to use the basis of self-adjoint (or real) elements

$$\{\tilde{\pi}(\delta_{\sigma_\alpha}) \,|\, \alpha \in R\}$$

even though each basis element is indexed by exactly two elements in R and there is an extra imaginary factor because of the definition of $\tilde{\pi}$. Here is this important definition:

Definition 14.3. *Let the linear map* $\tilde{\lambda} : {}_{inv}\Gamma \to \Omega^1(P)$ *be defined on the basis* $\{\tilde{\pi}(\delta_{\sigma_\alpha}) \,|\, \alpha \in R\}$ *of* $_{inv}\Gamma$ *by*

$$\tilde{\lambda}(\tilde{\pi}(\delta_{\sigma_\alpha}))(x) := h_\alpha(x)\,\alpha \in \mathfrak{D}^1 \cong \mathfrak{D}^1 \otimes 1 \subset \Omega^1(P)$$

for all $x \in P$, *where* $h_\alpha : P \to \mathbb{R}$ *is a family of* C^∞-*functions indexed by* $\alpha \in R$ *and satisfying these conditions for all* $\alpha \in R$ *and all* $x \in P$:

1. $h_{-\alpha}(x) = -h_\alpha(x)$.
2. $h_{\alpha g}(x) = h_\alpha(xg^{-1})$.

(Note that the equation in [31] corresponding to Eq. 1 here inadvertently omits the minus sign on the right side.)

Then we say that $\tilde{\lambda}$ is a Dunkl quantum connection displacement *or, more briefly,* a Dunkl QCD.

This terminology is justified in the next result.

Proposition 14.2. *Let* $\tilde{\lambda}$ *be a Dunkl QCD. Then* $\tilde{\lambda}$ *is a QCD.*

Proof. Condition 1 guarantees that $\tilde{\lambda}$ is well defined, since even though every $s \in S$ has exactly two representations, $s = \delta_{\sigma_\alpha} = \delta_{\sigma_{-\alpha}}$, they satisfy

$$\tilde{\lambda}(\tilde{\pi}(\delta_{\sigma_{-\alpha}}))(x) = h_{-\alpha}(x)\,(-\alpha) = (-1)^2 h_\alpha(x)\,\alpha = \tilde{\lambda}(\tilde{\pi}(\delta_{\sigma_\alpha}))(x).$$

Next, since h_α is real-valued and α is also real, we see that

$$\tilde{\lambda}(\tilde{\pi}(\delta_{\sigma_\alpha})^*) = \tilde{\lambda}(\tilde{\pi}(\delta_{\sigma_\alpha})) = h_\alpha\,\alpha = (h_\alpha\,\alpha)^* = \left(\tilde{\lambda}(\tilde{\pi}(\delta_{\sigma_\alpha}))\right)^*,$$

which shows that $\tilde{\lambda}$ is a $*$-morphism.

It remains to show the second defining property of a QCD, namely, that

$$\Omega_{(P)}\Psi(\tilde{\lambda}(\theta)) = (\tilde{\lambda} \otimes 1_A)\mathrm{ad}(\theta)$$

for all $\theta \in {}_{inv}\Gamma$. But it suffices to prove this for the basis elements $\theta = \pi(\delta_{\sigma_\alpha})$ for $\alpha \in R$. Expanding the right side, we see that

$$
\begin{aligned}
(\tilde{\lambda} \otimes id_A)\mathrm{ad}(\theta) &= (\tilde{\lambda} \otimes id_A)\mathrm{ad}(\pi(\delta_{\sigma_\alpha})) \\
&= (\tilde{\lambda} \otimes id_A)(\pi \otimes id_A)\mathrm{ad}(\delta_{\sigma_\alpha}) \\
&= \sum_{g \in G} (\tilde{\lambda} \otimes id_A)(\pi \otimes id_A)(\delta_{g\sigma_\alpha g^{-1}} \otimes \delta_g) \\
&= \sum_{g \in G} \tilde{\lambda}\pi(\delta_{g\sigma_\alpha g^{-1}}) \otimes \delta_g \\
&= \sum_{g \in G} \tilde{\lambda}\pi(\delta_{\sigma_{\alpha g^{-1}}}) \otimes \delta_g \\
&= \sum_{g \in G} i\tilde{\lambda}\tilde{\pi}(\delta_{\sigma_{\alpha g^{-1}}}) \otimes \delta_g \\
&= \sum_{g \in G} ih_{\alpha g^{-1}}(\alpha g^{-1}) \otimes \delta_g.
\end{aligned}
$$

To understand the right side, we first need to compute that

$$
g \cdot (\tilde{\lambda}\tilde{\pi}(\delta_{\sigma_\alpha})) = g \cdot (h_\alpha\, \alpha) = (g \cdot h_\alpha)(\alpha g^{-1})
$$

using the covariance of forms. Therefore, for $x \in P$, we have

$$
g \cdot (\tilde{\lambda}\tilde{\pi}(\delta_{\sigma_\alpha}))(x) = (g \cdot h_\alpha)(x)(\alpha g^{-1}) = h_\alpha(xg)(\alpha g^{-1}) = h_{\alpha g^{-1}}(x)(\alpha g^{-1}),
$$

where in the last step we used condition 2. So

$$
\begin{aligned}
\Omega_{(P)}\Psi(\tilde{\lambda}(\theta)) &= \Omega_{(P)}\Psi(\tilde{\lambda}\pi(\delta_{\sigma_\alpha})) \\
&= i\,\Omega_{(P)}\Psi(\tilde{\lambda}\tilde{\pi}(\delta_{\sigma_\alpha})) \\
&= \sum_{g \in G} ig \cdot \tilde{\lambda}\tilde{\pi}(\delta_{\sigma_\alpha}) \otimes \delta_g \\
&= \sum_{g \in G} ih_{\alpha g^{-1}}(\alpha g^{-1}) \otimes \delta_g.
\end{aligned}
$$

And that completes the proof. ∎

Definition 14.4. *Suppose* $\tilde{\lambda} : {}_{inv}\Gamma \to \Omega^1(P)$ *is a Dunkl quantum connection displacement. Then the quantum connection defined by*

$$
\omega := \tilde{\omega} + \lambda,
$$

where $\lambda := -i\tilde{\lambda}$, *is called a* Dunkl quantum connection.

Next, we give a rather large class of examples of Dunkl QCDs. Of course, each
one gives a corresponding Dunkl quantum connection. Suppose that $\psi_\alpha : \mathbb{R}\setminus\{0\} \to$
$\mathbb{R}$ are odd C^∞-functions for each $\alpha \in R$ that satisfy $\psi_{\alpha g} = \psi_\alpha$ for all $g \in G$.
This means that ψ_α is really indexed by the G-orbits in R, that is, for each G-
orbit we pick an odd C^∞-function. Also, these odd functions are determined by
their restrictions to the half-line $(0, \infty)$. So these examples are in bijection with the
elements in the infinite-dimensional vector space

$$\prod_{\alpha \in O} C^\infty(0, \infty),$$

where O is the finite set of G-orbits in R. Then for a given choice of such functions
ψ_α, we define for each $\alpha \in R$ the function

$$h_\alpha(x) := \psi_\alpha(\langle \alpha, x \rangle)$$

for all $x \in P$. Notice that $x \in P$ implies that $\langle \alpha, x \rangle \neq 0$ for all $\alpha \in R$ so that each
$\psi_\alpha(\langle \alpha, x \rangle)$ is well defined. Moreover, since $\psi_{-\alpha} = \psi_{\alpha\sigma_\alpha} = \psi_\alpha$, we see that

$$h_{-\alpha}(x) = \psi_{-\alpha}(\langle -\alpha, x \rangle) = \psi_\alpha(\langle -\alpha, x \rangle) = -\psi_\alpha(\langle \alpha, x \rangle) = -h_\alpha(x),$$

which is condition 1 in the definition of a Dunkl QCD.
   Next, for $\alpha \in R, g \in G \subset O(E)$, and $x \in P$, we have

$$h_{\alpha g}(x) = \psi_{\alpha g}(\langle \alpha g, x \rangle) = \psi_\alpha(\langle \alpha g, x \rangle) = \psi_\alpha(\langle \alpha, x g^{-1} \rangle) = h_\alpha(x g^{-1}),$$

which is condition 2 in the definition of a Dunkl QCD.
   So this establishes that we have defined an ample class of Dunkl QCDs.
A particular case of this example is given by defining for all real $r \neq 0$

$$\psi_\alpha(r) := \frac{\mu(\alpha)}{r},$$

where $\mu : R \to \mathbb{R}$ is a multiplicity function (that is, it is G-invariant). Then the
corresponding functions h_α are given by

$$h_\alpha(x) = \psi_\alpha(\langle \alpha, x \rangle) = \frac{\mu(\alpha)}{\langle \alpha, x \rangle}. \tag{14.7}$$

Definition 14.5. *The Dunkl quantum connection corresponding to the choice
in (14.7) is called the* standard Dunkl quantum connection.

   We next are going to identify the covariant derivative of Dunkl quantum
connections and, in particular, of standard Dunkl quantum connections. We start
from (14.6) to get the covariant derivative

$$D_{\tilde{\omega}+\lambda}(\varphi) = D_{\tilde{\omega}}(\varphi) + (-1)^k \sum_{s \in S} (\varphi - \varphi_s)\lambda(\pi(\delta_s))$$

$$= D(\varphi) + (-1)^k \sum_{\alpha \in \mathcal{R}^+} (\varphi - \varphi_{\sigma_\alpha})\tilde{\lambda}(\tilde{\pi}(\delta_{\sigma_\alpha}))$$

$$= D(\varphi) + (-1)^k \sum_{\alpha \in \mathcal{R}^+} (\varphi - \sigma_\alpha \cdot \varphi)\tilde{\lambda}(\tilde{\pi}(\delta_{\sigma_\alpha}))$$

$$= D(\varphi) + (-1)^k \sum_{\alpha \in \mathcal{R}^+} h_\alpha(\varphi - \sigma_\alpha \cdot \varphi)\alpha$$

for a Dunkl quantum connection $\tilde{\omega} + \lambda$ with a family of functions h_α as above and $\varphi \in \mathfrak{D}^k = \mathfrak{hor}^k(P)$. Here we also used $\varphi_g = g \cdot \varphi$ [see (14.4)]. Notice that in this case the quantum connection $\tilde{\omega} + \lambda$ is the sum of a *-morphism $\tilde{\omega}$ and a *-antimorphism λ.

In particular, for a standard Dunkl quantum connection with

$$h_\alpha = \frac{\mu(\alpha)}{\langle \alpha, \cdot \rangle},$$

we get the more specific formula

$$D_{\tilde{\omega}+\lambda}(\varphi) = D(\varphi) + (-1)^k \sum_{\alpha \in \mathcal{R}^+} \mu(\alpha)\frac{(\varphi - \sigma_\alpha \cdot \varphi)}{\langle \alpha, \cdot \rangle}\alpha.$$

Again, this is for $\varphi \in \mathfrak{D}^k = \mathfrak{hor}^k(P)$. The Dunkl operators in [19] are then obtained as the case $k = 0$, as we state in the following.

Theorem 14.5 ([31]). *For every $f \in C^\infty(P)$, the covariant derivative Tf of the standard Dunkl quantum connection associated with the multiplicity function μ is given for all $x \in P$ by*

$$Tf(x) := D_{\tilde{\omega}+\lambda}f(x) = Df(x) + \sum_{\alpha \in \mathcal{R}^+} \mu(\alpha)\frac{f(x) - f(x\sigma_\alpha)}{\langle \alpha, x \rangle}\alpha,$$

which is the Dunkl operator in vector form. If we take the jth component for $j = 1, \ldots, n = \dim E$ with respect to some orthonormal basis $\{\varepsilon_j\}$ of E, this becomes

$$T_j f(x) = \frac{\partial f}{\partial x_j} + \sum_{\alpha \in \mathcal{R}^+} \mu(\alpha)\frac{f(x) - f(x\,\sigma_\alpha)}{\langle \alpha, x \rangle}\alpha_j,$$

which are the Dunkl operators in their standard coordinate form. Here we write $\alpha = \sum_j \alpha_j \varepsilon_j$, and $\partial f/\partial x_j$ denotes the directional derivative associated to the unit vector ε_j.

This theorem is the main result of [31]. It establishes a concrete link between Dunkl operators, which originally were introduced in [19] in the context of harmonic analysis, and covariant derivatives of specific quantum connections of certain QPBs with differential calculus in the theory of noncommutative geometry.

Exercise 14.2. *Show that the standard Dunkl quantum connection* $\omega = \tilde{\omega} + \lambda$ *in Theorem 14.5 is not regular.* **Hint:** *Think about the graded Leibniz rule.*

This construction of the Dunkl operators gives in one fell swoop their action on horizontal k-forms on P for every $k \geq 0$. Of course, the extension from 0-forms to arbitrary classical k-forms is both natural and straightforward, starting with the theory as presented in [19]. So this aspect of the present theory has only its little bit of elegance and not anything much in the way of novelty. But for the 0-forms, we should note that in this theory they are the functions in $C^\infty(P)$ and P is only a dense, open set in E. Of course, we can extend the domain of the Dunkl operators to $C^\infty(E)$ in a trivial way by

$$T_{\mu,\xi} f := T_{\mu,\xi}(f|_P) \in C^\infty(P) \quad \text{for } f \in C^\infty(E).$$

However, in analysis the formula for an operator is often more important than the domain (or more typically, domains) on which that formula gives a well-defined operator. In this sense the present theory, which does produce the defining formula for Dunkl operators, provides a new way of viewing them, namely, as covariant derivatives in a noncommutative geometric setting.

As a secondary result, the authors of [31] included a proof of the well-known result of Theorem 14.2. This result is shown to be a consequence of the standard Dunkl quantum connection having curvature zero. So Theorem 14.2 now has a noncommutative geometric interpretation in terms of curvature. That proof can be found in [31].

The following points underline in detail why this construction of the Dunkl operators uses in a fundamental way the theory of QPBs, and not the classical theory.

1. The zero-dimensional structure group G is given a quantum, nontrivial differential calculus.
2. The total space P is given a quantum differential calculus that is not its classical differential calculus.
3. The quantum connection introduced on the QPB with differential calculus is not a classical connection. It is not even a regular connection, unlike classical connections.
4. The concepts of covariant derivative and horizontal form are defined within the general setting of noncommutative geometry. The classical definitions of these concepts are not sufficient in order to realize this construction of Dunkl operators. In particular, covariant derivatives in classical differential geometry are local operators, while the Dunkl operators are not local operators.

14.4 Notes

This chapter is based on some of the material presented in [31]. Micho
Đurđevich and I had many years of discussions about various topics in my research,
mostly about their possible relation to the ideas of noncommutative geometry. These
were brainstorming sessions. We just threw around ideas and tried and tried to get
something to work. I owe a lot—and I really mean a lot—to Micho's continued
patience as we discarded one thing after another. When I started to work on the
generalized Segal–Bargmann spaces associated to Coxeter groups and their Dunkl
operators (especially the associated Dunkl Laplacian and its heat kernel in [75]),
Micho became fascinated by their defining formulas. He wanted to know whether I
knew from where they came. "From the head of Charles Dunkl" was not the right
answer even though we are truly indebted to him for using his head so well! Micho
wanted to know their underlying structure. But neither of us was eager to retrace
the original path taken by Dunkl himself. (By the way, that story is told in [20].)
The point was whether we could understand Dunkl operators in our own way. What
are the Dunkl operators anyway? And more to the point, do they have something to
do with noncommutative geometry? As often happens in research, finding the right
question is a key step. But the devil, as always, lies in the details. And this time the
details worked out.

Chapter 15
What Next?

After plowing through 14 chapters of mathematics, readers face the question of "what next?" What would one want to do with this information? How could one use it?

The long-term goal of any theory is to turn itself into a language. Take calculus, for example. Nowadays calculus is often labeled "trivial." But it is the engrained, traditional language of an enormous amount of science. It is indispensable in physics and much of mathematics. It permeates chemistry and biology and even reaches the far shores of economics. We cannot do science without it!

But calculus is one of many other useful languages. Some of these are older languages such as algebra and the language of equations. (Yes, there was mathematics before there were equations and algebraic notation!) And some of these are newer languages that are derived from calculus and are not so widely spoken. One such language is stochastic analysis, which includes the Ito integral together with other extensions of the language of calculus. Another modern language has arisen from category theory. This is now a complement to the older language of set theory.

Also, recall that the "differential" in differential geometry is the same as the "differential" in differential calculus. Of course, differential geometry is a specialized geometric theory that continues to thrive in current research. But it also has become a language for expressing many physical theories. It is not as widely spoken as the language of calculus, but it has its community of "native speakers."

The theory presented in this book, as well as the alternative approaches forwarded by various well-established researchers, are probably just the first steps toward establishing yet another language derived from calculus. It is not easy to become a fluent speaker in a second language, especially when that language is in the process of formation. But eventually that could be farther down the path that is merely introduced in this book. Again, as I remarked in the preface, this book is an introduction, really nothing more.

Explicitly, readers might want to read more about this theory and variants of it. To do this, they could start by consulting the necessarily incomplete bibliography given here as well as what is available on arxiv.org and other resources on the

© Springer International Publishing Switzerland 2015
S.B. Sontz, *Principal Bundles*, Universitext, DOI 10.1007/978-3-319-15829-7_15

Internet and elsewhere. It should not come as a surprise that I have found the papers of Đurđevich and of Woronowicz to be of the utmost utility in this regard. As readers learn the theory in ever more detail and are able to apply it in their own areas of research, it will facilitate the formation of a new language, this time derived from the language of differential geometry though now in a noncommutative setting. Whether such a language will form and take hold is unforeseeable. If it does, it most likely will be as a language spoken in physics since that is where I expect the important applications to be found. So that is my answer to the question of "what next?" And I do hope to find many physicists in my audience.

Appendix A
m Is a Bimodule Morphism

This appendix presents a detailed demonstration that for any associative algebra $\mathcal{A}$ with unit 1, the multiplication map $m = m_{\mathcal{A}} : \mathcal{A} \otimes \mathcal{A} \to \mathcal{A}$ is an $\mathcal{A}$-bimodule map. This is the sort of elementary result that the reader should be able to produce with only a hint at most, but we present it in detail to show the reader what is entailed in a rather complete, thorough proof. Other results of this sort will not always be written out in such great detail.

The first step is to define canonical $\mathcal{A}$-bimodule structures on $\mathcal{A}$ and on $\mathcal{A} \otimes \mathcal{A}$. We start with $\mathcal{A}$. The left action of $\mathcal{A}$ on $\mathcal{A}$ is simply the multiplication $m : \mathcal{A} \otimes \mathcal{A} \to \mathcal{A}$. That this is a left action follows from $a(bx) = (ab)x$ and $1x = x$ for all $a, b, x \in \mathcal{A}$, namely, the associativity of multiplication in $\mathcal{A}$ and the fact that 1 is a left unit. Also, m is linear, an essential though trivial observation that we will not repeat hereafter. This left action gives a canonical left $\mathcal{A}$-module structure to $\mathcal{A}$. (Canonical here really means functorial, but going into that would involve too many trivial details.) The right action of $\mathcal{A}$ on $\mathcal{A}$ is also the multiplication $m : \mathcal{A} \otimes \mathcal{A} \to \mathcal{A}$. That this is a right action follows from $(xa)b = x(ab)$ and $x1 = x$ for all $a, b, x \in \mathcal{A}$. We are now using the fact that 1 is a right unit for the multiplication. This right action gives a canonical right $\mathcal{A}$-module structure to $\mathcal{A}$. But we have not yet proved enough to conclude that these two lateral $\mathcal{A}$-module structures give an $\mathcal{A}$-bimodule structure. We also have to show that they are compatible, namely, that $a(xb) = (ax)b$ for all $a, b, x \in \mathcal{A}$. But this is just the associativity again, and so we have a canonical $\mathcal{A}$-bimodule structure on $\mathcal{A}$.

We next define the $\mathcal{A}$-bimodule structure on $\mathcal{A} \otimes \mathcal{A}$. The left action of $\mathcal{A}$ on $\mathcal{A} \otimes \mathcal{A}$ is given by left multiplication on the first factor: $a(x \otimes y) := ax \otimes y$ for all $a, x, y \in \mathcal{A}$. The diagram for this is

$$\mathcal{A} \otimes \mathcal{A} \otimes \mathcal{A} \xrightarrow{m \otimes id} \mathcal{A} \otimes \mathcal{A}. \tag{A.1}$$

© Springer International Publishing Switzerland 2015
S.B. Sontz, *Principal Bundles*, Universitext, DOI 10.1007/978-3-319-15829-7

The right action of $\mathcal{A}$ on $\mathcal{A} \otimes \mathcal{A}$ is given by right multiplication on the second factor: $(x \otimes y)a := x \otimes ya$ for all $a, x, y \in \mathcal{A}$. Its diagram is

$$\mathcal{A} \otimes \mathcal{A} \otimes \mathcal{A} \xrightarrow{id \otimes m} \mathcal{A} \otimes \mathcal{A}. \tag{A.2}$$

Exercise A.1. *Prove:*

- *$m \otimes id$ gives $\mathcal{A} \otimes \mathcal{A}$ the structure of a left $\mathcal{A}$-module.*
- *$id \otimes m$ gives $\mathcal{A} \otimes \mathcal{A}$ the structure of a right $\mathcal{A}$-module.*
- *$m \otimes id$ and $id \otimes m$ give $\mathcal{A} \otimes \mathcal{A}$ the structure of an $\mathcal{A}$-bimodule.*

Proposition A.1. *The multiplication map $m : \mathcal{A} \otimes \mathcal{A} \to \mathcal{A}$ is an $\mathcal{A}$-bimodule map.*

Proof. First, we show that m commutes with the left actions of $\mathcal{A}$ on $\mathcal{A} \otimes \mathcal{A}$ and on $\mathcal{A}$, respectively. This comes down to the following diagram chase:

$$\begin{array}{ccc} a \otimes x \otimes y & \xrightarrow{l.a.} & ax \otimes y \\ {\scriptstyle id \otimes m} \downarrow & & \downarrow {\scriptstyle m} \\ a \otimes xy & \xrightarrow{l.a.} & a(xy) = (ax)y, \end{array} \tag{A.3}$$

where the horizontal arrows are the appropriate left actions. The proof that m commutes with the right actions of $\mathcal{A}$ on $\mathcal{A} \otimes \mathcal{A}$ and on $\mathcal{A}$ is equally straightforward. ∎

Corollary A.1. $\mathcal{A}^2 := \mathrm{Ker}\, m$ *is an $\mathcal{A}$-sub-bimodule of $\mathcal{A} \otimes \mathcal{A}$.*

This corollary is an essential ingredient in the discussion in the main text of universal first-order differential calculi (FODCs).

Appendix B
Hopf Algebras, an Overview

To make this presentation more self-contained, we present a quick introduction to Hopf algebras. The idea is to derive—or at least mention—some basic facts that are used throughout the text. For further details, there are an amazing number of tomes dedicated to this topic. Among these, Kassel's book [45] seems to be the most leisurely. Other standard references are [1] and [78].

Unfortunately, this overview is long on definitions and short on examples. a typical failing of expositions of this sort. Anyway, the reader has been forewarned.

B.1 Basic Properties and Identities

Definition B.1. *An* algebra *with unit over the field of complex numbers* $\mathbb{C}$ *is a triple* $\langle A, \mu, 1_A \rangle$, *where* A *is a vector space over* $\mathbb{C}$ *together with a multiplication, that is, a bilinear map*

$$\mu : A \times A \to A$$

and with a unit element $1_A \in A$. *We denote* $\mu(a, b)$, *the multiplication of elements* $a, b \in A$, *simply as* ab. *And when context permits the ambiguity in notation, we write* 1 *instead of* 1_A. *These satisfy*

1. *(Associativity)* $a(bc) = (ab)c$ *for all* $a, b, c \in A$.
2. *(Unit property)* $1_A a = a 1_A = a$ *for all* $a \in A$.

Remarks. The vector space A need not be finite dimensional. The bilinear multiplication induces a unique linear map, also denoted by μ:

$$\mu : A \otimes A \to A.$$

© Springer International Publishing Switzerland 2015
S.B. Sontz, *Principal Bundles*, Universitext, DOI 10.1007/978-3-319-15829-7

The tensor product here is the *algebraic* tensor product. The unit element also induces a unique linear map

$$\eta : \mathbb{C} \to \mathcal{A}$$

defined by $\eta(\lambda) := \lambda \, 1_{\mathcal{A}}$ for all $\lambda \in \mathbb{C}$. We often simply say that $\mathcal{A}$ is an algebra without explicitly mentioning the multiplication and unit. But at the other notational extreme, we sometimes write the multiplication and unit as $\mu_{\mathcal{A}}$ and $\eta_{\mathcal{A}}$, respectively, in order to emphasize that these structures pertain to a particular vector space $\mathcal{A}$.

This definition can be written equivalently in terms of the vector space $\mathcal{A}$ and the two linear maps μ and η. In that case, the two defining properties are translated into the requirement that these two diagrams commute:

$$
\begin{array}{ccc}
\mathcal{A} \otimes \mathcal{A} \otimes \mathcal{A} & \xrightarrow{\;id \otimes \mu\;} & \mathcal{A} \otimes \mathcal{A} \\
{\scriptstyle \mu \otimes id}\Big\downarrow & & \Big\downarrow{\scriptstyle \mu} \\
\mathcal{A} \otimes \mathcal{A} & \xrightarrow{\;\mu\;} & \mathcal{A},
\end{array}
$$

$$
\begin{array}{ccccc}
\mathcal{A} \otimes \mathbb{C} & \xrightarrow{\;id \otimes \eta\;} & \mathcal{A} \otimes \mathcal{A} & \xleftarrow{\;\eta \otimes id\;} & \mathbb{C} \otimes \mathcal{A} \\
{\scriptstyle \cong}\Big\downarrow & & \Big\downarrow{\scriptstyle \mu} & & \Big\downarrow{\scriptstyle \cong} \\
\mathcal{A} & \xrightarrow{\;id\;} & \mathcal{A} & \xleftarrow{\;id\;} & \mathcal{A}.
\end{array}
$$

The vertical arrow on the left of the last diagram is the canonical isomorphism $\mathcal{A} \otimes \mathbb{C} \to \mathcal{A}$ given by $a \otimes \lambda \mapsto \lambda \, a$. A similar comment holds for the vertical arrow on the right.

Definition B.2. *Suppose that $\mathcal{A}_1$ and $\mathcal{A}_2$ are algebras. A morphism f from $\mathcal{A}_1$ to $\mathcal{A}_2$, denoted $f : \mathcal{A}_1 \to \mathcal{A}_2$, is a linear map f from the vector space $\mathcal{A}_1$ to the vector space $\mathcal{A}_2$ satisfying preservation of multiplication, namely, the commutativity of the diagram*

$$
\begin{array}{ccc}
\mathcal{A}_1 \otimes \mathcal{A}_1 & \xrightarrow{\;f \otimes f\;} & \mathcal{A}_2 \otimes \mathcal{A}_2 \\
{\scriptstyle \mu_{\mathcal{A}_1}}\Big\downarrow & & \Big\downarrow{\scriptstyle \mu_{\mathcal{A}_2}} \\
\mathcal{A}_1 & \xrightarrow{\;f\;} & \mathcal{A}_2,
\end{array}
$$

and preservation of unit, namely, the commutativity of

$$
\begin{array}{ccc}
\mathbb{C} & \xrightarrow{\;id\;} & \mathbb{C} \\
{\scriptstyle \eta_{\mathcal{A}_1}}\Big\downarrow & & \Big\downarrow{\scriptstyle \eta_{\mathcal{A}_2}} \\
\mathcal{A}_1 & \xrightarrow{\;f\;} & \mathcal{A}_2.
\end{array}
$$

These diagrams can also be written as $f(ab) = f(a)f(b)$ *for all* $a, b \in \mathcal{A}_1$ *and* $f(1) = 1$.

Corollary B.1. *The collection whose objects are all the algebras over* $\mathbb{C}$ *and whose arrows are all the morphisms between algebras is a category.*

We leave to the reader the definition of the composition of morphisms as well as the remaining details in the proof of the result.

So all the usual terminology of categories can be applied here. In particular, we have monomorphisms, epimorphisms, and isomorphisms of algebras.

The dual structure is defined next. This is done from the outset using tensor products and diagrams.

Definition B.3. *A* co-algebra *over the field of complex numbers* $\mathbb{C}$ *is a triple* $\langle C, \phi, \varepsilon \rangle$, *where* C *is a vector space over* $\mathbb{C}$ *together with a* co-multiplication, *which is a linear map*

$$\phi : C \to C \otimes C,$$

and with a co-unit, *a linear map,*

$$\varepsilon : C \to \mathbb{C}$$

such that the following two diagrams commute. First, we require that the co-commutativity diagram be commutative:

$$
\begin{array}{ccc}
C \otimes C \otimes C & \xleftarrow{\; id \otimes \phi \;} & C \otimes C \\
{\scriptstyle \phi \otimes id}\big\uparrow & & \big\uparrow{\scriptstyle \phi} \\
C \otimes C & \xleftarrow{\;\;\phi\;\;} & C.
\end{array}
$$

Notice that this diagram is dual to the commutativity diagram for algebras in the sense that all arrows have been reversed and so multiplication has been replaced by co-multiplication. This condition is equivalent to the formula

$$(\phi \otimes id)\phi = (id \otimes \phi)\phi. \tag{B.1}$$

Second, we require that the co-unit diagram be commutative:

$$
\begin{array}{ccccc}
C \otimes \mathbb{C} & \xleftarrow{\; id \otimes \varepsilon \;} & C \otimes C & \xrightarrow{\; \varepsilon \otimes id \;} & \mathbb{C} \otimes \mathcal{A} \\
{\scriptstyle \cong}\big\uparrow & & {\scriptstyle \phi}\big\uparrow & & \big\uparrow{\scriptstyle \cong} \\
C & \xleftarrow{\;\; id \;\;} & C & \xrightarrow{\;\; id \;\;} & C.
\end{array}
$$

This is equivalent to

$$(\varepsilon \otimes id)\phi \cong id, \tag{B.2}$$

$$(id \otimes \varepsilon)\phi \cong id. \tag{B.3}$$

Continuing as in the case of algebras, we now define morphisms. (For the curious reader, the concept of a co-morphism, namely, the dual of a morphism, is identical with that of a morphism.)

Definition B.4. *Suppose that C_1 and C_2 are co-algebras. A morphism f from C_1 to C_2, denoted $f : C_1 \to C_2$, is a linear map f from the vector space C_1 to the vector space C_2 satisfying preservation of co-multiplication, namely, the commutativity of*

$$
\begin{array}{ccc}
C_1 \otimes C_1 & \xrightarrow{f \otimes f} & C_2 \otimes C_2 \\
\phi_{C_1} \uparrow & & \uparrow \phi_{C_2} \\
C_1 & \xrightarrow{f} & C_2,
\end{array}
$$

and preservation of co-unit, namely, the commutativity of

$$
\begin{array}{ccc}
\mathbb{C} & \xrightarrow{id} & \mathbb{C} \\
\varepsilon_{C_1} \uparrow & & \uparrow \varepsilon_{C_2} \\
C_1 & \xrightarrow{f} & C_2.
\end{array}
$$

Corollary B.2. *The collection whose objects are all the co-algebras over $\mathbb{C}$ and whose arrows are all the morphisms between co-algebras is a category.*

Next, we put these two structures on one and the same vector space.

Definition B.5. *A bi-algebra over the complex numbers $\mathbb{C}$ is $\langle B, \mu, \eta, \phi, \varepsilon \rangle$ such that $\langle B, \mu, \eta \rangle$ is an algebra, $\langle B, \phi, \varepsilon \rangle$ is a co-algebra and these two structures are compatible in the sense that ϕ and ε are morphisms of algebras.*

The condition on ε means that

$$\varepsilon(ab) = \varepsilon(a)\varepsilon(b), \tag{B.4}$$

$$\varepsilon(1) = 1. \tag{B.5}$$

The condition on ϕ means that

$$\phi(ab) = \phi(a)\phi(b), \tag{B.6}$$

$$\phi(1) = 1 \otimes 1. \tag{B.7}$$

We will come back and explain later in more detail the condition on ϕ in terms of Sweedler's notation.

Exercise B.1. *Define morphism of bi-algebras and define the corresponding category.*

After the definitions of these three mathematical structures, we are finally ready to define the structure of interest in this appendix. First, we recall that the definition of the *twist map* $\sigma_A : A \otimes A \to A \otimes A$ is given by $\sigma_A(a \otimes b) := b \otimes a$.

Definition B.6. *A* Hopf *algebra is* $\langle A, \mu, \eta, \phi, \varepsilon, \kappa \rangle$, *where* $\langle A, \mu, \eta, \phi, \varepsilon \rangle$ *is a bi-algebra and* $\kappa : A \to A$ *is a co-inverse (or* antipode*) for the bi-algebra. This means that* κ *is an* antimultiplicative *morphism, namely,*

$$\kappa\mu = \mu(\kappa \otimes \kappa)\sigma_A, \tag{B.8}$$

$$\kappa(1) = 1 \quad \text{or} \quad \kappa\eta = \eta, \tag{B.9}$$

and that κ *is an* anti-co-multiplicative *morphism, namely,*

$$\phi\kappa = \sigma_A(\kappa \otimes \kappa)\phi \tag{B.10}$$

$$\varepsilon\kappa = \varepsilon, \tag{B.11}$$

and that κ *makes these diagrams commute:*

$$
\begin{array}{ccccccc}
A & \xrightarrow{\phi} & A \otimes A & \xrightarrow{\kappa \otimes id} & A \otimes A & \xrightarrow{\mu} & A \\
\downarrow{=} & & & & & & \downarrow{=} \\
A & & \xrightarrow{\varepsilon} & \mathbb{C} & \xrightarrow{\eta} & & A,
\end{array}
$$

$$
\begin{array}{ccccccc}
A & \xrightarrow{\phi} & A \otimes A & \xrightarrow{id \otimes \kappa} & A \otimes A & \xrightarrow{\mu} & A \\
\downarrow{=} & & & & & & \downarrow{=} \\
A & & \xrightarrow{\varepsilon} & \mathbb{C} & \xrightarrow{\eta} & & A.
\end{array}
$$

The formulas equivalent to these diagrams are

$$\mu(\kappa \otimes id)\phi = \eta\varepsilon \tag{B.12}$$

$$\mu(id \otimes \kappa)\phi = \eta\varepsilon \tag{B.13}$$

The experts will realize that this definition is not the most efficient since the identities (B.12) and (B.13) suffice to prove the other identities in the definition. See [45], Theorem III.3.4, for the details. In the same theorem in [45] it is also shown that $\kappa^2 = id_A$ provided that either the multiplication of A is commutative or the co-multiplication of A is co-commutative.

Another formula we use is

$$\phi(\kappa(a^{(1)}))(a^{(2)} \otimes 1) = 1 \otimes \kappa(a), \tag{B.14}$$

where we use Sweedler's notation $\phi(a) = a^{(1)} \otimes a^{(2)}$ as discussed below. To prove (B.14), we note that by (B.10) we have

$$\phi\kappa(a^1) = \sigma_A(\kappa \otimes \kappa)\phi(a^1)$$

$$= \sigma_A(\kappa \otimes \kappa)(a^{11} \otimes a^{12})$$

$$= \kappa(a^{12}) \otimes \kappa(a^{11}).$$

Then we complete the calculation as follows:

$$\phi\kappa(a^1)(a^2 \otimes 1) = (\kappa(a^{12}) \otimes \kappa(a^{11}))(a^2 \otimes 1)$$

$$= (\kappa(a^{21}) \otimes \kappa(a^1))(a^{22} \otimes 1)$$

$$= \kappa(a^{21})\, a^{22} \otimes \kappa(a^1)$$

$$= \eta\, \varepsilon(a^2) \otimes \kappa(a^1)$$

$$= \varepsilon(a^2)\, 1 \otimes \kappa(a^1)$$

$$= 1 \otimes \kappa(\varepsilon(a^2)\, a^1)$$

$$= 1 \otimes \kappa(a),$$

where we used the co-associativity (B.1), (B.12), and (B.2). So that finishes the proof of (B.14).

One more identity that we use is

$$(\eta\varepsilon \otimes id)\phi(a) = 1 \otimes a. \tag{B.15}$$

Again using Sweedler's notation introduced below, this is so because

$$(\eta\varepsilon \otimes id)\phi(a) = (\eta\varepsilon \otimes id)(a^{(1)} \otimes a^{(2)})$$

$$= \eta\varepsilon(a^{(1)}) \otimes a^{(2)}$$

$$= 1 \otimes \eta\varepsilon(a^{(1)})a^{(2)}$$

$$= 1 \otimes \varepsilon(a^{(1)})a^{(2)}1$$

$$= 1 \otimes a.$$

Exercise B.2. *Define morphisms of Hopf algebras and the category of Hopf algebras over* $\mathbb{C}$.

The co-inverse κ is not necessarily bijective though that is true in many examples. In general, a given bi-algebra may have no co-inverse. But if a co-inverse does exist, then it is known to be unique. For example, see [45] for these and other details.

There are a number of different types of convolution associated with Hopf algebras. The definition of the *convolution* of a linear functional $g : \mathcal{A} \to \mathbb{C}$ and an element $a \in \mathcal{A}$ is defined to be

$$g * a := (id \otimes g)\phi(a) = a^{(1)} \otimes g(a^{(2)}) \in \mathcal{A} \otimes \mathbb{C}$$

using Sweedler's notation. Or alternatively, using $\mathcal{A} \otimes \mathbb{C} \cong \mathcal{A}$,

$$g * a := g(a^{(2)})a^{(1)} \in \mathcal{A}.$$

Using this definition, we see that

$$\varepsilon(g * a) = \varepsilon(g(a^{(2)})a^{(1)}) = g(a^{(2)})\varepsilon(a^{(1)}) = g(\varepsilon(a^{(1)})a^{(2)}) = g(a). \qquad \text{(B.16)}$$

Similarly, but with the opposite order, we define the *convolution* of an element $b \in \mathcal{A}$ and a linear functional $f : \mathcal{A} \to \mathbb{C}$ by

$$b * f := (f \otimes id)\phi(b) = f(b^{(1)})b^{(2)}.$$

As in (B.16), it immediately follows that

$$\varepsilon(b * f) = f(b). \qquad \text{(B.17)}$$

Moreover, we have the associativity identity

$$f * (c * g) = (f * c) * g \qquad \text{(B.18)}$$

for all linear functionals $f, g : \mathcal{A} \to \mathbb{C}$ and all $a \in \mathcal{A}$. To see this, we first expand the left side, getting

$$f * (c * g) = f * (g(c^{(1)})c^{(2)}) = g(c^{(1)})(f * c^{(2)}) = g(c^{(1)})f(c^{(22)})c^{(21)}.$$

On the other hand, the right side gives us

$$(f * c) * g = f(c^{(2)})c^{(1)} * g = f(c^{(2)})g(c^{(11)})c^{(12)}.$$

Then the co-associativity of the co-multiplication proves that the two sides of (B.18) are equal.

We next quote some identities for the bijections r and s that are given in [86]. These follow readily from the definitions of these maps, and so the proofs will be left to the reader. In the following, $q \in \mathcal{A} \otimes \mathcal{A}$ and $a, b \in \mathcal{A}$.

$$r((a \otimes 1)q) = (a \otimes 1)r(q) \qquad \text{(B.19)}$$

$$r(q(1 \otimes b)) = r(q)\phi(b) \qquad \text{(B.20)}$$

$$s\big((a \otimes 1)q\big) = (1 \otimes a)s(q) \tag{B.21}$$

$$s\big(q(1 \otimes b)\big) = s(q)\phi(b) \tag{B.22}$$

$$\mu r^{-1} = id \otimes \varepsilon \tag{B.23}$$

$$\mu s^{-1} = \varepsilon \otimes id \tag{B.24}$$

$$(\varepsilon \otimes id)r^{-1}(a \otimes b) = \varepsilon(a)b \tag{B.25}$$

$$(\varepsilon \otimes id)s^{-1}(a \otimes b) = \varepsilon(b)a \tag{B.26}$$

Next, we consider the dual space $\mathcal{A}'$ of a Hopf algebra $\mathcal{A}$, where

$$\mathcal{A}' := \{f : \mathcal{A} \to \mathbb{C} \mid f \text{ is linear}\}.$$

We would like to use the pullbacks of the operations in $\mathcal{A}$ to define a Hopf algebra structure on $\mathcal{A}'$. But in general this does not quite work out. Here are some details on this topic. First, we take $f \in \mathcal{A}'$ and define

$$\phi^o(f) := \mu_{\mathcal{A}}^* f = f \circ \mu_{\mathcal{A}} : \mathcal{A} \otimes \mathcal{A} \to \mathbb{C},$$

where $\mu_{\mathcal{A}} : \mathcal{A} \otimes \mathcal{A} \to \mathcal{A}$ denotes the linear map induced by the bilinear multiplication map of $\mathcal{A}$, and $\mu_{\mathcal{A}}^*$ is its pullback. So the composition $f \circ \mu_{\mathcal{A}}$ is linear; that is, $\phi^o(f) \in (\mathcal{A} \otimes \mathcal{A})'$. We would like to say that ϕ^o is a co-multiplication on $\mathcal{A}'$, but such a co-multiplication acts as a map

$$\mathcal{A}' \to \mathcal{A}' \otimes \mathcal{A}'.$$

And while $\mathcal{A}' \otimes \mathcal{A}' \subset (\mathcal{A} \otimes \mathcal{A})'$ this inclusion is often proper. Therefore we have no guarantee in general that the range of ϕ^o is contained in the smaller space $\mathcal{A}' \otimes \mathcal{A}'$. This may help motivate in part the following definition.

Definition B.7. $\mathcal{A}^o := \{f \in \mathcal{A}' \mid \phi^o(f) \in \mathcal{A}' \otimes \mathcal{A}'\}.$

Then the main result is that $\mathcal{A}^o$ can be made into a Hopf algebra in a natural way. We have already defined the co-multiplication ϕ^o. Here are the definitions of the remaining operators. The multiplication μ^o in $\mathcal{A}^o$ is

$$\mu^o(f \otimes g) := (f \otimes g)\phi : \mathcal{A} \to \mathbb{C} \otimes \mathbb{C} \cong \mathbb{C}.$$

To define the unit η^o, we first note that the co-unit $\varepsilon : \mathcal{A} \to \mathbb{C}$ is an element in $\mathcal{A}^o$. We define $1_{\mathcal{A}^o} := \varepsilon$. Then we define the unit $\eta^o : \mathbb{C} \to \mathcal{A}^o$ by

$$\eta^o(\lambda) := \lambda 1_{\mathcal{A}^o} \quad \text{for } \lambda \in \mathbb{C}.$$

The co-unit $\varepsilon^o : \mathcal{A}^o \to \mathbb{C}$ is defined for $f \in \mathcal{A}^o$ by

$$\varepsilon^o(f) := f(1_{\mathcal{A}}) \in \mathbb{C}.$$

The co-inverse $\kappa^o : \mathcal{A}^o \to \mathcal{A}^o$ is defined for $f \in \mathcal{A}^o$ by the composition

$$\kappa^o(f) := f \circ \kappa : \mathcal{A} \to \mathbb{C}.$$

Notice that each of these five operations on $\mathcal{A}^o$ is the pullback of its "dual" operation on $\mathcal{A}$.

Exercise B.3. *Show that $\langle \mathcal{A}^o, \mu^o, \eta^o, \phi^o, \varepsilon^o, \kappa^o \rangle$ is a Hopf algebra.*

We now briefly discuss dual pairings of Hopf algebras.

Definition B.8. *Let $\langle \mathcal{A}_j, \mu_j, \eta_j, \phi_j, \varepsilon_j, \kappa_j \rangle$ be Hopf algebras for $j = 1, 2$. Then a bilinear mapping $\mathcal{A}_1 \times \mathcal{A}_2 \to \mathbb{C}$, denoted by $(f, a) \mapsto \langle f, a \rangle$, is said to be a* dual pairing *if for all $f, g \in \mathcal{A}_1$ and $a, b \in \mathcal{A}_2$, we have*

$$\langle \phi_1(f), a \otimes b \rangle = \langle f, ab \rangle = \langle f, \mu_2(a \otimes b) \rangle,$$

$$\langle \mu_1(f \otimes g), a \rangle = \langle fg, a \rangle = \langle f \otimes g, \phi_2(a) \rangle,$$

$$\varepsilon_1(f) = \langle f, 1_{\mathcal{A}_2} \rangle,$$

$$\varepsilon_2(a) = \langle 1_{\mathcal{A}_1}, a \rangle,$$

$$\langle \kappa_1(f), a \rangle = \langle f, \kappa_2(a) \rangle.$$

For example, the map $\mathcal{A}^o \times \mathcal{A} \ni (f, a) \mapsto f(a)$ is a dual pairing of Hopf algebras.

Exercise B.4. *Show that the last condition on the co-inverses is a consequence of the previous four conditions.*

We say that the pairing is *nondegenerate* if the usual conditions hold, namely, (a) $\langle f, a \rangle = 0$ for all $f \in \mathcal{A}_1$ implies $a = 0$ and (b) $\langle f, a \rangle = 0$ for all $a \in \mathcal{A}_2$ implies $f = 0$.

B.2 Sweedler's Notation

To introduce Sweedler's notation (see [78]), we take the primary example of a co-multiplication map

$$\phi : \mathcal{A} \to \mathcal{A} \otimes \mathcal{A}.$$

For any $a \in \mathcal{A}$, we can always write

$$\phi(a) = \sum_{k \in K} b_k \otimes c_k,$$

where the elements $b_k, c_k \in \mathcal{A}$ and k runs over some finite index set K. This is how one writes the co-product of a in standard notation. This is rather bulky, so we look for a more compact, manageable notation. One standard shortcut is to omit the index set K from the notation, and so one writes

$$\phi(a) = \sum_{k} b_k \otimes c_k.$$

The reader now has to understand that k runs over some unnamed finite index set. The Einstein convention carries this one step further by dropping the summation sign. So in that convention one writes

$$\phi(a) = b_k \otimes c_k.$$

Now it is the presence of the repeated index k that indicates a summation should be taken with respect to it. But Sweedler's notation is different from Einstein's. The idea is that both b_k and c_k depend on $a \in \mathcal{A}$, and so the notation should reflect this fact, even though b_k and c_k are not *uniquely* determined by a. By the way, the index set K is also not uniquely determined. So in Sweedler's notation we write $a^{(1)}$ instead of b_k and $a^{(2)}$ instead of c_k. Then we have Sweedler's notation for the co-product,

$$\phi(a) = a^{(1)} \otimes a^{(2)}.$$

The 1 in the notation $a^{(1)}$ indicates that it is the first factor (actually, a finite set of factors since there is an implicit summation here), while the 2 in $a^{(2)}$ indicates that it is the second factor. The summation is implicit, as in the case of Einstein's notation, but now the way we know this is by the repetition of the symbol a on the right side. Notice that the symbol a appears on both sides of this equation. This "balancing" property is characteristic of Sweedler's notation. The super-indices are placed in parentheses so that they are not confused with powers of a. Some authors put these as subindices and so they write

$$\phi(a) = a_{(1)} \otimes a_{(2)}.$$

This form of Sweedler's notation is not used in this book. Even other forms of Sweedler's notation maintain the summation sign, but we will also not use any of that. In [45] there is even another version of Sweedler's notation that uses primes instead of superindices.

However, when doing rough calculations, the reader may prefer not to write the parentheses in the superindices since that can slow down the mental processes needed to do the calculation. But they must be placed in any document meant for public circulation since otherwise people will easily confuse them for powers.

Why is Sweedler's notation more convenient? In the above very simple example of the co-multiplication, we gain very little, if anything at all. But in large complicated expressions with multiple occurrences of the co-product ϕ this notation provides enormous benefits. To get a hint of this, let's consider a "double" co-product

$$(\phi \otimes id_A)\phi : A \to A \otimes A \otimes A.$$

Calculating the action of this on an element $a \in A$, we obtain

$$(\phi \otimes id_A)\phi = (\phi \otimes id_A)(a^{(1)} \otimes a^{(2)})$$
$$= \phi(a^{(1)}) \otimes a^{(2)}$$
$$= a^{(11)} \otimes a^{(12)} \otimes a^{(2)},$$

where there is an iterated use of Sweedler's notation in the last equality but again in a compact form. More specifically, we write

$$\phi(a^{(1)}) = a^{(11)} \otimes a^{(12)} \tag{B.27}$$

instead of the bulkier, but totally correct form

$$\phi(a^{(1)}) = (a^{(1)})^{(1)} \otimes (a^{(1)})^{(2)},$$

which nobody wants to use! I call the notation of the right side of (B.27) the *iterated Sweedler's notation*, but I doubt that this is standard terminology. This form of Sweedler's notation is quite natural and easy to use for anyone with a recursive bent of mind.

Now let's consider the other "double" co-product

$$(id_A \otimes \phi)\phi : A \to A \otimes A \otimes A.$$

Calculating the action of this on an element $a \in A$, we obtain

$$(id_A \otimes \phi)\phi = (id_A \otimes \phi)(a^{(1)} \otimes a^{(2)})$$
$$= a^{(1)} \otimes \phi(a^{(2)})$$
$$= a^{(1)} \otimes a^{(21)} \otimes a^{(22)},$$

where we now are using the iterated Sweedler's notation to write out $\phi(a^{(2)})$.

Now, we generally work with co-multiplications that are co-associative, which means that

$$(\phi \otimes id_A) \phi = (id_A \otimes \phi) \phi.$$

In that case, of course, we have the identity

$$a^{(11)} \otimes a^{(12)} \otimes a^{(2)} = a^{(1)} \otimes a^{(21)} \otimes a^{(22)}$$

in the iterated Sweedler's notation. But be careful! We do not necessarily have $a^{(11)} = a^{(1)}$ and so forth. Each side of the equation here has an implicit double summation, and the index sets on the two sides need not be equal. And even if they are equal, this "element-by-element" type of equality will not hold in general.

However, in this case there is yet another variant of Sweedler's notation that is used in this book. We write

$$a^{(1)} \otimes a^{(2)} \otimes a^{(3)} := a^{(11)} \otimes a^{(12)} \otimes a^{(2)} = a^{(1)} \otimes a^{(21)} \otimes a^{(22)}.$$

I call $a^{(1)} \otimes a^{(2)} \otimes a^{(3)}$ the *single-digit Sweedler's notation* although, again, I doubt that this is standard terminology. Also, it is a bit of a misnomer when one comes to an expression with $a^{(k)}$ and $k \geq 10$ in it, but that does not happen in usual calculations. Of course, the 22 in $a^{(22)}$ is not twenty-two! As the reader can see, the meaning of a simple expression such as $a^{(1)}$ depends on the context of the formula where it is being used. Its correct interpretation, including the number of implicit summations involved, depends on the other occurrences of the symbol a in Sweedler's notation in the same expression. As a quick thought exercise, the reader might review the various uses of $a^{(1)}$ and $a^{(2)}$ in this discussion and identify their exact meanings.

In this book I use both the iterated and the single-digit versions of Sweedler's notations, always without summation signs. (But some authors include the summation sign in their versions of Sweedler's notation!) And sometimes I use both notations in the same expression, for example, in

$$(id \otimes \phi \otimes id)(a^{(1)} \otimes a^{(2)} \otimes a^{(3)}) = a^{(1)} \otimes a^{(21)} \otimes a^{(22)} \otimes a^{(3)}.$$

This is actually a dangerous practice since ambiguities can easily arise that can confuse one rather thoroughly. At least, that is my experience! But, as the saying goes, it is okay if you know what you are doing.

Of course, I could also use here only the single-digit version of Sweedler's notation by writing

$$(id \otimes \phi \otimes id)(a^{(1)} \otimes a^{(2)} \otimes a^{(3)}) = a^{(1)} \otimes a^{(2)} \otimes a^{(3)} \otimes a^{(4)}.$$

But then the meanings of the $a^{(k)}$'s are quite different for each side of the equation. And there is no "audit trail" as there is with the iterated notation. What I mean is that by the co-associativity of ϕ, we also have

$$(\phi \otimes id \otimes id)(a^{(1)} \otimes a^{(2)} \otimes a^{(3)}) = a^{(1)} \otimes a^{(2)} \otimes a^{(3)} \otimes a^{(4)}$$

and $$(id \otimes id \otimes \phi)(a^{(1)} \otimes a^{(2)} \otimes a^{(3)}) = a^{(1)} \otimes a^{(2)} \otimes a^{(3)} \otimes a^{(4)},$$

so that one cannot tell by just examining the expression on the right side, namely, $a^{(1)} \otimes a^{(2)} \otimes a^{(3)} \otimes a^{(4)}$, how it was produced. In fact, the single-digit Sweedler's notation only makes sense if co-associativity holds. Notice that there is an algorithm that produced the superindices when going from the expression on the left side to that on the right. This algorithm erases the factor $a^{(k)}$ on which ϕ acts, replaces it with the factor $a^{(k)} \otimes a^{(k+1)}$, and also replaces each factor $a^{(l)}$ with $l > k$ in the original expression with $a^{(l+1)}$.

The single-digit Sweedler's notation has an algorithm as well for reducing the number of factors. Consider

$$(id \otimes id \otimes \varepsilon \otimes id)(a^{(1)} \otimes a^{(2)} \otimes a^{(3)} \otimes a^{(4)}) = a^{(1)} \otimes a^{(2)} \otimes \varepsilon(a^{(3)}) \otimes a^{(4)}$$

$$= a^{(1)} \otimes \varepsilon(a^{(3)})a^{(2)} \otimes 1_A \otimes a^{(4)}.$$

Now we can apply the identity (B.3) to $\varepsilon(a^{(3)})a^{(2)}$ and this will yield the element denoted by $a^{(2)}$, but with respect to another single-digit Sweedler's notation, namely, the one with three factors and with the digits 1, 2, and 3. So we end up with

$$(id \otimes id \otimes \varepsilon \otimes id)(a^{(1)} \otimes a^{(2)} \otimes a^{(3)} \otimes a^{(4)}) = a^{(1)} \otimes a^{(2)} \otimes 1_A \otimes a^{(3)}.$$

So now the algorithm replaces the occurrences of $a^{(k)}$ and $a^{(k+1)}$ with an occurrence of $a^{(k)}$ and also replaces each factor $a^{(l)}$ with $l > k + 1$ in the original expression with $a^{(l-1)}$. After a while one gets the hang of these algorithms, and the single-digit Sweedler's notation becomes rather convenient. As is usual in these sorts of matters, a bit of practice is quite essential.

Another virtue of the single-digit Sweedler's notation is that each side of an equation satisfies a consistency condition. If the number of occurrences of $a^{(k)}$ is n on one side of the equation, then necessarily the index k takes each of the values in $\{1, 2, \ldots, n\}$ (or in $\{0, 1, \ldots, n-1\}$) exactly once on that side. Thus it helps explain how it happens that the symbols $a^{(k)}$ can have different meanings on the two sides of one equation, since the number of occurrences often depends on which side you are considering. Notice that the algorithms for introducing and removing occurrences of $a^{(k)}$ are compatible with this consistency check in the sense that they transform an expression satisfying the condition into another expression also satisfying the condition. This consistency check can help you both in finding correct expressions as well as in correcting silly errors. Of course, all errors are silly.

We now come back to a detail not fully explained earlier, namely, what does it mean to say that the co-multiplication ϕ is an algebra morphism? Quickly, this means that $\phi(ab) = \phi(a)\phi(b)$ as well as $\phi(1) = 1 \otimes 1$. To be more explicit, we write these in Sweedler's notation as

$$(ab)^{(1)} \otimes (ab)^{(2)} = a^{(1)}b^{(1)} \otimes a^{(2)}b^{(2)},$$

$$1^{(1)} \otimes 1^{(2)} = 1 \otimes 1.$$

So far we have only considered Sweedler's notation for a co-multiplication. The reader is invited to think about using the same notation for any linear map $V_1 \to V_2 \otimes V_3$, where each V_j is a vector space. In particular, we can and will use both versions of Sweedler's notation for right and left co-actions. The only difference will be that the superindices (in both the iterated and the single-digit notations) will often start from 0 instead of 1. I try to use the superindices $1, 2, \ldots$ for the factors that are repeated and the superindex 0 for the one factor that occurs exactly once. (However, excuse me, kind reader, if I sometimes forget to do this.) For example, if $\Phi : V \to V \otimes \mathcal{A}$ is a right co-action of $\mathcal{A}$ on the vector space V, then the notation is

$$\Phi(v) = v^{(0)} \otimes v^{(1)}.$$

And for the iterated co-action $(\Phi \otimes id_{\mathcal{A}})\Phi : V \to V \otimes \mathcal{A} \otimes \mathcal{A}$, the notation is

$$(\Phi \otimes id_{\mathcal{A}})\Phi(v) = \Phi(v^{(0)}) \otimes v^{(1)} = v^{(0)} \otimes v^{(1)} \otimes v^{(2)}$$

in the single-digit Sweedler's notation. So here $v^{(0)} \in V$, the factor that always occurs exactly once, while for $j \geq 1$ we have that $v^{(j)} \in \mathcal{A}$, the (possibly) repeated factor. Notice that, as before, the exact meaning of an expression such as $v^{(0)}$ or $v^{(1)}$ depends on the context. Using the identity $(\Phi \otimes id_{\mathcal{A}})\Phi = (id_V \otimes \phi)\Phi$ for a co-action, we also can write

$$(\Phi \otimes id_{\mathcal{A}})\Phi(v) = (id_V \otimes \phi)\Phi(v)$$

$$= (id_V \otimes \phi)(v^{(0)} \otimes v^{(1)})$$

$$= v^{(0)} \otimes v^{(1)} \otimes v^{(2)},$$

where we have used the single-digit Sweedler's notation for ϕ. This is not distinguishable with the previous expansion of this expression. And again we do not have an audit trail. Also, this notation requires that the co-action identity holds.

Using the iterated Sweedler's notation instead, we have

$$(\Phi \otimes id_{\mathcal{A}})\Phi(v) = \Phi(v^{(0)}) \otimes v^{(1)} = v^{(00)} \otimes v^{(01)} \otimes v^{(1)}$$

and

$$(id_V \otimes \phi)\Phi(v) = v^{(0)} \otimes v^{(11)} \otimes v^{(12)}.$$

So in this version of Sweedler's notation, the co-action identity becomes

$$v^{(00)} \otimes v^{(01)} \otimes v^{(1)} = v^{(0)} \otimes v^{(11)} \otimes v^{(12)}.$$

Bibliography

1. E. Abe, *Hopf Algebras*, Cambridge Tracts in Mathematics, **74**, Cambridge University Press, 1980.
2. E. Artin, Theorie de Zöpfe, Abh. Math. Sem. Univ. Hamburg **4** (1925) 47–72.
3. P.F. Baum, P.M Hajac, R. Matthes, and W. Szymański, Noncommutative geometry approach to principal and associated bundles, to appear in Quantum Symmetries in Noncommutative Geometry (P. M. Hajac, ed.) 2007. arXiv:math/0701033.
4. P.F. Baum, K. De Commer, and P.M. Hajac, Free Actions of Compact Quantum Groups on Unital C^*-algebras, arXiv:1304.2812v1
5. B. Blackadar, *Operator Algebras*, Springer, 2006.
6. T. Brzeziński and S. Majid, Quantum group gauge theory on quantum spaces, Commun. Math. Phys. **157** (1993) 591–638. Erratum: Commun. Math. Phys. **167** (1995) 235.
7. T. Brzeziński, Quantum Fibre Bundles. An Introduction. (1995) arXiv:q-alg/9508008.
8. T. Brzeziński, Translation map in quantum principal bundles, J. Geom. Phys. **20** (1996) 349. arXiv:hep-th/9407145v2.
9. T. Brzeziński and P.M. Hajac, The Chern–Galois character, C. R. Acad. Sci. Paris, Ser. I **338** (2004) 113–116.
10. T. Brzezinski and P.M. Hajac, Galois-Type Extensions and Equivariant Projectivity, arXiv:0901.0141
11. V. Chari and A. Pressley, *A Guide to Quantum Groups*, Cambridge University Press, 1994.
12. A. Connes, Noncommutative differential geometry, Inst. Hautes Études Sci. Publ. Math. **62** (1985) 257–360.
13. A. Connes, *Noncommutative Geometry*, Academic Press, 1994.
14. C. Davis and E.W. Ellers, eds., *The Coxeter Legacy, Reflections and Projections*, Am. Math. Soc., 2006.
15. V.G. Drinfeld, Hopf algebras and the quantum Yang–Baxter equation, Soviet Math. Dokl. **32** (1985) 254–258.
16. M. Dubois-Violette, Dérivations et calcul differentiel non commutatif, C.R. Acad. Sci. Paris, **307** I (1988) 403–408.
17. M. Dubois-Violette, R. Kerner, and J. Madore, Noncommutative differential geometry of matrix algebras, J. Math. Phys. **31** (1990) 316–322.
18. M. Dubois-Violette, R. Kerner, and J. Madore, Noncommutative differential geometry and new models of gauge theory, J. Math. Phys. 31 (1990) 323–330.
19. C.F. Dunkl, Differential difference operators associated to reflection groups, Trans. Am. Math. Soc. **311** (1989) 167–183.
20. C.F. Dunkl, Reflection groups for analysis and applications, Japan. J. Math. **3** (2008) 215–246. DOI: 10.1007/s11537-008-0819-3

© Springer International Publishing Switzerland 2015

S.B. Sontz, *Principal Bundles*, Universitext, DOI 10.1007/978-3-319-15829-7

21. M. Đurđevich, Geometry of quantum principal bundles I, Commun. Math. Phys. **175** (1996) 457–520.
22. M. Đurđevich, Geometry of quantum principal bundles II, extended version, Rev. Math. Phys. **9**, No. 5 (1997) 531–607.
23. M. Đurđevich, Differential structures on quantum principal bundles, Rep. Math. Phys. **41** (1998) 91–115.
24. M. Đurđevich, Quantum principal bundles as Hopf–Galois extensions, Acad. Res. **23** (2009) 41–49.
25. M. Đurđevich, Characteristic classes of quantum principal bundles, Alg. Groups Geom. **26** (2009) 241–341.
26. M. Đurđevich, Geometry of quantum principal bundles III, Alg. Groups Geom. **27** (2010) 247–336.
27. M. Đurđevich, Quantum classifying spaces and universal quantum characteristic classes, Banach Center Publ. **40** (1997) 315–327.
28. M. Đurđevich, General frame structures on quantum principal bundles, Rep. Math. Phys. **44** (1999) 53–70.
29. M. Đurđevich, Quantum principal bundles and corresponding gauge theories, J. Physics A: Math. Gen. **30** (1997) 2027–2054.
30. M. Đurđevich, Quantum gauge transformations and braided structure on quantum principal bundles, Misc. Alg. **5** (2001) 5–30.
31. M. Đurđevich and S.B. Sontz, Dunkl operators as covariant derivatives in a quantum principal bundle, SIGMA **9** (2013), 040, 29 pages. http://dx.doi.org/10.3842/SIGMA.2013.040
32. D. Eisenbud, *Commutative Algebra: With a View Towards Algebraic Geometry*, Springer, 1995.
33. P. Etingof, *Calogero–Moser Systems and Representation Theory*, Euro. Math. Soc., 2007.
34. R.P. Feynman, Feynman's Office: The Last Blackboards, Phys. Today, **42**, No. 2, (1989) p. 88.
35. G.B. Folland, *Quantum Field Theory, A Tourist Guide for Mathematicians*, Math. Surv. Mono. **149**, Am. Math. Soc. (2008).
36. L.C. Grove and C.T. Benson, *Finite Reflection Groups*, Springer, 1985.
37. P.M. Hajac, Strong connections on quantum principal bundles, Commun. Math. Phys. **182** (1996) 579–617.
38. P.M. Hajac and S. Majid, Projective module description of the q-monopole, Commun. Math. Phys. **206** (1999) 247–264.
39. S. Helgason, *Differential Geometry and Symmetric Spaces*, Academic Press, 1962.
40. J.E. Humphreys, *Reflection Groups and Coxeter Groups*, Cambridge, 1990.
41. T.W. Hungerford, *Algebra*, Springer, 1974.
42. J.C. Jantzen, *Lectures on Quantum Groups*, Am. Math. Soc., 1996.
43. M. Jimbo, A q-difference analogue of $U(\mathfrak{g})$ and the Yang–Baxter equation, Lett. Math. Phys. **10** (1985) 63–69.
44. M. Jimbo, A q-analog of $U(\mathfrak{gl}(N+1))$, Hecke algebras and the Yang–Baxter equation, Lett. Math. Phys. **11** (1986) 247–252.
45. C. Kassel, *Quantum Groups*, Springer, 1995.
46. C. Kassel and V. Turaev, *Braid Groups*, Springer, 2008.
47. M. Khalkhali, *Basic Noncommutative Geometry*, Eur. Math. Soc., 2009.
48. A. Klimyk and K. Schmudgen, *Quantum Groups and Their Representations*, Springer, 1997.
49. S. Kobayashi and K. Nomizu, *Foundations of Differential Geometry*, Vol. I, John Wiley & Sons, 1963.
50. L.I. Korogodski and Y.S. Soibelman, *Algebras of Functions on Quantum Groups: Part I*, Am. Math. Soc., 1998.
51. P.P. Kulish and N. Yu. Reshetikhin, Quantum linear problem for the sine–Gordon equation and higher representations, Zap. Nauch. Sem. LOMI **101** (1981) 101–110.
52. G. Landi, C. Pagani, and C. Reina, A Hopf bundle over a quantum four-sphere from the symplectic group, Commun. Math. Phys. **263** (2006) 65–88.
53. G. Lusztig, *Introduction to Quantum Groups*, Birkhäuser, 1993.

54. J. Madore, *An Introduction to Noncommutative Differential Geometry and Its Physical Applications*, Cambridge University Press, 1995.
55. S. Majid, *Foundations of Quantum Group Theory*, Cambridge University Press, 1995.
56. Yu. Manin, *Quantum Groups and Non-Commutative Geometry*, CRM, 1988.
57. Yu. Manin, *Topics in Noncommutative Geometry*, Princeton University Press, 1991.
58. E.H. Moore, Concerning the abstract group of order $k!$ and $\frac{1}{2}k!\ldots$, Proc. London Math. Soc. (1) **28** (1897) 357–366.
59. F.J. Murray and J. von Neumann, On rings of operators, Ann. Math. **37** (1936) 116–229.
60. F.J. Murray and J. von Neumann, On rings of operators 2, Trans. Amer. Math. Soc. **41** (1937) 208–248.
61. F.J. Murray and J. von Neumann, On rings of operators 4, Ann. Math. **44** (1943) 716–808.
62. J. von Neumann, On rings of operators III, Ann. Math. **41** (1940) 94–161.
63. M. Noumi and K. Mimachi, Quantum 2-spheres and big q-Jacobi polynomials, Commun. Math. Phys. **128** (1990) 521–531.
64. B. Parshall and J. Wang, Quantum linear groups, Mem. Am. Math. Soc. **439** 1991.
65. M. Pflaum, Quantum groups on fibre bundles, Commun. Math. Phys. **166** (1994) 279.
66. P. Podles, Quantum spheres, Lett. Math. Phys. **14** (1987) 193–202.
67. N.Yu. Reshetikhin, L.A. Takhtadjian, and L.D. Faddeev, Quantization of Lie groups and Lie algebras, Leningrad Math. J. **1** (1990) 193–225.
68. M. Rösler, Dunkl operators: theory and applications, in *Orthogonal Polynomials and Special Functions* (Leuven, 2002), Lecture Notes in Math., Vol. 1817, Springer, Berlin, 2003, 93–135. math.CA/0210366.
69. M. Rösler, Generalized Hermite polynomials and the heat equation for Dunkl operators, Commum. Math. Phys. **192** (1998) 519–542. q-alg/9703006.
70. M. Rösler and M. Voit, Markov processes related with Dunkl operators, Adv. in Appl. Math. **21** (1998) 575–643.
71. H.J. Schneider, Principal homogeneous spaces for arbitrary Hopf algebras, Isr. J. Math. **72** (1990) 167–195.
72. H.J. Schneider, Hopf–Galois extensions, cross products, and Clifford theory, Eds. J. Bergen et al., Advances in Hopf Algebras. Conference, August 10–14, 1992, Chicago. New York: Marcel Dekker. Lect. Notes Pure Appl. Math. **158** (1994) 267–297.
73. S. Shnider and S. Sternberg, *Quantum Groups: From Coalgebras to Drinfeld Algebras*, International Press, 1993.
74. E.K. Sklyanin, Some algebraic structures connected with the Yang–Baxter equation, Funct. Anal. Appl. **17** (1983) 273–284.
75. S.B. Sontz, On Segal-Bargmann analysis for finite Coxeter groups and its heat kernel, Math. Z. **269** (2011) 9–28.
76. S.B. Sontz, *Principal Bundles: The Classical Case*, in preparation.
77. B. Sutherland, Exact results for a quantum many-body problem in one-dimension. II, Phys. Rev. A **5** (1972) 1372–1376.
78. M.E. Sweedler, *Hopf Algebras*, W.A. Benjamin, New York, 1969.
79. M. Takeuchi, Quantum orthogonal and symplectic groups and their embedding into quantum GL, Proc. Japan Acad. Ser. A Math. Sci. **65** (1989) 55–58.
80. T. Timmermann, *An Invitation to Quantum Groups and Duality*, Euro. Math. Soc., 2008.
81. L.L. Vaksman and Y.S. Soibelman, Algebra of functions on the quantum group $SU(2)$, Funct. Anal. Appl. **22** (1988) 170–181.
82. J.C. Varilly, *An Introduction to Noncommutative Geometry*, Eur. Math. Soc., 2006.
83. S.L. Woronowicz, Pseudospaces, pseudogroups, and Pontryagin duality, in: *Proceedings of the International Conference on Mathematics and Physics*, Lausanne 1979, Lecture Notes in Physics, Vol. 116, pp. 407–412, Springer, 1980.

84. S.L. Woronowicz, Compact matrix pseudogroups, Commun. Math. Phys. **111** (1987) 613–665.
85. S.L. Woronowicz, Twisted $SU(2)$ group. an example of a non-commutative differential calculus, Publ. RIMS., Kyoto Univ. **23** (1987) 117–181.
86. S.L. Woronowicz, Differential calculus on compact matrix pseudogroups (quantum groups), Commun. Math. Phys. **122** (1989) 125–170.

Index

Printed in the United States
By Bookmasters

Printed in the United States
By Bookmasters